Graduate Texts in Physics

Graduate Texts in Physics

Graduate Texts in Physics publishes core learning/teaching material for graduate- and advanced-level undergraduate courses on topics of current and emerging fields within physics, both pure and applied. These textbooks serve students at the MS- or PhD-level and their instructors as comprehensive sources of principles, definitions, derivations, experiments and applications (as relevant) for their mastery and teaching, respectively. International in scope and relevance, the textbooks correspond to course syllabi sufficiently to serve as required reading. Their didactic style, comprehensiveness and coverage of fundamental material also make them suitable as introductions or references for scientists entering, or requiring timely knowledge of, a research field.

More information about this series at http://www.springer.com/series/8431

Mário J. de Oliveira

Equilibrium Thermodynamics

Second Edition

Mário J. de Oliveira
Institute of Physics
University of São Paulo
São Paulo, Brazil

ISSN 1868-4513 ISSN 1868-4521 (electronic)
Graduate Texts in Physics
ISBN 978-3-662-57121-7 ISBN 978-3-662-53207-2 (eBook)
DOI 10.1007/978-3-662-53207-2

Printed on acid-free paper

This Springer imprint is published by Springer Nature
The registered company is Springer-Verlag GmbH Germany
The registered company address is: Heidelberger Platz 3, 14197 Berlin, Germany

To my beloved wife
Tânia

Preface

If a birth date could be assigned to the science of thermodynamics, it would certainly be the day of publication of the Clausius article on the first and second law of thermodynamics, which occurred in 1850. Independently, these two laws were also laid by Kelvin. Both authors arrived at the two fundamental laws of thermodynamics relying on the ideas of Carnot on the operation of heat engines and on the work of Mayer and Joule on the mechanical equivalent of heat. Subsequently, the thermodynamic theory received contributions from other authors including Maxwell and Helmholtz. We point out the fundamental contribution of Gibbs whose formulation of the thermodynamics on the basis of the convexity properties of the thermodynamic potentials is that we follow in this book starting from Chap. 3.

By its origin, thermodynamics is closely related to the study of heat engines and thermodynamic processes. However, the science of thermodynamics should also be understood as the study of thermodynamic properties of substances. In this book, we adopt the point of view according to which thermodynamics is concerned with the study of macroscopic properties obtained from macroscopic laws. This does not mean that the laws of thermodynamics cannot be obtained from microscopic laws. We just understand that this task belongs to the domain of another discipline, statistical mechanics.

In the first two chapters of this book, we use the thermodynamic processes to introduce the first law of thermodynamics, which we call Joule principle, and the first part of the second law of thermodynamics, which we call Carnot principle, which allows us to define the absolute temperature and entropy. In Chap. 3 we present the principle of maximum entropy as introduced by Gibbs, which corresponds to the second part of the second law of thermodynamics. This principle leads to the convexity of thermodynamic potentials, studied in Chap. 4 along with the Legendre transformations. Chapter 5 is reserved for the analysis of the consistency of equations of state and Maxwell relations. The Nernst-Planck principle, which is the third law of thermodynamics, is presented in Chap. 6.

This book treats not only the fundamental principles of thermodynamics but also the thermodynamics of phase transitions and critical phenomena observed in various types of systems. Initially, we study the transitions between ordinary phases, solid, liquid, and vapor, in pure substances and in binary mixtures. We then examine the transitions of the order-disorder type in binary alloys and then pass to the study of phase transitions in magnetic systems, which include the ferromagnetic and antiferromagnetic materials, and in dielectrics.

The chapter on the Gibbs phase rule and the structure of the phase diagrams of multicomponent systems is based on the work of Griffiths and Wheeler. The study of criticality is done in accordance with the view that the systems have universal behavior in the critical region and can be described by the scaling theory introduced by Widom. In the study of phase transitions and criticality, we use equations of state that are introduced ad hoc and whose most notorious example is the van der Waals equation. Such equations, in spite of being different, describe the same critical behavior if they have the same symmetries. This universal behavior can be understood by the Landau theory of phase transitions, also described in this book.

Other topics studied in this book include the thermodynamics of solids, liquid crystals, thermal radiation, and thermochemistry. We examine the structural phase transitions and the phase transitions occurring between the mesophases of liquid crystals. In the study of thermal radiation, we arrive at the Planck distribution by a thermodynamic reasoning. The last chapter concerns the equilibrium thermodynamics of chemical reactions.

São Paulo, Brazil Mário J. de Oliveira

Contents

Chapter 1
Joule Principle

1.1 Introduction

Thermal Phenomena

The thermal phenomena are events that occur around us every day and so are part of our common experience. We feel the changes in temperature throughout the day and perceive climate changes caused by variations in atmospheric pressure and temperature. We notice that the fire is the most striking manifestation of the daily thermal processes. We learn that it can be used to heat water and make it boil and discover its enormous potentiality to generate motion and useful work.

In the absence of fire, the scalding water cools spontaneously to room temperature. In this process of spontaneous cooling, we find that the energy dissipates irreversibly in the form of heat transferred to the environment. However, it is possible to avoid the dissipation if the heat generated by fire or by other means is used to obtain useful work. In fact, this is achieved by means of a heat engine which transforms part of the heat into work that can be used directly or converted into another form of energy.

We also notice that a body does not cool spontaneously, unless it is placed in contact with a colder body. On contact, the hotter body cools because it gives off heat spontaneously to the colder body. However, it is possible to cool a body in the absence of a colder one. In fact, this can be achieved by the refrigerator, which does the opposite of the spontaneous process, that is, it removes heat from a colder body by the consumption of work and gives heat to the warmer body.

The operation principles of heat engines and refrigerators are directly related to the fundamental laws of thermodynamics, which are laws that govern how the heat is transformed into work and vice versa. These fundamental laws also govern the thermal behavior of bodies such as the thermal expansion, the decrease in volume by compression, the increase in temperature by heat absorption, etc.

M.J. de Oliveira, *Equilibrium Thermodynamics*, Graduate Texts in Physics,
DOI 10.1007/978-3-662-53207-2_1

In general, the bodies expand when heated, as happens to a stem of iron or to mercury inside a thermometer. Some materials, by contrast, contract when heated, as happens to rubber. We mention also the peculiar behavior of water when heated, which expands if it is above four degrees Celsius and contracts if it is below this temperature. The bodies also contract by the increase of the pressure exerted on them. Some bodies can be compressed more easily than others. In general, the gases have large variations in volume when compressed, contrary to what happens with the liquids and solids which have a small decrease in volume.

The amount of heat that the bodies can receive is a fundamental thermodynamic property. Some bodies can receive a large amount of heat with a small increase in temperature. The variation in temperature can also be zero, as in a system comprising two or more thermodynamic phases in coexistence. A trivial example is the system formed by water in coexistence with ice. If this system receives heat, a certain amount of ice will melt becoming liquid, but the temperature remains unchanged.

Viewpoint

We adopt the point of view according to which the science of thermodynamics concerns the study of the macroscopic properties of the bodies from macroscopic fundamental laws. This does not mean that the laws of thermodynamics can not be obtained from the laws governing the microscopic motion of matter. We just understand that this task is beyond the scope of thermodynamics and belongs to the field of statistical mechanics.

According to our point of view, the thermodynamic laws and properties do without reference to the microscopic constitution of matter. However, they acquire greater understanding if we can resort on the atomic constitution of matter. After all, the big breakthrough in the study of thermal phenomena, which gave rise to the science of thermodynamics, was due to the understanding of heat as the agitation of matter at the microscopic level.

Thermodynamic Equilibrium

The existence of thermodynamic equilibrium states is a fundamental postulate of thermodynamics. The thermodynamic equilibrium comprises the thermal equilibrium and other forms of equilibrium, depending on the type of thermodynamic system under study. A mechanical system in thermodynamic equilibrium is in thermal equilibrium and in mechanical equilibrium.

When a hot body is placed in contact with a cold body, a flow of heat is established from the first to the second up to the moment they reach a situation in which the flow ceases. From that moment on the two bodies are in thermal equilibrium. The thermal equilibrium can be checked by measuring the temperature of the bodies. If the two bodies have the same temperatures then they are in thermal equilibrium. If several bodies are in thermal contact with each other, they will be in thermal equilibrium if they have the same temperature.

We should also consider the internal thermal equilibrium of a thermodynamic system. In this case it is useful to understand the system as consisting of several parts that make up a set of multiple bodies in contact. The internal equilibrium means that there is no heat flows between the various parts of the system so that they all must have the same temperature. In summary, all parts of a system in thermal equilibrium have the same temperature.

If the object of study is a mechanical system, the thermodynamic equilibrium will also comprise mechanical equilibrium. For example, two gases in a closed vessel and separated by a movable wall will be in mechanical equilibrium if the wall is at rest. In this situation, the pressure is the same on both sides of the wall. Generally speaking, the mechanical equilibrium of a fluid, in the absence of external forces, requires no pressure gradient inside the fluid. Thus all parts of a system in mechanical equilibrium have the same pressure, in the absence of external forces.

Thermodynamic Process

The thermodynamic state of a gas can be changed in several ways depending on the type of wall which delimits the container which encloses the gas. If the walls are rigid but allow heat exchange, diathermal walls, the state of the gas can be changed simply by contact of the container, for example, with a warmer body. In this case, the gas receives heat and the pressure increases while the volume remains unchanged. If, on the other hand, the walls of the container do not allow the exchange of heat, adiabatic walls, but are mobile, the state may be modified by movements of the walls. A compression decreases the volume of the gas increasing its pressure. If the walls are rigid and adiabatic, then the gas is completely isolated from the outside and its equilibrium state remains unchanged.

If a system has its thermodynamic state modified, we say that it undergoes a thermodynamic process. In the study of equilibrium thermodynamics we are particularly interested in processes whose initial and final states are thermodynamic equilibrium states. The intermediate states cannot be all equilibrium states because it is necessary to remove the system from its initial equilibrium state for the process to take place. This is what happens with the free expansion of a gas. In a container with two compartments, a gas occupies one of the two compartments while the other is empty. At a certain moment the wall separating the compartments is removed, the gas expands freely and after a certain time the equilibrium is restored in a state in which the gas fills all the container.

We can also imagine, on the other hand, that the intermediate states, although out of equilibrium, are very close to equilibrium so that the thermodynamic process can be considered as composed by a succession of equilibrium states, intercalated between the initial and final states. To this end, small perturbations are applied to the system at regular intervals of time. The interval between two disturbances must be large enough to enable the system to relax to a new equilibrium state before it will be disturbed again. If the disturbance is too small, then the equilibrium state occurring after each disturbance will be very close to the disturbed state so that the time interval between two disturbance may be reduced. In the limit where the disturbances are arbitrarily small and infinitely large in number, the successive equilibrium states form a continuous sequence of equilibrium states that defines a thermodynamic process which we call quasi-static. If the disturbance is made by the contact with the external environment, then the system and the external environment should be close to equilibrium. This implies, for instance, that the temperature of the external environment and the system must be very close. Notice that this does not happen with the free expansion of a gas, even if the gas flow from one compartment to the other is arbitrarily small.

The increase of energy of a system has two contributions. One is the work done on the system and the other is the heat received. This second contribution can also be understood as work done on the system, but differs from the former because it occurs at the microscopic level. However, this distinctiveness between heat and work is not useful from the thermodynamic point of view. The thermodynamic distinctiveness between them is made by postulating the existence of adiabatic walls whose fundamental property is preventing the passage of heat. Later on, we will see how the use of adiabatic walls will allow us to distinguish between heat and work.

1.2 Work

Representation of Quasi-Static Processes

Consider a quantity of gas confined in a container. Moving the container walls, we can adjust the volume V of the gas at will. Introducing or removing heat, we can vary the gas pressure p until it reaches the value we wish. Once the gas is in its equilibrium state it is characterized by the volume V and the pressure p and therefore may be represented by a point in pressure versus volume diagram, called Clapeyron diagram.

Imagine that the gas undergoes a quasi-static process that takes it from an initial equilibrium state A to a final equilibrium state B. These two states are represented by two points in the Clapeyron diagram. The intermediate stages of the process are also states of equilibrium, and therefore have a Clapeyron diagram representation. Thus, the quasi-static process corresponds to a succession of points in the Clapeyron diagram, forming a continuous path that connects the points A and B, as can be seen in Fig. 1.1.

Notice that out of equilibrium processes, that is, such that the gas is not in equilibrium, do not have representation in the Clapeyron diagram. Suppose, for

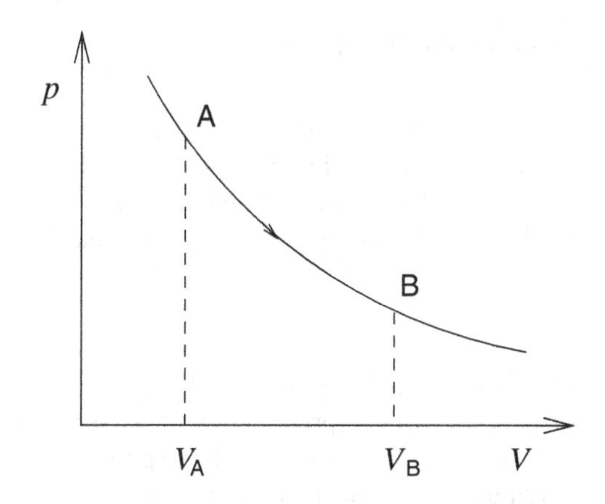

Fig. 1.1 Representation of a quasi-static process connecting the states A and B, in the Clapeyron diagram. The area under the curve between the points A and B equals the work done

example, that a gas undergoes a free expansion. Before the expansion, the gas is in equilibrium and its state is represented by a point in the Clapeyron diagram. After the expansion and as soon it enters again in equilibrium, the state of the gas will be represented by another point which is distinct from the first. However, these two points are not connected by a trajectory because the free expansion, a process that is out of equilibrium, has no representation in the Clapeyron diagram, although the initial and final states have it.

Isochoric and Isobaric Processes

The simplest quasi-static process that can be carried out is the one in which the volume of the gas remains unchanged, called isochoric process. The corresponding trajectory is an isochoric line, which in the Clapeyron diagram is simply a straight line parallel to the p-axis. Introducing heat into the gas slowly while the walls of the container are kept unchanged, we will be facing an isochoric process.

The quasi-static process in which the pressure remains unchanged is called isobaric process. The corresponding trajectory is an isobaric line or isobar, which in the Clapeyron diagram is merely a straight line parallel to the V-axis. Let us imagine that the gas is confined in a cylindrical container having rigid base and side walls and whose top can move freely. Over the top there is a block and the system composed by the block, the container, and the gas are in mechanical equilibrium. Suppose that from a certain instant the gas becomes subjected to a continuous flow of heat, for example, through the base of the container. Suppose further that the heat transfer is slow enough so that the system can be considered to be in mechanical equilibrium at each instant. Due to the continuous flow of heat, the gas expands and the block rises. Since the block has a fixed weight, the gas pressure remains constant throughout the expansion and we are faced with an isobaric process.

Mechanical Work

Consider the isobaric process just described. If we denote by A the area of the top, then the force exerted by the gas on the block is $F = pA$. If the top rises from the height ℓ_1 to the height ℓ_2 then the work W done by the gas in the isobaric process will be $W = F(\ell_2 - \ell_1) = pA(\ell_2 - \ell_1)$, that is,

$$W = p(V_2 - V_1), \tag{1.1}$$

where $V_1 = A\ell_1$ is the initial volume and $V_2 = A\ell_2$ is the final volume. The work done by the gas is represented in the Clapeyron diagram by the area under the line segment that represents the isobaric process.

Let us imagine next an arbitrary quasi-static process connecting the states A and B, like the one shown in Fig. 1.1. This process can be accomplished by conveniently changing the mass of the block which rests on the top of the container as well as the heat flow. The process is then approximated by a succession of isobaric subprocesses such that the volume change in each one of them is ΔV. Denoting by p_i the pressure corresponding to the starting point of each subprocess, then the work done is given by the sum of the works corresponding to each isobaric subprocess,

$$W = \sum_i p_i \Delta V. \tag{1.2}$$

In the limit where the number of subprocesses increases without limit, $\Delta V \to 0$ and the work becomes equal to the integral

$$W = \int p dV. \tag{1.3}$$

The area under the path is equal to the work performed by the gas. When a gas does work, increasing its volume, W is positive because the pressure of the gas is positive. When a gas consumes work, reducing its volume, W is negative. Note that in an isochoric process, the work done by the gas is zero.

As an example of the calculation of work, suppose that a gas undergoes a process such that its pressure is inversely proportional to the volume, that is,

$$p = \frac{B}{V} \tag{1.4}$$

along the path, where B is constant. The work done by the gas when it expands from a volume V_1 to a volume V_2 will be

$$W = \int_{V_1}^{V_2} p dV = B \int_{V_1}^{V_2} \frac{1}{V} dV = B \ln \frac{V_2}{V_1}. \tag{1.5}$$

As a second example, consider another process such that the gas pressure varies according to the equation

$$p = \frac{A}{V^{\gamma}}, \tag{1.6}$$

where A is a constant and γ is a numeric constant greater than one. If the gas expands from an initial volume V_1 to a final volume V_2, then the work done by the gas will be

$$W = \int_{V_1}^{V_2} p dV = A \int_{V_1}^{V_2} V^{-\gamma} dV = \frac{A}{-\gamma + 1} \{V_2^{-\gamma+1} - V_1^{-\gamma+1}\}. \tag{1.7}$$

Using the notation $p_1 = AV_1^{-\gamma}$ and $p_2 = AV_2^{-\gamma}$, then

$$W = \frac{1}{\gamma - 1}(p_1 V_1 - p_2 V_2). \tag{1.8}$$

In the International System of Units, the volume is measured in m^3 and the pressure in Pascal (Pa), defined as 1 $\mathrm{N/m}^2$. Another very common unit for pressure, but not belonging to the International Unit System is the atmosphere (atm), defined as 101,325 Pa. The unit of work in the International Unit System is the joule (J), defined as 1 N m.

1.3 Heat

Measure of Heat

By putting a body in contact with a warmer body, there will be heat transfer from the second to the first. However, in this procedure we will not know how much heat is transferred from one body to the other. To transfer a determined amount of heat to a system we proceed as in the following example.

A system comprising a fluid confined in a container with adiabatic walls has in its interior a propeller shaft which is wounded by an inextensible wire. This wire passes through a small pulley and is attached at its end to a block of mass m. Initially, the block is at rest and propeller immobile. Then the block is released and falls slowly with a constant speed v due to friction of the helix with the fluid. As the speed is constant the power dissipated by the helix is equal to the work performed by the force of gravity per unit time mgv, where g is the acceleration of gravity. In a time interval Δt the fluid receives an amount of heat $Q = mgv\Delta t$. Such a procedure in which a measurable amount of heat is introduced into a system we call Joule procedure.

Another way of performing the Joule procedure is to put within the system a resistor connected to a battery. If the circuit is closed during a time interval Δt, the heat received by the system equals $Q = (E^2/R)\Delta t$, where R is the resistance of the resistor and E is the battery emf.

If a system is brought from a state A to a state B, through an arbitrary quasi-static process in which it receives heat, the Joule procedure allows the direct determination of the amount of heat received by the system. If on the other hand, the system experiences a quasi-static process from a state C to a state D, where it yields heat, the Joule procedure can not be applied directly because this procedure can only be used to transfer heat to the system. We note, however, that in the inverse process the system receives heat and this heat can be measured by the Joule procedure. The amount of heat given off by the system in direct process CD is the same amount of heat received in the reverse process DC.

Conventionally, the heat received by the system under consideration is positive and the heat given off by the system is negative. Denoting by Q_{AB} the heat exchanged in the process AB, then according to this convention $Q_{\mathrm{AB}} > 0$. Similarly, $Q_{\mathrm{CD}} < 0$, $Q_{\mathrm{DC}} > 0$ and also $Q_{\mathrm{CD}} = -Q_{\mathrm{DC}}$.

Calorie

The heat has the same unit as work. In the International System of Units both quantities are measured in joule (J). Another unit often used for heat, but not belonging to the International System of Units, is the calorie (cal) defined as the amount of heat required to raise the temperature of 1 g of water by 1 °C. This definition is directly related to the definition of specific heat. However, the specific heat of water is not constant but varies with temperature, which makes the definition of calorie inaccurate unless one declares what temperature is taken as reference. Thus, many calories may be defined such as the thermochemical calorie, equivalent to 4.184 J; the IT calorie, equivalent to 4.1868 J; the average calorie, equivalent to 4.1900 J; the 15 °C calorie, equivalent to 4.1858 J, and the 20 °C calorie, equivalent to 4.1819 J.

The origin of the calorie as a unit of heat occurred before the development of thermodynamics, when the heat measurements were not related to the measures of work. The most common method to measure the heat developed in a process is to compare it with that necessary to raise the temperature by one unit a certain quantity of water. Another method for measuring heat is the one used by Lavoisier and Laplace in which the heat developed by a system is determined by the amount of ice that the heat can melt. It is worth mentioning that the first method is linked to the specific heat of water and the second to the latent heat of melting ice.

The first values for calorie are found in the Joule experiments for determining the mechanical equivalent of heat, that is, the work required to raise the temperature of a unit a mass of water by one unit of temperature. Joule showed that the heat required to raise by 1 °F the temperature of one pound of water is equivalent to the work of a body of 772 pounds falling from a height of one foot. In the International System of Units this is equal to 4.16 J. This value obtained by Joule for the mechanical equivalent of heat, and therefore for the calorie, is very close to the current values of the calorie. Before Joule the mechanical equivalent of heat was obtained by Mayer. According to Mayer, the heating of a certain a quantity of water by 1 °C corresponds to the falling of the same amount of water from a height of 365 m. This result is equivalent to 3.58 J.

Quasi-Static Adiabatic Process

Consider a gas confined to a cylindrical vessel with adiabatic walls, so that no heat can be exchanged between the environment and the system. Then imagine that the vessel top is raised slowly, leading to a gas expansion. This can be done by assuming that the top initially contain a certain amount of sand. The sand is then slowly removed, so that the process is quasi-static. The gas then undergoes a quasi-static process where no heat is exchanged with the environment, that is, a quasi-static adiabatic process. The corresponding trajectory is an adiabatic line.

1.4 Conservation of Energy

Internal Energy

To define the internal energy of a gas, we use as a reference one point in the Clapeyron diagram, labeling it by $\mathsf{O} = (V_0, p_0)$, and associate arbitrarily to this point an energy U_0. To determine the energy U_A of an arbitrary point $\mathsf{A} = (V_\mathsf{A}, p_\mathsf{A})$, we choose a process consisting of an adiabatic line followed by an isochoric process, as shown in Fig. 1.2. An adiabatic line connects the point O to an intermediate point $\mathsf{B} = (V_\mathsf{B}, p_\mathsf{B})$ while an isochoric line connects the point B to the point final A. If we denote by $W_{0\mathsf{B}}$ the work done by the gas in the adiabatic process and by Q_{BA} the heat received by the gas in the isochoric process, then the energy U_A is defined by

$$U_\mathsf{A} = U_0 - W_{0\mathsf{B}} + Q_{\mathsf{BA}}, \tag{1.9}$$

in which the work is determined by the formula (1.3) while the heat is measured by the Joule procedure. It may happen, however, that Q_{BA} is negative, resulting in the inability to use the Joule procedure. In this case, we use the inverse process and determine U_A by

$$U_0 = U_\mathsf{A} - W_{\mathsf{B}0} + Q_{\mathsf{AB}}, \tag{1.10}$$

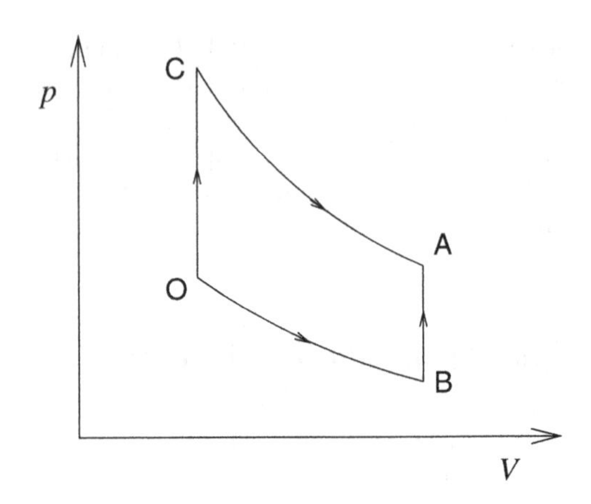

Fig. 1.2 The point A is connected to the reference point O in two ways. The first consists of an adiabatic line (OB) followed by an isochoric line (BA). The second comprises an isochoric line (OC) followed by an adiabatic line (CA)

since $Q_{\mathsf{AB}} = -Q_{\mathsf{BA}} > 0$. Since *any two states can be connected by an adiabatic line followed by an isochoric line*, then the internal energy of gas is defined for any state.

The Principle of Conservation of Energy

Suppose we choose another path to reach the point A from the point O. For example, the path OCA, shown in Fig. 1.2, comprising an isochoric line followed by an adiabatic line. In this case, the work and heat are different from the previous case. However, the Joule principle, or principle of conservation of energy states that *the energy of an equilibrium state is independent of the process used to reach it*. Thus the energy of the point A is the same regardless of the path used to reach the point A from the reference point O. We may write then, for any path,

$$U_{\mathsf{A}} = U_0 + Q - W, \tag{1.11}$$

where Q is the heat exchanged and W is the work done by the system along the chosen path.

Consider now any two points A and B. The energy difference $\Delta U = U_{\mathsf{B}} - U_{\mathsf{A}}$ between these two points will then be

$$\Delta U = Q - W, \tag{1.12}$$

where Q is the heat exchanged and W is the work done by the system along any path between any A and B. Remember that heat exchanged is positive when the system receives heat and negative when the systems gives off heat; the work actually performed by the system is positive while the work consumed is negative.

First Law of Thermodynamics

The Joule principle, or principle of conservation of energy, is the first law of thermodynamics. It was established by several scientists, but mostly by Mayer and Joule, who assumed that the various forms of work could be converted into one another and, moreover, that all of them could be dissipated as heat. Although it seems quite obvious that the work can be transformed into heat, for example, by friction, one can not conclude that the work dissipated always produce the same amount of heat. It is necessary to experimentally verify such a law, which is summarized in the determination of the mechanical equivalent of heat. This determination was in fact performed by Mayer and exhaustively by Joule by numerous experiments. By assuming that a certain amount of work always turns into the same amount of heat, they were adopting the conservation of energy.

From the microscopic point of view, the principle of conservation of energy was advanced by Helmholtz. Assuming the atomic constitution of matter, Helmholtz extended the theorem of conservation of mechanical energy to the microscopic motion of atoms. That is, he assumed that the sum of the kinetic and potential energy of atoms is constant and is therefore the internal energy of a body. When any form of work dissipates as heat, this means to say, from the microscopic point of view, that atoms gain energy. Therefore, according to Helmholtz and other scientists that have established the kinetic theory of matter, heat is associated to the microscopic motion of atoms.

Problems

1.1 A gas undergoes a quasi-static process expanding from a state A characterized by a volume V_0 and a pressure p_0 to a state B corresponding to a volume V_1. In this expansion the pressure varies with volume according $p = p_0 V_0^{5/3} V^{-5/3}$. Determine the pressure p_1 corresponding to state B. Calculate the work done by the gas when it expands from state A to the state B. Assuming the expansion to be an adiabatic line, what is the change in internal energy? Is the energy of the gas increased or decreased?

1.2 A gas goes through the process described in the previous problem and then undergoes an isochoric compression until a final state C whose pressure is p_0, the same pressure of the state A. In this process the gas receives an amount of heat Q. Suppose now that the gas undergoes an isobaric process from A to C. What is the amount of heat received in this process?

1.3 For a given gas, the internal energy depends on volume and pressure according $U = (3/2)pV$. Determine the work done by the gas when it is expanded from a state A to a state B, both points A and B belong to the same adiabatic line. Determine the heat received by the gas when it undergoes an isochoric process from state B to a state C such that C has the same energy as A. Suppose now that the gas undergoes a process leading from A to C by a quasi-static process in which the energy is constant. Determine the work and heat during this process. Given: V_A, p_A and V_B, p_B.

1.4 Suppose that the same gas of the previous problem undergoes a quasi-static adiabatic expansion starting from a reference state (V_0, p_0) of the Clapeyron diagram. Along the adiabatic line the heat exchanged is zero, so that the work W carried by the gas to a generic point (V, p) is equal to the energy variation, that is,

$$-\int_{V_0}^{V} p dV = \frac{3}{2}pV - \frac{3}{2}p_0 V_0.$$

Use this equation to determine the equation of the adiabatic curve which passes through the reference point. Hint: derive both members of this equation with respect to V to find a differential equation for $p(V)$.

1.5 For a given gas the equation of the adiabatic line that passes through a reference point (V_0, p_0) is $pV^{5/3} = p_0 V_0^{5/3}$. Furthermore, the heat introduced in a quasi-static process at constant volume, between the points (V, p_1) and (V, p) is $Q_v = (3/2)(p - p_1)V$. Determine the internal energy as a function of V and p.

Chapter 2
Carnot Principle

2.1 Temperature

Isothermal Process

When two bodies are placed in thermal contact, the hotter body gives off heat to the colder body. As long as the temperatures are different, there will be a flow of heat between them. After a while, the temperature becomes equal and the heat flow ceases. From that moment the bodies are in thermal equilibrium.

Once a body is in thermal equilibrium its temperature does not change anymore, unless it is disturbed. To verify possible changes in temperature, we used a tool comprising a substance having a thermometric property, that is, a property which varies with temperature. We call this instrument thermoscope and not thermometer, considering that we did not introduce a temperature scale yet.

Consider a gas confined in a cylindrical container whose base allows the exchange of heat and whose top can move freely. Over the top we place a certain amount of sand. Starting from a certain pressure adjusted by the quantity of sand on the top, we lead the gas through a process along which the temperature remains constant by monitoring the temperature of the gas by means of the thermoscope. We place the gas in the initial state and mark the value of the thermometric property. Then we allow the gas to expand by introducing a small amount of heat. If the temperature has changed, we vary the pressure by the removal of sand, until the temperature returns to the same value. This operation is then repeated many times. If the amount of heat introduced into each stage is small enough, the procedure approaches a quasi-static process carried out at constant temperature, or an isothermal process. The corresponding trajectory is an isotherm.

Another way to achieve an isothermal process is as follows. It is an experimental fact that can be verified by the use of thermoscope that a pure substance, such as pure water, has its temperature unchanged while it boils at constant pressure. If we place the cylinder containing the gas in contact with boiling water, the gas temperature

M.J. de Oliveira, *Equilibrium Thermodynamics*, Graduate Texts in Physics,
DOI 10.1007/978-3-662-53207-2_2

is remains constant. Removing sand from the top in small quantities, we will be conducting an isothermal process.

The above procedures allow a system to go through an isothermal process. However, we do not know the temperature of the system since we have not defined a temperature scale. In the following we define a temperature scale, called absolute temperature T, which allows us to determine the temperature of equilibrium systems.

Carnot Cycle

Suppose that a system, comprising a fluid contained in vessel, runs a cycle consisting of two isothermal processes and two adiabatic process. Starting from a state A, the system undergoes an isothermal expansion to a state B, then an adiabatic expansion to a state C, after an isothermal compression to a state D and finally an adiabatic compression back to the initial state A. This cycle, shown in Fig. 2.1, is referred to as the Carnot cycle. In the isothermal expansion, the system receives an amount of heat Q_1 from a heat reservoir at a temperature T_1 and, in the isothermal compression, gives off an amount of heat Q_2 to a heat reservoir at temperature T_2. The Carnot principle states that *the ratio of the work and heat received by a system which operates according to a Carnot cycle depends only on the temperature of the reservoirs*. Denoting by W the work done by the Carnot cycle then W/Q_1 depends only on T_1 and T_2. This principle is universal and is therefore independent of the substance that comprises the system that undergoes the cycle.

By the principle of conservation of energy, the work done in a closed cycle equals the heat received minus the heat given off, $W = Q_1 - |Q_2|$, so $W/Q_1 = 1 - |Q_2|/Q_1$. We conclude from the Carnot principle that the ratio of heat given off $|Q_2|$

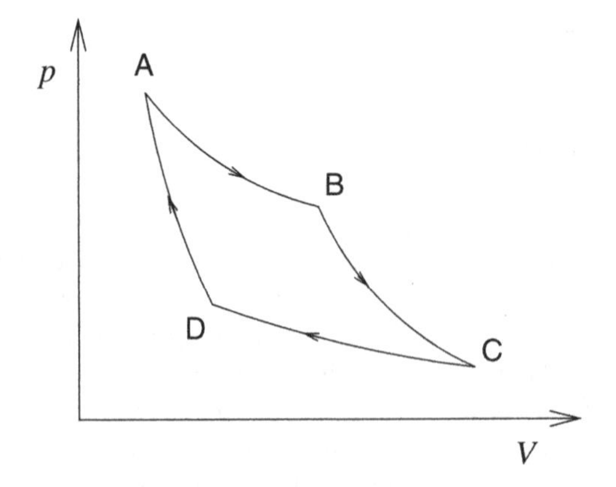

Fig. 2.1 Representation of the Carnot Cycle in the Clapeyron diagram. AB represents an isothermal expansion, BC an adiabatic expansion, CD an isothermal compression and DA an adiabatic compression

and the heat received Q_1 depends only on the temperatures of the reservoirs, that is

$$\frac{|Q_2|}{Q_1} = f(T_1, T_2). \tag{2.1}$$

If now we use another system which runs a Carnot cycle whose isotherms correspond to temperatures T_2 and T_3 then

$$\frac{|Q_3|}{Q_2'} = f(T_2, T_3), \tag{2.2}$$

where Q_2' is the heat received and $|Q_3|$ is the heat given off by the system running this second cycle. We conveniently choose the second cycle so that the heat given off by the first system is received by the second system so that $Q_2' = |Q_2|$. Multiplying the two equations (2.1) and (2.2) and taking into account that $Q_2' = |Q_2|$, we get

$$\frac{|Q_3|}{Q_1} = f(T_1, T_2)f(T_2, T_3). \tag{2.3}$$

But Q_1 and Q_3, appearing on the left side, can be considered respectively as the heat received and heat given off by a Carnot cycle operating between the temperatures T_1 and T_3. Therefore,

$$\frac{|Q_3|}{Q_1} = f(T_1, T_3), \tag{2.4}$$

so that

$$f(T_1, T_2)f(T_2, T_3) = f(T_1, T_3). \tag{2.5}$$

A function that satisfies this equation is of the form

$$f(T, T') = \frac{\phi(T')}{\phi(T)}, \tag{2.6}$$

from which we conclude that

$$\frac{|Q_2|}{Q_1} = \frac{\phi(T_2)}{\phi(T_1)}, \tag{2.7}$$

where $\phi(T)$ is a function to be defined but which is independent of the substances which is used to run the cycle.

Next we use the Carnot principle, in the form given by (2.7), to define the absolute scale of temperature. Following the prescription of Kelvin, we define absolute temperature T as the quantity such that $\phi(T) \propto T$ or, equivalently,

$$\frac{|Q_2|}{Q_1} = \frac{T_2}{T_1}. \tag{2.8}$$

Thus, to construct an absolute temperature scale, it suffices to assign a numerical value to the temperature of a reference state. Any other temperature is determined in accordance with (2.8). For the Kelvin scale (K) one uses the triple point of water as a reference state whose temperature we assign the exact value

$$T_0 = 273.16\,\text{K}. \tag{2.9}$$

It is worth mentioning that in the Kelvin scale the temperature of the melting of ice at a pressure of 1 atm is $T_{\text{sl}} = 273.15\,\text{K}$ and that the temperature of boiling water at the same pressure of 1 atm is $T_{\text{lv}} = 373.15\,\text{K}$.

The Celsius scale (°C) of temperature θ is defined by

$$\theta = T - 273.15\,°\text{C}, \tag{2.10}$$

where T is given in kelvins. In the Celsius scale, the temperature of the triple point of water is $\theta_0 = 0.01\,°\text{C}$, whereas the temperature of the melting ice and boiling water at the pressure of 1 atm are $\theta_{\text{sl}} = 0\,°\text{C}$ and $\theta_{\text{lv}} = 100\,°\text{C}$, respectively.

2.2 Entropy

Definition of Entropy

The definition of temperature allowed us to determine the temperature of a gas for any values of p and volume V. Thus, to each point of the Clapeyron diagram is associated a temperature. The points having the same temperature form an isotherm. The isotherms are a set of lines that never intersect, as can be seen in Fig. 2.2.

We can also plot on the same Clapeyron diagram, adiabatic lines corresponding to the same gas. To set up an adiabatic line, we confine the gas in a cylindrical container with adiabatic walls and slowly raise the top of the cylinder, while measuring the volume and pressure. Repeating this procedure several times, we get several adiabatic lines. The adiabatic lines form a set of lines which do not intersect, as can be seen in Fig. 2.2.

Just as there is a thermodynamic quantity associated with the isotherms, which is the temperature, already defined, we also associate to the adiabatic lines a thermodynamic quantity. This quantity should be invariant along an adiabatic line,

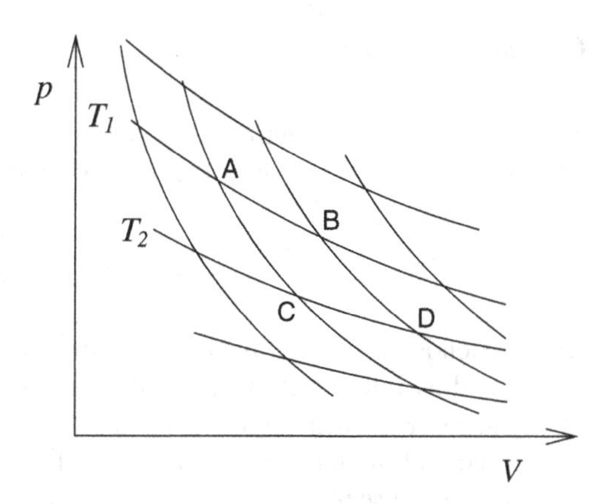

Fig. 2.2 Set of isotherms and adiabatic lines in the Clapeyron diagram. The lines AB and CD are isotherms corresponding to temperatures T_1 and T_2, respectively. The lines AC and BD are adiabatic lines

but this is not sufficient to establish its value for each adiabatic line. To set this magnitude, called entropy, we proceed as follows.

Consider any two adiabatic lines constructed in the manner shown above. Consider a point A on the first adiabatic line and a point B on the second adiabatic line such that A and B are in the same isotherm, corresponding to a temperature T_1, as shown in Fig. 2.2. If S_A is the entropy of the first adiabatic line, then the entropy S_B of the second adiabatic line is defined by

$$S_\mathsf{B} = S_\mathsf{A} + \frac{Q_1}{T_1}, \tag{2.11}$$

where Q_1 is the amount of heat received by the system from A to B along the isotherm T_1.

Suppose that another pair of points C and D are chosen on the two adiabatic lines, C over the first adiabatic and D over the second, but belonging to another isotherm, corresponding to a temperature T_2, as shown in Fig. 2.2. Using the same definition we also have

$$S_\mathsf{B} = S_\mathsf{A} + \frac{Q_2}{T_2}, \tag{2.12}$$

where Q_2 is the amount of heat received by the system from C to D along the isotherm T_2. At first sight, expressions (2.11) and (2.12) seem to be inconsistent. But by the Carnot principle $Q_1/T_1 = Q_2/T_2$ so that (2.11) and (2.12) are in fact consistent and the entropy of the second adiabatic line becomes independent of the choice of the isotherms.

Thus, taking an adiabatic line as reference for entropy, all other adiabatic lines will be associated with well defined values of entropy. To each point of the Clapeyron diagram is associated a certain value of entropy. As the adiabatic lines connect points with the same entropy value, they are also called isentropic lines. The quasi-static adiabatic processes may equivalently be called isentropic processes.

Clausius Integral

Next we show that the entropy difference between any two points A and B in the Clapeyron diagram is given by the integral Clausius

$$S_{\mathsf{B}} - S_{\mathsf{A}} = \int \frac{dQ}{T}, \tag{2.13}$$

which extends over *any path* connecting the points A and B. This integral should be understood as follows. A path chosen from A to B is partitioned into a number of segments. For each segment, we determine the ratio Q_i/T_i between the heat exchanged and the temperature at which the system is found along the segment. However, as the segment is not necessarily an isotherm, the temperature may vary along the segment. To avoid ambiguity we set T_i as the temperature of the system at the beginning of the segment. Thus, we calculate the sum

$$\sum_i \frac{Q_i}{T_i} \tag{2.14}$$

over all segments of the path. The integral in (2.13) should then be understood as the limit of this sum when the number of segments grows without bound.

Let us demonstrate now that the sum in (2.14) is an approximation to the difference in entropy $S_{\mathsf{B}} - S_{\mathsf{A}}$, while the number of sections is finite, but which becomes exact in the limit as the number of segments becomes infinite. To this end, we chose a smooth and monotonic decreasing path. In the extreme points of the segments, we draw adiabatic curves. We focus next in a segment between two consecutive adiabatic lines as the one shown in Fig. 2.3. The initial and final point of the segment are I and F. The curves IG and HF are the two adiabatic lines and GH is the isotherm built in such a way that the work W_{IGHF} done by the system along the path IGHF equals the work W_{IF} along the path IF. Thus, the heat Q_{IGHF} received by the system along the path IGHF equals the heat Q_{IF} received along the segment IF. Indeed, from the energy difference between the states F and I,

$$U_{\mathsf{F}} - U_{\mathsf{I}} = Q_{\mathsf{IF}} - W_{\mathsf{IF}} = Q_{\mathsf{IGHF}} - W_{\mathsf{IGHF}}, \tag{2.15}$$

calculated along the paths IF and IGHF, we conclude that $Q_{\mathsf{IGHF}} = Q_{\mathsf{IF}}$ because $W_{\mathsf{IF}} = W_{\mathsf{IGHF}}$. But $Q_{\mathsf{IGHF}} = Q_{\mathsf{GH}}$, since IG and HF are adiabatic lines, so that

$$Q_{\mathsf{GH}} = Q_{\mathsf{IF}}. \tag{2.16}$$

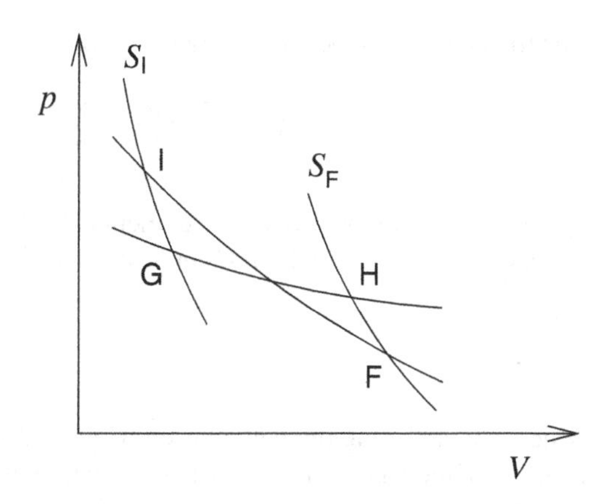

Fig. 2.3 The curve IF is a segment of a path in the Clapeyron diagram. The curves IG and HF are the adiabatic lines corresponding to the entropies S_I and S_F, respectively. The curve GH is an isotherm built in such a way that the works done by the system along the paths IGHF and IF are equal

From the definition of entropy, we see that the difference $S_F - S_I$ between the points F and I is given by

$$S_F - S_I = \frac{Q_{GH}}{T_G}, \tag{2.17}$$

where T_G is the temperature corresponding to the isotherm GH. Therefore,

$$S_F - S_I = \frac{Q_{IF}}{T_G}. \tag{2.18}$$

Thus the entropy difference $S_B - S_A$ will be

$$S_B - S_A = \sum_i (S_F^i - S_I^i) = \sum_i \frac{Q_{IF}^i}{T_G^i}, \tag{2.19}$$

where the summation is over all segments and the index i indicates that the quantities refer to the i-th segment. By construction, the expression (2.19) is exact, even for a finite number of segments. However, it cannot be identified with (2.14) because T_G^i and T_i are not equal, although $Q_{IF}^i = Q_i$. However, in the limit in which the number of segments increase indefinitely and the size of the segments decreases, T_G^i approaches $T_I^i = T_i$, and the two summations become equal in this limit and they are identified as the integral of expression (2.13).

Conservation of Energy in Differential Form

According to (2.18), the heat introduced along the segment IF is related to the entropy difference between the extreme points by

$$Q_{IF} = T_G(S_F - S_I). \tag{2.20}$$

The total heat Q introduced along the whole path is therefore given by

$$Q = \sum_i Q^i_{\mathsf{IF}} = \sum_i T^i_{\mathsf{G}}(S^i_{\mathsf{F}} - S^i_{\mathsf{I}}), \tag{2.21}$$

where the summation is over all segments of the path. In the limit where the number of segments growths without bounds, the summation becomes the integral

$$Q = \int TdS, \tag{2.22}$$

which extends along the path chosen between A and B. This equation says that the heat exchanged along a process is identified with the area under the corresponding trajectory in the T-S diagram.

Using (2.22) and (1.3), then the principle of conservation of energy can be written in integral form

$$\Delta U = \int TdS - \int pdV, \tag{2.23}$$

where ΔU is the internal energy difference between the points A and B, on the diagram S-V, and the integrals extends over any trajectory that connects these two points. Equivalently,

$$dU = TdS - pdV, \tag{2.24}$$

which is the equation of energy conservation in differential form. From this equation, we get

$$T = \left(\frac{\partial U}{\partial S}\right)_V \quad \text{and} \quad p = -\left(\frac{\partial U}{\partial V}\right)_S, \tag{2.25}$$

where the derivatives should be carried out keeping constant the variable that appears as an index.

2.3 Ideal Gas

Equation of State

All gases at sufficiently low densities behave similarly, irrespective of the type of molecules that they are composed. This behavior is called the ideal behavior and the gas is called an ideal gas. Therefore, an ideal gas must be understood as a shorthand for a gas in the regime of low densities.

If the temperature is kept constant, the pressure of an ideal gas is inversely proportional to the volume, that is,

$$p = \frac{A}{V}, \tag{2.26}$$

which is the Boyle law. Experimentally, one finds that the constant A is proportional to the temperature, which is the Gay-Lussac Law, so that we may write

$$p = \frac{BT}{V}. \tag{2.27}$$

The internal energy U of a gas which follows the law (2.27) is independent of the volume, depending only on the temperature. This independence on the volume was demonstrated experimentally by Joule by means of the free expansion of an ideal gas. In many cases the energy is linear with temperature, so that

$$U = CT, \tag{2.28}$$

where we arbitrarily set $U = 0$ for $T = 0$. If another behavior of the internal energy of an ideal gas is not explicitly stated, this means that we will be assuming the behavior given by (2.28).

To demonstrate that U depends only on the temperature, we use (2.24) in the form

$$dS = \frac{1}{T}dU + \frac{p}{T}dV. \tag{2.29}$$

Since S is a state function, then dS is an exact differential and, therefore,

$$\frac{\partial}{\partial V}\left(\frac{1}{T}\right) = \frac{\partial}{\partial U}\left(\frac{p}{T}\right). \tag{2.30}$$

The right hand side of this equation vanishes because $p/T = B/V$ depends on V but not on U. Therefore, $1/T$ does not depend on V but only on U or, in other words, U depends on T but not on V.

Next, we examine the consequences of assuming that U depends only on temperature. This is to say that the left side of (2.30) vanishes. Therefore, p/T depends only on V but not U. The equation of state (2.27) is therefore compatible with this result although not the most general equation.

Avogadro Hypothesis

Consider two vessels with distinct volumes containing different amounts of the same gas under the same temperature and pressure. Since the conditions are the same, then the ratio between the numbers of molecules in the containers must be equal to the ratio between the volumes. In other words, the number of molecules should be proportional to the volume. Bearing in mind that, from (2.27), B is proportional to the volume at T and p fixed, we then conclude that B is proportional to the number of molecules.

Using the standard for the number of molecules called mole, we may write $B = NR$ where N is the number of moles and R is the constant of proportionality. Therefore, for the gas we are considering, the following equation holds

$$p = \frac{NRT}{V}. \tag{2.31}$$

So far R is a constant which in principle should be different for each type of ideal gas. However, the Avogadro hypothesis implies that the constant R must be universal, that is, it must be the same for any gas at low densities. In fact, the Avogadro hypotheses states that equal volumes of distinct ideal gases under the same temperature and pressure contains the same number of molecules. This hypothesis implies, for instance, that the ratio NT/pV should be the same for any ideal gas. But according to (2.31), this ratio is a constant R, which therefore, must be universal. It is called the universal gas constant and its value is

$$R = 8.314510\,\text{J/mol K}. \tag{2.32}$$

Under normal conditions of temperature and pressure, that is, under the pressure of 1 atm (101,325 Pa) and temperature of 0 °C, one mole of gas occupies a volume equal to $22.41410\,\text{dm}^3$.

An argument similar to that used above leads us to conclude that the constant C that appears in (2.28) is also proportional to the number of moles, so that

$$U = NcT. \tag{2.33}$$

However, the constant c is not universal. For a monoatomic gas $c = (3/2)R$, for a diatomic gas $c = (5/2)R$.

An alternative form of presenting the internal energy of ideal gases that obey (2.33) is

$$U = \frac{c}{R}pV, \tag{2.34}$$

obtained form (2.31) and (2.33).

Adiabatic Curve

To determine the equation of an adiabatic curve, or isentropic line, in the Clapeyron diagram, we first calculate the differential dU from (2.34),

$$dU = \frac{c}{R}(pdV + Vdp). \tag{2.35}$$

Substituting (2.35) in (2.29) we obtain

$$dS = \frac{c+R}{RT}pdV + \frac{c}{RT}Vdp, \tag{2.36}$$

or

$$dS = \frac{c}{RT}(\gamma pdV + Vdp), \tag{2.37}$$

where the constant γ is defined by

$$\gamma = \frac{c+R}{c}. \tag{2.38}$$

Along the isentropic line, S is constant so that

$$\gamma pdV + Vdp = 0. \tag{2.39}$$

Integrating this differential equation, we get

$$pV^{\gamma} = \text{const.}, \tag{2.40}$$

that gives the dependence of p on V along the adiabatic line of an ideal gas. The constant γ varies from one gas to another. For a monatomic gas, $\gamma = 5/3$ because $c = (3/2)R$ and for a diatomic gas, $\gamma = 7/5$ because $c = (5/2)R$.

Entropy

To determine the entropy of an ideal gas corresponding to a point in the Clapeyron diagram, we use the equation of state (2.31) to eliminate T in (2.37). Then we obtain the following equation

$$dS = Nc(\frac{\gamma}{V}dV + \frac{1}{p}dp). \tag{2.41}$$

Integrating this equation along a path formed by an isobaric line and by an isochoric line, starting from a reference point $O = (V_0, p_0)$, we find

$$S - S_0 = Nc\gamma \int_{V_0}^{V} \frac{1}{V} dV + Nc \int_{p_0}^{p} \frac{1}{p} dp, \tag{2.42}$$

or

$$S = S_0 + Nc\gamma \ln \frac{V}{V_0} + Nc \ln \frac{p}{p_0}, \tag{2.43}$$

where S_0 is the entropy of the reference point. But $V_0 = Nv_0$ and $S_0 = Ns_0$ are proportional to the number of moles N so that

$$S = Ns_0 + Nc\gamma \ln \frac{V}{Nv_0} + Nc \ln \frac{p}{p_0}, \tag{2.44}$$

equation that gives the entropy of an ideal gas as a function of volume, pressure and number of moles.

2.4 Cyclic Processes

Heat Engines

Consider a system composed of a fluid confined in a cylindrical vessel and suppose that initially it is a particular state. The fluid undergoes a process and returns to its initial state. Suppose that the cyclic process is such that it performs some positive work, $W > 0$. In the Clapeyron diagram, this work corresponds to the area of the internal region of the cycle. As the fluid is back to the initial state, the change of internal energy is zero so that the work equals the total heat Q. This total heat is composed of the heat Q_R effectively received by gas and the heat Q_C given way by the gas, that is, $Q = Q_R + Q_C$. But $Q_R > 0$ and $Q_C < 0$, so that $Q = Q_R - |Q_C|$. Therefore,

$$W = Q_R - |Q_C|. \tag{2.45}$$

Since $W > 0$, the heat received by the system is greater than the heat given away. Therefore, in this case, we can say that part of the heat received by the system becomes work and the other part is transferred to the outside. A heat engine works in this way, converting heat into useful work.

Because not all the heat effectively received by the system becomes work, it is important to know what fraction η of the heat received turns into work. This fraction is called the efficiency of the heat engine and is defined by

$$\eta = \frac{W}{Q_R}, \tag{2.46}$$

or

$$\eta = \frac{Q_R - |Q_C|}{Q_R} = 1 - \frac{|Q_C|}{Q_R}. \tag{2.47}$$

In general the efficiency is distinct for different cyclic processes and depends on the substance which runs the cyclic process.

The simplest heat engine is that operating according to a Carnot cycle. As we have seen, the Carnot cycle consists of four steps: (a) an isothermal expansion, at a temperature T_1, in which the system receives a heat $Q_1 > 0$; (b) an adiabatic expansion; (c) an isothermal compression, at a temperature T_2, in which the system yields a heat $Q_2 < 0$; and (d) an adiabatic compression. The system returns to its initial state, so that the work done $W = Q_1 + Q_2 = Q_1 - |Q_2|$. The temperature T_1 is greater than T_2. Thus, in this cycle, the system receives an amount of heat Q_1 from a heat source at a higher temperature and gives off a smaller amount of heat to a heat sink at a lower temperature.

According to the Carnot principle, the ratio $|Q_2|/Q_1 = T_2/T_1$ is independent of the substance that undergoes the Carnot cycle. Therefore, the efficiency is

$$\eta = 1 - \frac{T_2}{T_1} = \frac{T_1 - T_2}{T_1}, \tag{2.48}$$

which is independent of the substance and depends only on the temperatures of the heat source and heat sink.

Otto Cycle

Let us examine a machine that operates with an ideal gas in a cycle called Otto cycle. It consists of four processes: (a) an isochoric compression from A to B, (b) an adiabatic expansion from B to C, (c) an isochoric decompression from C to D, and (d) an adiabatic contraction from D to the starting point A. In the calculations below we use the equation for the internal energy of an ideal gas in the form given by (2.34).

In the first process the gas receives an amount of heat Q_{AB}. As the gas does not perform work, then the heat is equal to the change of internal energy, that is,

$$Q_{\mathsf{AB}} = U_{\mathsf{B}} - U_{\mathsf{A}} = \frac{c}{R} p_{\mathsf{B}} V_{\mathsf{B}} - \frac{c}{R} p_{\mathsf{A}} V_{\mathsf{A}}. \tag{2.49}$$

In the second process the gas performs a work W_{BC}. Because there is no heat exchange, then the work is equal to the decrease in energy, that is,

$$W_{\mathrm{BC}} = U_{\mathrm{B}} - U_{\mathrm{C}} = \frac{c}{R} p_{\mathrm{B}} V_{\mathrm{B}} - \frac{c}{R} p_{\mathrm{C}} V_{\mathrm{C}}. \tag{2.50}$$

In the third process there is not work done, so that the heat in this process is given by

$$Q_{\mathrm{CD}} = U_{\mathrm{D}} - U_{\mathrm{C}} = \frac{c}{R} p_{\mathrm{D}} V_{\mathrm{D}} - \frac{c}{R} p_{\mathrm{C}} V_{\mathrm{C}}. \tag{2.51}$$

Finally, in the last process there is no heat exchanged so that the work is

$$W_{\mathrm{DA}} = U_{\mathrm{D}} - U_{\mathrm{A}} = \frac{c}{R} p_{\mathrm{D}} V_{\mathrm{D}} - \frac{c}{R} p_{\mathrm{A}} V_{\mathrm{A}}. \tag{2.52}$$

The efficiency will be then

$$\eta = \frac{W_{\mathrm{BC}} + W_{\mathrm{DA}}}{Q_{\mathrm{AB}}} = \frac{p_{\mathrm{B}} V_{\mathrm{B}} - p_{\mathrm{C}} V_{\mathrm{C}} + p_{\mathrm{D}} V_{\mathrm{D}} - p_{\mathrm{A}} V_{\mathrm{A}}}{p_{\mathrm{B}} V_{\mathrm{B}} - p_{\mathrm{A}} V_{\mathrm{A}}}, \tag{2.53}$$

or

$$\eta = 1 - \frac{p_{\mathrm{C}} V_{\mathrm{C}} - p_{\mathrm{D}} V_{\mathrm{D}}}{p_{\mathrm{B}} V_{\mathrm{B}} - p_{\mathrm{A}} V_{\mathrm{A}}}. \tag{2.54}$$

Recalling that $V_{\mathrm{B}} = V_{\mathrm{A}}$ and $V_{\mathrm{D}} = V_{\mathrm{C}}$ then

$$\eta = 1 - \frac{V_{\mathrm{C}}}{V_{\mathrm{A}}} \frac{p_{\mathrm{C}} - p_{\mathrm{D}}}{p_{\mathrm{B}} - p_{\mathrm{A}}}. \tag{2.55}$$

Now, the points A and D are on the same adiabatic line so that $p_{\mathrm{A}} V_{\mathrm{A}}^{\gamma} = p_{\mathrm{D}} V_{\mathrm{D}}^{\gamma}$. Likewise, the points B and C are over the same adiabatic line so that $p_{\mathrm{B}} V_{\mathrm{B}}^{\gamma} = p_{\mathrm{C}} V_{\mathrm{C}}^{\gamma}$. Recalling that $V_{\mathrm{B}} = V_{\mathrm{A}}$ and $V_{\mathrm{D}} = V_{\mathrm{C}}$ and combining these two relations we get

$$\frac{p_{\mathrm{D}}}{p_{\mathrm{A}}} = \frac{p_{\mathrm{C}}}{p_{\mathrm{B}}} = \left(\frac{V_{\mathrm{A}}}{V_{\mathrm{C}}}\right)^{\gamma}. \tag{2.56}$$

Using these relations we obtain

$$\eta = 1 - \frac{V_{\mathrm{C}}}{V_{\mathrm{A}}} \frac{p_{\mathrm{C}}}{p_{\mathrm{B}}} = 1 - \left(\frac{V_{\mathrm{A}}}{V_{\mathrm{C}}}\right)^{\gamma - 1}. \tag{2.57}$$

Refrigerators

If we run a heat engine in the reverse sense, then we are creating a refrigerator. For a better understanding, let us examine a Carnot cycle operated in the opposite direction. We start from a state corresponding to a temperature T_1. First we perform an adiabatic expansion in which the system cools to a temperature T_2 smaller than T_1. Then we go through an isothermal expansion at the temperature T_2, the end of which the system will have received a heat $Q_2 > 0$. Then, the system is compressed adiabatically to the temperature T_1. Finally, it is compressed isothermally at the temperature T_1 to the starting point. In this last stage it gives way a heat $Q_1 < 0$. The total work W is negative, which means that the system actually consumes work.

As the total energy change is zero, then $W = Q_1 + Q_2$, which can be written as

$$|W| = |Q_1| - Q_2, \tag{2.58}$$

so that the heat given way is greater than the heat received. However, the important point is that the system takes a certain amount of heat from a body at a lower temperature and reject a greater amount of heat to a body at a higher temperature.

The purpose of the refrigerator is to remove as much heat as possible for a given consumption of work. It is therefore convenient to define a performance coefficient ω of the refrigerator by

$$\omega = \frac{Q_2}{|W|} = \frac{Q_2}{|Q_1| - Q_2}, \tag{2.59}$$

where Q_2 is the heat extract from the colder body and W is the work consumed. Notice that Q_2 can be greater than $|W|$. Taking into account that for the Carnot cycle $|Q_1|/Q_2 = T_1/T_2$ then the performance coefficient for the cycle is

$$\omega = \frac{T_2}{T_1 - T_2}. \tag{2.60}$$

Paradoxically, a refrigerator may also function as a very efficient heater because the Q_1 rejected to the body at the higher temperature may be very large. It is at least greater than the work consumed.

Problems

2.1 Determine the work done, the absorbed heat, the energy variation and the change in entropy of an ideal gas undergoing the following processes. (a) Isothermal expansion at a temperature T between two states of volumes V_1 and V_2. (b) adiabatic expansion from a state of volume V_1 and pressure p_1 to a state of volume V_2. (c)

Isobaric expansion at the pressure p between two states of volumes V_1 and V_2. (d) Isochoric compression at the volume V between two states of pressures p_1 and p_2.

2.2 An ideal gas goes from an initial state A whose pressure is p_0 and volume is V_0 to a final state B whose pressure is $2p_0$ and volume is $2V_0$, through two different processes comprising: (a) isothermal expansion to a point C followed by an isochoric increase of pressure, (b) an isothermal compression to a point D followed by an isobaric expansion. Represent the two processes in the Clapeyron diagram. Find the volume and the pressure of the point C and the point D. For each process, calculate for each section the work, the heat absorbed and the change in internal energy.

2.3 An ideal gas undergoes a Carnot cycle composed by (a) an isothermal expansion (AB) at a temperature T_1, (b) an adiabatic expansion (BC), (c) an isothermal compression (CD) at a temperature T_2 and (d) an adiabatic compression (DA). Shown that the volumes and the pressures of the four states A, B, C and D, satisfies the relations $V_{\mathrm{C}}/V_{\mathrm{D}} = V_{\mathrm{B}}/V_{\mathrm{A}}$ and $p_{\mathrm{C}}/p_{\mathrm{D}} = p_{\mathrm{B}}/p_{\mathrm{A}}$. Determine the work done and the heat exchanged in each section of the cycle. Show explicitly that the efficiency is given by $\eta = 1 - T_2/T_1$.

2.4 An ideal gas undergoes a cyclic process formed by an isobaric line (AB), an isochoric line (BC) and an isotherm (CA). Make a sketch of the possible cycles in the Clapeyron diagram. Determine the efficiency of a heat engine working according to the cycle such that the temperature T_{B} of state B is greater than the temperature T_{A} of state A, $T_{\mathrm{B}} > T_{\mathrm{A}}$.

2.5 Determine the efficiency of a heat engine that operates with an ideal gas in accordance with the Brayton-Joule cycle consisting of two adiabatic processes and two isobaric processes. Make a sketch of the cycle in the Clapeyron diagram.

2.6 Determine the efficiency of a heat engine that works with an ideal gas in accordance with the Diesel cycle compose by (a) an isobaric expansion, (b) an adiabatic expansion, (c) an isochoric decompression, and (d) an adiabatic compression. Make a sketch of the cycle in the Clapeyron diagram.

2.7 An ideal gas undergoes the cyclic transformation ABCA composed by an isochoric process (AB), an adiabatic process (BC) and an isobaric process (CA). Represent the cycle in the Clapeyron diagram given that the temperature T_{B} of B is greater than the temperature T_{A} of A, $T_{\mathrm{B}} > T_{\mathrm{A}}$. Show that the temperature T_{C} of C is related to T_{A} and T_{B} by

$$T_{\mathrm{C}}^{\gamma} = T_{\mathrm{B}} T_{\mathrm{A}}^{\gamma-1}.$$

Determine the heat exchanged, the work done, the variation of the internal energy in each of the processes as well as the efficiency of a heat engine operating according to this cycle. Give the answers in terms of the temperatures of the three points.

2.8 An ideal gas undergoes a free expansion from a volume V_0 to a volume $2V_0$. Determine the variation of the entropy in this process.

2.9 Consider a straight line in the Clapeyron diagram passing through the point A and having a slope equal to $-\alpha < 0$. Suppose that an ideal gas undergoes an expansion along this straight line starting from point A. Determine the heat Q received by the gas as a function of the gas volume V. Plot Q versus V, indicating the interval in which the heat increases with volume and the interval in which it decreases with volume. Do the same with entropy.

2.10 In the Clapeyron diagram, consider the segment of a straight line connecting two points A and B found in the same adiabatic curve. The volume of A is smaller than the volume of B, that is, $V_\mathsf{A} < V_\mathsf{B}$. An ideal gas undergoes the cycle formed by the segment and the adiabatic curve. Initially, the gas expands from A to B along the straight line. Then, it s compressed adiabatically going back to the point A. Determine the efficiency of a heat engine operating according to this cycle.

2.11 Show that the efficiency of a heat engine working between two temperatures T_1 and T_2 in any cyclic quasi-static process is always smaller than the one operating according to a Carnot cycle between the same temperatures. The temperatures T_1 an T_2 must be understood as the maximum and minimum temperatures attained by the substances undergoing the cycle. Hint: plot the cycle in a T-S diagram and compare it with the Carnot cycle. Remember that in this diagram the heat exchanged is identified as the area under the trajectory and that the Carnot cycle is a rectangle.

Chapter 3
Clausius-Gibbs Principle

3.1 Thermodynamic Coefficients

Introduction

A system in thermodynamic equilibrium must be stable. This means to say that small perturbations do not remove the system from its equilibrium. In a mechanical system in stable equilibrium described by a potential energy $\mathscr{V}(x)$, the stability implies that the coefficient $d^2\mathscr{V}/dx^2$ is non-negative. Similarly, a gas confined in a vessel with adiabatic walls, which can be viewed as a mechanical system, the mechanical stability implies that the coefficient $\partial^2 U/\partial V^2$ is nonnegative, where $U(V)$ is the internal energy and V the volume of the vessel. This quantity is directly related to the adiabatic compressibility κ_s, which will be defined later, by $\partial^2 U/\partial V^2 = 1/V\kappa_s$. Therefore, the mechanical stability of a gas means that the adiabatic compressibility is nonnegative.

The overall stability of a thermodynamic systems is related not only to the adiabatic compressibility, but also to other coefficients such as the isothermal compressibility and the heat capacities at constant volume and at constant pressure, as defined below. A system in thermodynamic equilibrium implies that these coefficients are nonnegative. The positivity of the thermodynamic coefficients, however, does not necessarily imply global stability of thermodynamic systems. It implies the local stability, that is, the stability with respect to small disturbances. The global stability, which implies the local stability, of the thermodynamic systems is guaranteed by the Clausius-Gibbs principle.

M.J. de Oliveira, *Equilibrium Thermodynamics*, Graduate Texts in Physics,
DOI 10.1007/978-3-662-53207-2_3

Heat Capacity

The heat capacity of a body is defined as the ratio $Q/\Delta T$ between the heat received Q and the corresponding increase in temperature ΔT. More precisely, as the limit of this ratio when $\Delta T \to 0$. Because the amount of heat is small, then $Q = T\Delta S$, where ΔS is the increase in entropy. Therefore, the heat capacity is given by TdS/dT. We however should discriminate the way in which the heat is received by the system. If done at constant volume, we have the isochoric heat capacity C_v, given by

$$C_v = T\left(\frac{\partial S}{\partial T}\right)_V. \quad (3.1)$$

If done at constant pressure, we have the isobaric heat capacity C_p, given by

$$C_p = T\left(\frac{\partial S}{\partial T}\right)_p. \quad (3.2)$$

Figure 3.1 shows the heat capacity per unit mass, or specific heat, of water for temperatures between 0 and 100 °C, at the pressure of 1 atm.

In the first case, being the heat introduced at constant volume, it will be equal to the variation of the internal energy so that we may write in an equivalent manner

$$C_v = \left(\frac{\partial U}{\partial T}\right)_V. \quad (3.3)$$

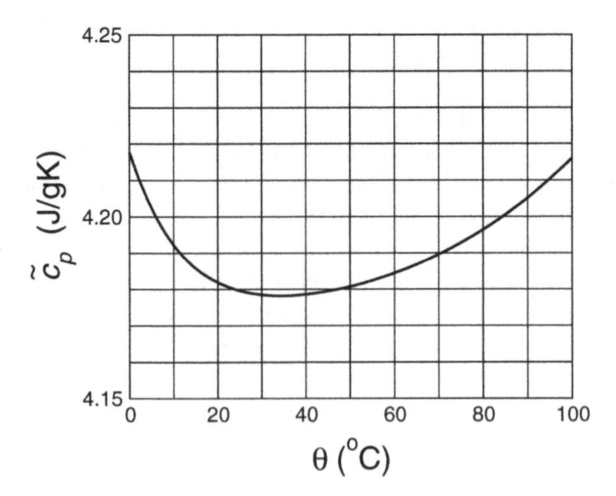

Fig. 3.1 Isobaric specific heat $\tilde{c}_p$ of water as a function of temperature at the pressure of 1 atm. At 15 °C, $\tilde{c}_p$ is 4.1858 J/gK. *Source*: AIP

In the second case, the heat introduced is equal to the variation of energy added to the work done by the system. Therefore, in this case

$$C_p = \left(\frac{\partial U}{\partial T}\right)_p + p\left(\frac{\partial V}{\partial T}\right)_p. \tag{3.4}$$

Using the equations of the ideal gas, we obtain

$$C_v = Nc, \qquad C_p = Nc + NR. \tag{3.5}$$

Therefore, the constant γ defined by (2.38), which appears in (2.40) of the adiabatic line of an ideal gas, is equal to the ratio between the two heat capacities of the ideal gas, that is, $\gamma = C_p/C_v$.

Particularly useful formulas for obtaining entropy from the heat capacities are as follows. Along an isochoric process, we integrate both members of (3.1), after dividing by the temperature, to get

$$S = \int_{(\mathrm{V})} \frac{C_v}{T} dT, \tag{3.6}$$

where the symbol (V) indicates that the integral must be performed along a path in which the volume remains constant. Along an isobaric process, we integrate both member of (3.2), after dividing by the temperature, to get

$$S = \int_{(\mathrm{p})} \frac{C_p}{T} dT, \tag{3.7}$$

where the symbol (p) indicates that the integral must be done along a trajectory in which the pressure is kept constant.

Coefficient of Thermal Expansion

The coefficient of thermal expansion α is related to the ratio between the increase in the volume and the increase in the temperature in a process at constant pressure. More precisely, α is defined by

$$\alpha = \frac{1}{V}\left(\frac{\partial V}{\partial T}\right)_p. \tag{3.8}$$

For the case of an idea gas, it is given by

$$\alpha = \frac{NR}{pV} = \frac{1}{T}. \tag{3.9}$$

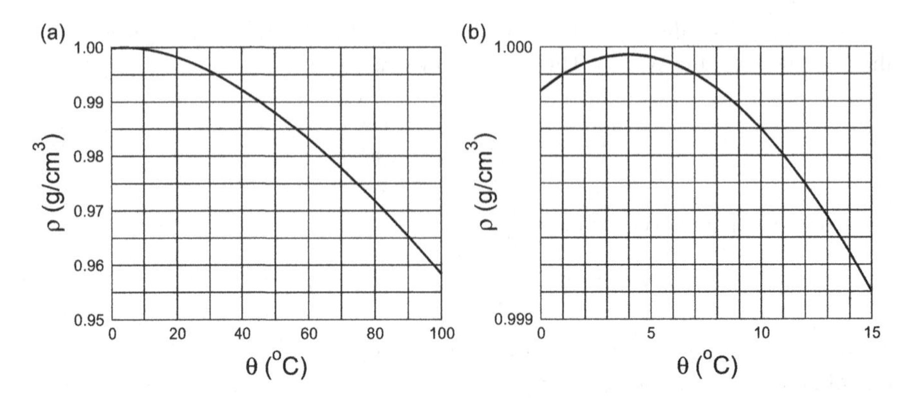

Fig. 3.2 Density of water ρ as a function of temperature at the pressure of 1 atm. The maximum value of density 0.999972 g/cm^3 occurs when $\theta = 4.0\,°\text{C}$. *Source*: LB

The coefficient of thermal expansion can be positive, zero or negative. For example, at normal pressure, the water has a positive value of α for temperature above 4 °C and a negative value of α, below 4 °C. Equivalently, we may say that above 4 °C the density of water decreases with temperature and below 4 °C the density increases with temperature. At 4 °C the water has its maximum density as seen in Fig. 3.2.

Compressibility

The compressibility is related to the ratio between the decrease in the volume and the increase in the pressure. There are two types of compressibility. The isothermal compressibility, denoted by κ_T, and the adiabatic compressibility, denoted by κ_s. The first is defined by

$$\kappa_T = -\frac{1}{V}\left(\frac{\partial V}{\partial p}\right)_T, \tag{3.10}$$

whereas the second is defined by

$$\kappa_s = -\frac{1}{V}\left(\frac{\partial V}{\partial p}\right)_S. \tag{3.11}$$

In the Clapeyron diagram, the isothermal compressibility is related to the slope of the isotherm whereas the adiabatic compressibility is related to the slope of the adiabatic curve.

Sometimes, it is useful to employ the bulk modulus, defined as the reciprocal of the compressibility. We have the bulk modulus at constant temperature $B_T = 1/\kappa_T$ and the adiabatic bulk modulus $B_s = 1/\kappa_s$.

Using the equation of state of the ideal gas, $V = NRT/p$, we get

$$\kappa_T = \frac{1}{p}, \qquad B_T = p. \tag{3.12}$$

Using the equation $pV^\gamma = \text{const.}$, valid for the ideal gas along an adiabatic line, we get

$$\kappa_s = \frac{1}{\gamma p}, \qquad B_s = \gamma p. \tag{3.13}$$

3.2 Thermodynamic Stability

Principle of Maximum Entropy

We focus our attention on two distinct states of an isolated system composed of several bodies, which can be interpreted as the various parts of the same system. In the first state, the bodies are separated and each one is in thermodynamic equilibrium at a given temperature and a given pressure. The separation between them is imposed by the existence of one or more constraints of various types. A rigid wall, for example, is a mechanical constraint. An adiabatic wall is a thermal constraint. In this state, the system as a whole is in a state of constrained equilibrium. In the second state, the constraints are absent, the bodies are in thermal and mechanical contact with each other forming a system in an unconstrained equilibrium, all at the same temperature and same pressure.

Next, we imagine a thermodynamic process that takes the system from a state of constrained equilibrium to a state of unconstrained equilibrium, which is accomplished simply by the withdrawal of the constraints, by placing the bodies in thermal and mechanical contact. In this process, the energy and the volume remain invariant. Under these conditions, the maximum entropy principle states that the sum of the entropies of the body increases or remains constant. In other words, this principle states that *the entropy of the unconstrained equilibrium state is greater than or equal to that of the constrained equilibrium, as long as the two states have the same energy and the same volume.*

Now we use the principle of maximum entropy to determine the conditions of thermodynamic stability of a gas in contact with a thermal reservoir at temperature T_0 and a mechanical reservoir at a pressure p_0. The gas is confined in a vertical cylindrical vessel having side walls that are rigid and adiabatic and whose base is in contact with the thermal reservoir. This is an auxiliary system which exchanges only heat without change in temperature. Because the temperature is constant, the variation of its entropy ΔS_{RT} and the heat received Q_{RT} are related by $\Delta S_{\mathrm{RT}} = Q_{\mathrm{RT}}/T_0$.

The top of the cylinder is adiabatic and can move freely. Over the top there is a block which together with the surrounding air, act as a mechanical reservoir, maintaining a constant pressure on the gas, no matter what the height of the top. The pressure p_0 on the gas is equal to the ratio between the weight of the block and the area of top plus the pressure of the atmospheric air. Because the pressure remains constant, the work W_{RM} done by the mechanical reservoir and the variation ΔV_{RM} of its volume are related by $W_{\mathrm{RM}} = p_0 \Delta V_{\mathrm{RM}}$.

Next, we consider the constrained and unconstrained equilibrium states. In the first, the gas is in equilibrium but isolated from the two reservoirs. In the second, the gas is in contact with the two reservoirs forming a composite system in equilibrium. In the state of constrained equilibrium we denote by T_1 and p_1 the temperature and the pressure of the gas. We suppose that in this state its volume is V_1, its energy is U_1 and its entropy is S_1. The variation of the total entropy $\Delta S_{\mathrm{total}}$, between the unconstrained and the constrained equilibrium, is given by

$$\Delta S_{\mathrm{total}} = S_0 - S_1 + \Delta S_{\mathrm{RT}} = S_0 - S_1 + \frac{Q_{\mathrm{RT}}}{T_0}, \tag{3.14}$$

since the mechanical reservoir does not exchange heat. In addition, the variation of the total energy $\Delta U_{\mathrm{total}}$ between the two states is

$$\Delta U_{\mathrm{total}} = U_0 - U_1 + Q_{\mathrm{RT}} - W_{\mathrm{RM}} = U_0 - U_1 + Q_{\mathrm{RT}} + p_0(V_0 - V_1), \tag{3.15}$$

since $W_{\mathrm{RM}} = p_0 \Delta V_{\mathrm{RM}}$ and $\Delta V_{\mathrm{RM}} = (V_1 - V_0)$.

If the composite system is brought from the constrained equilibrium to the unconstrained equilibrium in such a way that the total energy is constant, then we have $\Delta U_{\mathrm{total}} = 0$ and, according to the principle of maximum entropy, $\Delta S_{\mathrm{total}} \geq 0$. In other terms, for

$$U_0 - U_1 + p_0(V_0 - V_1) + Q_{\mathrm{RT}} = 0, \tag{3.16}$$

we should have

$$S_0 - S_1 + \frac{Q_{\mathrm{RT}}}{T_0} \geq 0. \tag{3.17}$$

Replacing Q_{RT}, obtained from (3.16), into (3.17), we arrive at the principle o maximum entropy in the form

$$(S_1 - S_0) - \frac{1}{T_0}(U_1 - U_0) - \frac{p_0}{T_0}(V_1 - V_0) \leq 0, \tag{3.18}$$

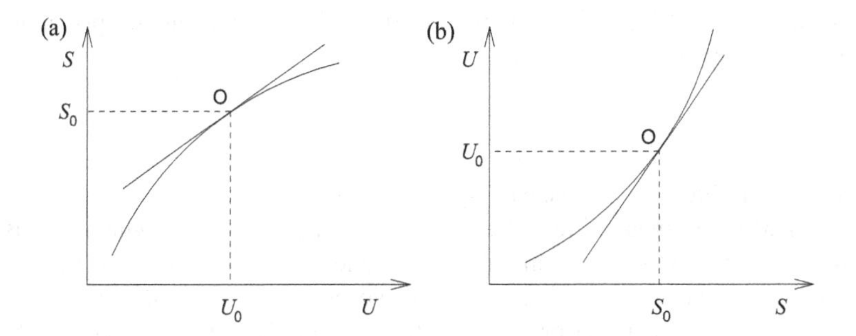

Fig. 3.3 Illustration of the equivalence between the principle of maximum entropy and the principle of minimum energy. (**a**) The curve $S(U)$ satisfies the inequality $S(U) \leq S_0 + (1/T_0)(U - U_0)$, whose right hand side describes the tangent straight line at O. (**b**) The curve $U(S)$ satisfies the inequality $U(S) \geq U_0 + T_0(S - S_0)$, whose right hand side describes the tangent straight line at O. The functions $S(U)$ an $U(S)$ form a pair of inverse functions

which we call Clausius-Gibbs principle, as illustrated in Fig. 3.3a for the case $V_1 = V_0$.

Principle of Minimum Energy

Suppose again that the set of various bodies undergoes a thermodynamic process that leads the set from the state of constrained equilibrium to the state of unconstrained equilibrium. However, the process that interests us here is the one in which the total entropy is kept constant. For the occurrence of this situation, the total energy can not remain invariant. In fact, according to the principle of minimum energy, the total energy should decrease or remain the same. In other words, this principle states that *the energy of the state of unconstrained equilibrium is less than or equal to that of the state of constrained equilibrium, as long as these states have the same entropy*.

We consider again, the gas confined in a cylinder vessel, initially separated from the thermal and mechanical reservoirs. Then they are placed in contact and reach equilibrium by a trajectory such that the total variation of entropy is zero. According to the principle of minimum energy, the total energy should decrease or remain constant. That is, for $\Delta S_{\text{total}} = 0$, we should have $\Delta U_{\text{total}} \leq 0$. In other words, for

$$S_0 - S_1 + \frac{Q_{\text{RT}}}{T_0} = 0, \tag{3.19}$$

we should have

$$U_0 - U_1 + p_0(V_0 - V_1) + Q_{\text{RT}} \leq 0. \tag{3.20}$$

Replacing Q_{RT}, obtained form (3.19), into (3.20), we arrive at the principle of minimum energy in the form

$$(U_1 - U_0) + p_0(V_1 - V_0) - T_0(S_1 - S_0) \geq 0, \tag{3.21}$$

as illustrated in Fig. 3.3b for the case $V_1 = V_0$.

Importantly, the inequalities (3.18) and (3.21) are equivalent, which means to say that the principles of maximum entropy and minimum energy in the forms given by these two inequalities are in fact two expressions of one and the same principle, which we call the principle of Clausius-Gibbs. The equivalence of the two inequalities for the case where the volume remains constant becomes evident from Fig. 3.3.

Thermal Stability

We interpret the inequalities (3.18) and (3.21) as the general stability conditions of a thermodynamic system. To study specific stabilities, we consider special cases of these inequalities. We imagine that a gas is taken from a reference state with energy U, entropy S and volume V, and led to another state with energy U', entropy S' and volume V', through a particular thermodynamic process. Defining $\Delta U = U' - U$ as the change in energy, $\Delta S = S' - S$ as the change in entropy and $\Delta V = V' - V$ as the change in the volume, then the inequality (3.21) is written as

$$\Delta U + p\Delta V - T\Delta S \geq 0, \tag{3.22}$$

where T and p are the temperature and pressure of the reference state.

Stabilities of various kinds can be examined depending on the type of disturbance made on the system. First, we analyze the thermal stability, which corresponds to the case in which the gas exchanges heat without performing work. That is, the gas suffers an isochoric disturbance. In this case, $\Delta V = 0$, so that the inequality (3.22) becomes

$$\Delta U - T\Delta S \geq 0. \tag{3.23}$$

Up to terms of second order in ΔS, the variation in energy, along an isochoric line is calculated from

$$\Delta U = T\Delta S + \frac{1}{2}\left(\frac{\partial T}{\partial S}\right)_V (\Delta S)^2 \tag{3.24}$$

where we used the equality $(\partial U/\partial S)_V = T$. Replacing this result into the inequality (3.23) gives

$$\left(\frac{\partial T}{\partial S}\right)_V (\Delta S)^2 \geq 0, \tag{3.25}$$

from which we conclude that

$$C_v = T\left(\frac{\partial S}{\partial T}\right)_V \geq 0. \tag{3.26}$$

That is, the stability by isochoric perturbation implies that the isochoric heat capacity cannot be negative.

Mechanical Stability

Now we consider the mechanical stability, which corresponds to an adiabatic perturbation. In this case, $\Delta S = 0$, so that the inequality (3.22) is written as

$$\Delta U + p\Delta V \geq 0. \tag{3.27}$$

If we consider only small variation in the volume, then

$$\Delta U = -p\Delta V - \frac{1}{2}\left(\frac{\partial p}{\partial V}\right)_S (\Delta V)^2, \tag{3.28}$$

where we used the equality $(\partial U/\partial V)_S = -p$. Replacing this result into the inequality (3.27) gives

$$-\left(\frac{\partial p}{\partial V}\right)_S (\Delta V)^2 \geq 0, \tag{3.29}$$

from which we get

$$\kappa_s = -\frac{1}{V}\left(\frac{\partial V}{\partial p}\right)_S \geq 0. \tag{3.30}$$

That is, the stability by adiabatic perturbation implies that the adiabatic compressibility cannot be negative. In other words, the adiabatic expansion (increase of the volume) of a gas implies the decrease in pressure.

Isothermal Stability

Let us assume now that the disturbance is such that the temperature of the gas remains constant and equal to T, that is, an isothermal disturbance. In this case, the variation in the internal energy of the gas is $\Delta U = Q_{\mathrm{T}} - W_{\mathrm{T}}$, where $Q_{\mathrm{T}} = T\Delta S$ is the heat introduced isothermally and W_{T} is the work done by the gas along an isotherm at temperature T. Considering small changes in the volume, then the pressure varies according to

$$p' = p + \left(\frac{\partial p}{\partial V}\right)_T \Delta V. \tag{3.31}$$

From this expression we get the isothermal work, given by

$$W_{\mathrm{T}} = p\Delta V + \frac{1}{2}\left(\frac{\partial p}{\partial V}\right)_T (\Delta V)^2. \tag{3.32}$$

Therefore, for an isothermal perturbation

$$\Delta U = Q_{\mathrm{T}} - W_{\mathrm{T}} = T\Delta S - p\Delta V - \frac{1}{2}\left(\frac{\partial p}{\partial V}\right)_T (\Delta V)^2, \tag{3.33}$$

which, replaced in (3.22), gives

$$-\left(\frac{\partial p}{\partial V}\right)_T (\Delta V)^2 \geq 0, \tag{3.34}$$

from which we conclude that

$$\kappa_T = -\frac{1}{V}\left(\frac{\partial V}{\partial p}\right)_T \geq 0. \tag{3.35}$$

That is, the stability by an isothermal perturbation implies that the isothermal compressibility cannot be negative. In other words, the isothermal expansion of a gas implies a decrease in pressure.

Isobaric Stability

In this case, the disturbance is such that the gas pressure remains the same, which corresponds to an isobaric disturbance. The variation of internal energy of the gas will then be $\Delta U = Q_{\mathrm{p}} - W_{\mathrm{p}}$, where $W_{\mathrm{p}} = p\Delta V$ is the isobaric work done by the gas and Q_{p} is the heat introduced at constant pressure. For small values of heat introduced the temperature varies with entropy according to

$$T' = T + \left(\frac{\partial T}{\partial S}\right)_p \Delta S, \tag{3.36}$$

from which we obtain the result

$$Q_{\mathrm{p}} = T\Delta S + \frac{1}{2}\left(\frac{\partial T}{\partial S}\right)_p (\Delta S)^2. \tag{3.37}$$

Therefore, for an isobaric perturbation,

$$\Delta U = Q_{\mathrm{p}} - W_{\mathrm{p}} = T\Delta S + \frac{1}{2}\left(\frac{\partial T}{\partial S}\right)_p (\Delta S)^2 - p\Delta V, \tag{3.38}$$

which, replaced in (3.22), gives

$$\left(\frac{\partial T}{\partial S}\right)_p (\Delta S)^2 \geq 0, \tag{3.39}$$

from which follows the result

$$C_p = T\left(\frac{\partial S}{\partial T}\right)_p \geq 0. \tag{3.40}$$

That is, the stability by an isobaric perturbation implies that the isobaric heat capacity cannot be negative.

3.3 Second Law of Thermodynamics

Statements and Meanings

The second law of thermodynamics is the law that distinguishes thermodynamics from the other areas of physics. At the same time, it is a fundamental law of nature that has implications that go beyond the field of physics. Many are their statements and their meanings. It was introduced independently by Clausius and Kelvin from the ideas of Carnot on heat engines and cyclic processes. For this reason, the original statements of this fundamental law refers to the operation of heat engines and cyclic processes. The statement associated with the name of Kelvin says that no process is possible whose sole result is the conversion of heat, extracted from a thermal reservoir, into work. In other words, it is impossible to construct a heat engine that converts all the heat received into work. The statement associated with the name of Clausius states that no process is possible whose sole result is the transfer of heat from a cooler to a hotter body. In other words, it is impossible to build a refrigerator that operates without the consumption of work. The further development of thermodynamic theory led to the Clausius definition of entropy as a state function and the formulation of the second law as a principle of maximum entropy, which is another form of representation of the second law.

The second law of thermodynamics must be understood as consisting of three parts. The first one leads to the definition of absolute temperature and entropy. Combined with the principle of conservation of energy, it allows us to set up the thermodynamic space and introduce the fundamental relation of systems in equilibrium. The second part consists of the form postulated by Gibbs for the principle of maximum entropy. It leads us to the property of convexity of the entropy and the conditions of stability of thermodynamic systems in equilibrium. These two parts refer only to equilibrium states and are represented by the Carnot principle and by the Clausius-Gibbs principle, respectively. These two principles altogether form the second law of thermodynamics for equilibrium systems.

The third part of the second law of thermodynamics refers to the time evolution of thermodynamic systems and has to do with the growth of entropy in spontaneous and irreversible processes. It therefore corresponds to the dynamic aspect of the principle of maximum entropy. We emphasize, however, that this aspect is not used explicitly, since we treat only systems in thermodynamic equilibrium. In contrast, the static aspect of the maximum entropy principle is used explicitly in the form given by the Clausius-Gibbs principle. When the initial and final states of an irreversible process are made up of equilibrium states a close relationship exists between the two aspects, which we analyze below.

Spontaneous Processes

When two bodies at different temperatures are placed in thermal contact, a flow of heat spontaneously establishes from the hotter to the colder body until the temperatures become equal and the flow ceases. From that moment on, the bodies remain in thermal equilibrium. Before being placed in contact, each body separately is in thermodynamic equilibrium. The moment they are brought into contact, the two bodies are in an state out of equilibrium that evolves into a state of final equilibrium. It is important to note that we are considering bodies constituting an isolated system, which does not exchange heat or work, that is, such that the total energy and volume remain invariant.

We can imagine the more general case where several bodies, or various parts of the same system, are separated by the existence of thermal constraints, arising from adiabatic walls, or mechanical constraints, due to rigid walls. When the constraints are removed while the system as a whole is isolated, they evolve spontaneously to the unrestricted equilibrium. A spontaneous process is therefore a process occurring in an isolated system, whose initial state is a state of constrained equilibrium and whose final state is a state of unconstrained equilibrium. The initial and final states have the same energy and the same volume since the system is isolated.

The principle of maximum entropy leads to conclude that *in a spontaneous process the entropy increases*. As an example, consider a body in thermal contact with a reservoir at a temperature T_0. Suppose that the body finds itself initially at a temperature T_1 and the thermal contact takes place at constant volume. The variation of the total energy of the reservoir and body is given by

$$\Delta U_{\text{total}} = U_0 - U_1 + Q_{\text{RT}}, \tag{3.41}$$

where U_0 and U_1 are the final and initial energies of the body and Q_{RT} is the heat exchanged with the reservoir. The variation of the total entropy of the body and reservoir is given by

$$\Delta S_{\text{total}} = S_0 - S_1 + \frac{Q_{\text{RT}}}{T_0}, \tag{3.42}$$

where S_0 and S_1 are the final and initial entropies of the body and Q_{RT}/T_0 is the variation of the entropy of the reservoir. Eliminating Q_{RT} from these two equations, we reach the following relation

$$\Delta S_{\text{total}} = S_0 - S_1 - \frac{1}{T_0}(U_0 - U_1) + \frac{1}{T_0}\Delta U_{\text{total}}. \tag{3.43}$$

Along an isochoric line

$$U_0 - U_1 = \int_{T_1}^{T_0} C_v dT \tag{3.44}$$

and

$$S_0 - S_1 = \int_{T_1}^{T_0} \frac{C_v}{T} dT, \tag{3.45}$$

where C_v is the isochoric heat capacity of the body. Therefore,

$$\Delta S_{\text{total}} = \int_{T_1}^{T_0} (\frac{1}{T} - \frac{1}{T_0}) C_v dT + \frac{1}{T_0}\Delta U_{\text{total}}. \tag{3.46}$$

Considering that the total energy of the system composed by the body and the reservoir is constant, then $\Delta U_{\text{total}} = 0$ and therefore the total variation of entropy is given by

$$\Delta S_{\text{total}} = \int_{T_1}^{T_0} (\frac{1}{T} - \frac{1}{T_0}) C_v dT. \tag{3.47}$$

If $T_1 < T_0$, then the expression in parentheses is positive, so that the integral is positive because, since $C_v \geq 0$. If $T_1 > T_0$, then the expression in parentheses is negative. But in this case, the lower limit of the integral is greater than the upper limit, so that the integral is also positive. If $T_1 = T_0$ the integral vanishes. Therefore, in all cases $\Delta S_{\text{total}} \geq 0$. Notice that in the second case, $T_1 > T_0$, the energy and the entropy of the body decrease.

Maximum Work

Suppose now that the body reaches the final state by a process in which the total energy is not maintained constant. We imagine that the difference in total energy is transformed into work W_{total}, performed against the outside. Thus, the variation of the total energy can be nonzero and moreover $W_{\text{total}} = -\Delta U_{\text{total}}$. We may ask about the maximum work that the composite system can perform when the body is taken from the initial state to the final equilibrium state with the reservoir.

From (3.46), we obtain

$$W_{\text{total}} = -\Delta U_{\text{total}} = T_0 \int_{T_1}^{T_0} (\frac{1}{T} - \frac{1}{T_0}) C_v dT - T_0 \Delta S_{\text{total}}. \tag{3.48}$$

Taking into account that $\Delta S_{\text{total}} \geq 0$, then the maximum work is obtained when the process is such that $\Delta S_{\text{total}} = 0$. For this case the maximum work is

$$W_{\text{total}} = T_0 \int_{T_1}^{T_0} (\frac{1}{T} - \frac{1}{T_0}) C_v dT \tag{3.49}$$

and is clearly positive.

Microscopic Interpretation of Entropy

Entropy has a microscopic interpretation that is used as the foundation of statistical mechanics. According to Boltzmann entropy is related to the number of microscopic states available to the system. The higher the number W of accessible microscopic states the higher the entropy S. However, entropy does not grow linearly with the number of accessible states but grows logarithmically, that is,

$$S = k_B \ln W, \tag{3.50}$$

where k_B is a universal proportionality constant called Boltzmann constant.

Let us examine what happens to entropy when an ideal gas expands freely. Imagine a vessel consisting of two compartments separated by a rigid and adiabatic wall. An ideal gas at a temperature T occupies one compartment whereas the other is empty. In a given moment, the separation wall is removed and the gas expands freely occupying the whole vessel and reaching equilibrium. Because the system is isolated, there is no variation of energy, so that the final temperature will be the same as the energy of an ideal gas depends only on the temperature. We suppose that the volume V of the vessel is ℓ times greater than the volume V_1 of the compartment where the gas was, that is, $V = \ell V_1$. As the gas is ideal, the final pressure p will be ℓ times smaller than the initial pressure p_1, that is, $p = p_1/\ell$. Using (2.44) for the entropy of an ideal gas, we obtain the following variation of entropy

$$\Delta S = NR \ln \ell. \tag{3.51}$$

After the expansion, each molecule of the gas may wander in a space ℓ times greater which means that it has a number of possibilities, or microscopic states, ℓ times greater. The gas as a whole has, therefore, a number of microscopic states ℓ^n times greater, where n is the number of gas molecules. The ratio W/W_1 between the final number W and the initial number W_1 of microscopic states is, therefore, $W/W_1 = \ell^n$. Using the Boltzmann formula (3.50), we get for the variation of entropy the expression

$$\Delta S = k_B \ln W - k_B \ln W_1 = k_B \ln \frac{W}{W_1}, \tag{3.52}$$

or

$$\Delta S = k_B \ln \ell^n = n k_B \ln \ell, \tag{3.53}$$

which is identical to the expression (3.51).

The number of molecules divided by the number of moles is equal to the number of molecules in one mole N_A, called Avogadro constant. Comparing expressions (3.51) and (3.53), we see that the Boltzmann constant k_B and the universal gas constant R are related by

$$k_B = \frac{R}{N_A}. \tag{3.54}$$

The knowledge of the Avogadro constant allows the determination of the Boltzmann constant. The value of the Avogadro constant is

$$N_A = 6,0221367 \times 10^{23}\,\text{mol}^{-1} \tag{3.55}$$

and of the Boltzmann constant is

$$k_B = 1.380658 \times 10^{-23}\,\text{J/K}. \tag{3.56}$$

Problems

3.1 Determine C_v, C_p, κ_T, κ_S and α of an ideal gas.

3.2 The coefficient of linear thermal expansion of a body is defined by $\beta = (1/L)(\partial L/\partial T)_p$ where L is the length of the body along a certain direction. Show that for an isotropic body $\alpha = 3\beta$. If the length of a metallic stem at temperature T_0 is L_0, show that for small values of $\Delta T = T - T_0$ the length of the stem varies according to $L = L_0(1 + \beta\Delta T)$.

3.3 Determine the variation of the entropy of an ideal gas along an isochoric line and along an isobaric line from the isochoric and isobaric heat capacities, respectively.

3.4 Find the variation of entropy along an isobaric line between the temperatures T_1 and T_2 of a body whose heat capacity at constant pressure varies according to $C_p = A + BT$.

3.5 Determine the variation of the entropy of a body when it is placed in thermal contact with a heat reservoir at temperature T_0. Suppose that initially the body is at temperature T_1 and that the isochoric heat capacity is constant. Find the variation of the total entropy ΔS_{total} and show explicitly that $\Delta S_{\text{total}} \geq 0$. Determine also the maximum work that can be obtained when the body passes from the initial state to the equilibrium state with the reservoir.

3.6 Two identical bodies are found at temperatures T_1 and T_2. They are placed in thermal contact and eventually reach the state of equilibrium with each other. Determine the equilibrium temperature and the variation of the total entropy ΔS_{total}. Show explicitly that $\Delta S_{\text{total}} \geq 0$. Find next the maximum work that one can extract form these bodies and determine, in this case, the final equilibrium temperature. Which temperature is the biggest? Suppose that the isochoric heat capacities of the bodies are constant.

3.7 Solve the previous problem considering three identical bodies at temperatures T_1, T_2 and T_3.

3.8 A body initially at a temperature T_1 is placed successively in contact with a series of n reservoirs numbered from 1 to n. The last reservoir, of number n, has temperature T_0 and the temperature difference between successive reservoirs is $\Delta T = (T_1 - T_0)/n$. Determine the entropy variation of the body, of the reservoirs, and the total entropy variation in the limit $n \to \infty$. Suppose next that the body at temperature T_1 is placed in contact with the last reservoir. Determine in this case the entropy variation of the body, of the reservoir and the total entropy variation. Compare the results of both cases. Suppose that the isochoric heat capacity of the body is constant.

3.9 Use the principle of minimum energy to show that

$$U_{11}(\Delta S)^2 + 2U_{12}(\Delta S)(\Delta V) + U_{22}(\Delta V)^2 \geq 0,$$

where U_{11}, U_{12} and U_{22} are the second order derivatives of U with respect to S and V. From this inequality, show that $U_{11} \geq 0$, $U_{22} \geq 0$ and $U_{11}U_{22} - U_{12}^2 \geq 0$.

3.10 Use the principle of maximum entropy to show that

$$S_{11}(\Delta U)^2 + 2S_{12}(\Delta U)(\Delta V) + S_{22}(\Delta V)^2 \leq 0.$$

where S_{11}, S_{12} and S_{22} are the second order derivatives of S with respect to U and V. From this inequality, show that $S_{11} \leq 0$, $S_{22} \leq 0$ and $S_{11}S_{22} - S_{12}^2 \geq 0$.

Chapter 4
Thermodynamic Potentials

4.1 Fundamental Relation

Energy Representation

The thermodynamic properties of an isotropic fluid, confined in a vessel, involve several thermodynamic quantities such as pressure, temperature, volume, the internal energy and entropy. These quantities are not all independent. To determine how many and which may be treated as independent we start by examining the equation

$$dU = TdS - pdV. \tag{4.1}$$

We see that it is natural to imagine as U as dependent on S and V. Furthermore, if we understand the internal energy $U(S, V)$ as a function of S and V, then this equation allows us to obtain also the pressure and temperature as functions of S and V through the relations

$$T = \frac{\partial U}{\partial S}, \qquad p = -\frac{\partial U}{\partial V}. \tag{4.2}$$

The reasonings above were done by considering a given quantity of material contained in the vessel. However, the amount of substance that composes the fluid may also be changed. To completely specify the thermodynamic state of the fluid we should therefore indicate the number of moles N of the substance in the vessel in addition to S and V. The thermodynamic equilibrium state becomes defined by the independent variables S, V and N. In an equivalent manner, we may say that each state of thermodynamic equilibrium of an isotropic fluid composed by a single pure substance corresponds to a point in the thermodynamic space (S, V, N). Note that we are assuming the fluid as composed by a single pure substance. For a fluid

M.J. de Oliveira, *Equilibrium Thermodynamics*, Graduate Texts in Physics,
DOI 10.1007/978-3-662-53207-2_4

comprising several substances it would be necessary to indicate the number of moles of each substance.

Since U is also a function of N, it is convenient to define a quantity that is associated with the variation of energy when one makes an increase in the number of moles of the pure substance that makes up the fluid. This quantity is called chemical potential μ and is defined by

$$\mu = \frac{\partial U}{\partial N}, \tag{4.3}$$

so that we may write the conservation of energy in differential form as

$$dU = TdS - pdV + \mu dN. \tag{4.4}$$

The internal energy $U(S, V, N)$ as a function of the independent variables S, V and N is called the fundamental relation in the energy representation because from it we can obtain the thermodynamic properties of the system in consideration. Deriving the energy with respect to S, V, and N, we obtain equations (4.2) and (4.3), called equations of state in the energy representation. We assume as a postulate that the internal energy U and its derivatives T, p and μ are continuous functions or, equivalently, that the internal energy is a continuous and differentiable function of entropy, volume, and number of moles.

Entropy Representation

Inverting the fundamental relation $U(S, V, N)$ with respect to S, we get the entropy $S(U, V, N)$ as a function of the independent variables U, V, and N, which is the fundamental relation in the entropy representation because from it we can obtain the thermodynamic properties of the system in consideration. In this representation, the thermodynamic state is defined by a point in the space (U, V, N). From (4.4) we may write

$$dS = \frac{1}{T}dU + \frac{p}{T}dV - \frac{\mu}{T}dN, \tag{4.5}$$

from which follow the three equations of state

$$\frac{1}{T} = \frac{\partial S}{\partial U}, \qquad \frac{p}{T} = \frac{\partial S}{\partial V}, \qquad \frac{\mu}{T} = -\frac{\partial S}{\partial N}, \tag{4.6}$$

in the entropy representation.

Since $S(U)$ is the inverse of $U(S)$ and, being the energy continuous and differentiable in S, V and N, then the entropy is a continuous and differentiable function of the energy, volume, and number of moles.

4.2 Extensivity

Extensive Variables

The thermodynamic systems exhibit the property of extensivity. The main consequence of this property is a reduction in number of variables necessary for the description of a thermodynamic system. For the analysis of the property of extensivity, we should initially distinguish the variables that are extensible from those that we call thermodynamic fields, which are not extensive. The volume V, the internal energy U, the entropy S and the number of moles N are extensive variables. The pressure p, the temperature T and the chemical potential μ are thermodynamic fields. In a system in equilibrium, the thermodynamic fields have the same value in each part of the system, regardless of the size of each part. The extensive variables on the other hand take on values that depend on the size of each part.

Let us imagine two copies of the same system such that one of them is an enlargement of the other in the following sense. The entropies, the volumes, and the number of moles of the two copies are related by $S_2 = \lambda S_1$, $V_2 = \lambda V_1$ and $N_2 = \lambda N_1$, where λ is the enlargement factor. The system is described by the fundamental equation $U(S, V, N)$ so that the energies are given by $U_2 = U(S_2, V_2, N_2)$ and $U_1 = U(S_1, V_1, N_1)$. It exhibits the property of extensivity if the energies U_1 and U_2 are also related by $U_2 = \lambda U_1$. This means to say that the fundamental relation obeys the equation

$$\lambda U(S, V, N) = U(\lambda S, \lambda V, \lambda N), \tag{4.7}$$

for any positive values of λ. That is, U must be a homogeneous first order function of S, V, and N.

Deriving both sides of (4.7) with respect to λ and then setting $\lambda = 1$, we get

$$U = TS - pV + \mu N, \tag{4.8}$$

which is the Euler equation. From the fundamental relation, the equations of state are obtained by differentiation. If, however, we do not know the fundamental relation, but the three equations of state, that is, T, p, and μ as functions of S, V and N, then it can be obtained by integration or by replacing into the Euler equation. In fact, the knowledge of two equations of state plus the property of extensivity suffices to obtain the fundamental relation.

Molar Quantities

The property of extensivity, given by relation (4.7), enables the description of a thermodynamic system by a smaller number of independent variables. Such a description involves quantities called thermodynamic densities, defined as the ratio of extensive quantities. Although they are not extensive, the thermodynamic densities are a class of thermodynamic variables that must be distinguished from the class of thermodynamic fields. Particularly useful are the thermodynamic densities obtained by dividing the extensive quantities by the number of moles N, called molar quantities. Thus, we define the energy per mole, or molar energy, $u = U/N$, entropy per mole, or molar entropy, $s = S/N$ and the volume per mole, or molar volume, $v = V/N$.

If in the relation (4.7) we set $\lambda = 1/N$, then the left hand side becomes the molar energy u and the right hand side becomes a function of two variables only: the molar entropy s and molar volume v. We may conclude that $u(s, v)$ is a function s and v only and that

$$U(S, V, N) = Nu(\frac{S}{N}, \frac{V}{N}). \tag{4.9}$$

Therefore, if we know $u(s, v)$, the fundamental relation can immediately be recovered by the above formula. That is, $u(s, v)$ is equivalent to the fundamental relationship.

From (4.2) and (4.9) we conclude that

$$T = \frac{\partial u}{\partial s}, \qquad p = -\frac{\partial u}{\partial v}, \tag{4.10}$$

which allows to write the differential form for the molar energy,

$$du = Tds - pdv. \tag{4.11}$$

The extensivity of $S(U, V, N)$ allows to conclude also that the thermodynamic properties can be determined from the molar entropy $s(u, v)$ as a function of the molar energy u and the molar volume v. If we know $s(u, v)$, we can recover the fundamental relation in the entropy representation through

$$S(U, V, N) = Ns(\frac{U}{N}, \frac{V}{N}). \tag{4.12}$$

From this equation and (4.6), we may conclude that

$$\frac{1}{T} = \frac{\partial s}{\partial u}, \qquad \frac{p}{T} = \frac{\partial s}{\partial v}, \tag{4.13}$$

which allows us to write the differential form,

$$ds = \frac{1}{T}du + \frac{p}{T}dv, \tag{4.14}$$

for the molar entropy.

The molar heat capacities $c_v = C_v/N$ and $c_p = C_p/N$, or specific heats, as well as κ_T, κ_s and α, may be obtained from the molar quantities

$$c_v = T\left(\frac{\partial s}{\partial T}\right)_v, \qquad c_p = T\left(\frac{\partial s}{\partial T}\right)_p, \tag{4.15}$$

$$\kappa_T = -\frac{1}{v}\left(\frac{\partial v}{\partial p}\right)_T, \qquad \kappa_s = -\frac{1}{v}\left(\frac{\partial v}{\partial p}\right)_s, \qquad \alpha = \frac{1}{v}\left(\frac{\partial v}{\partial T}\right)_p. \tag{4.16}$$

Ideal Gas

Let us consider, as an example, an ideal gas whose equations of state $pv = RT$ and $u = cT$ are written in the form

$$\frac{1}{T} = \frac{c}{u}, \qquad \frac{p}{T} = \frac{R}{v}. \tag{4.17}$$

Therefore, according to (4.13), we obtain

$$\frac{\partial s}{\partial u} = \frac{c}{u}, \qquad \frac{\partial s}{\partial v} = \frac{R}{v}, \tag{4.18}$$

from which we obtain s by the path integral

$$s = s_0 + \int \{\frac{1}{T}du + \frac{p}{T}dv\} = s_0 + \int \{\frac{c}{u}du + \frac{R}{v}dv\}. \tag{4.19}$$

Choosing a any path that begins at the reference point (u_0, v_0), we get

$$s = s_0 + c\ln\frac{u}{u_0} + R\ln\frac{v}{v_0}. \tag{4.20}$$

Inverting this relation,

$$u = u_0 \exp\{\frac{1}{c}[s - s_0 - R\ln\frac{v}{v_0}]\}. \tag{4.21}$$

Therefore, the fundamental relations of an ideal gas in the entropy and energy representations are, respectively

$$S = N\{s_0 + c\ln\frac{U}{Nu_0} + R\ln\frac{V}{Nv_0}\} \tag{4.22}$$

and

$$U = Nu_0 \exp\{\frac{1}{c}[\frac{S}{N} - s_0 - R\ln\frac{V}{Nv_0}]\}. \tag{4.23}$$

Performing the derivatives of U with respect to S and V, we recover the equations of state of the ideal gas, $T = U/Nc$ and $p = RU/cV$. Indeed, the first gives $U = NcT$ which, substituted in the second, gives $p = RNT/V$.

van der Waals Fluid

As a second example, let us consider a van der Waals fluid, that is a fluid that obeys the van der Waals equation

$$p = \frac{RT}{v-b} - \frac{a}{v^2} \tag{4.24}$$

and the following equation for the molar internal energy

$$u = cT - \frac{a}{v}. \tag{4.25}$$

From (4.25), we get

$$\frac{1}{T} = \frac{cv}{uv+a}, \tag{4.26}$$

which can be considered an equation of state in the entropy representation, since the right hand side depend on u and v. From equation (4.24), we get

$$\frac{p}{T} = \frac{R}{v-b} - \frac{a}{v^2}\frac{1}{T}, \tag{4.27}$$

or, using equation (4.26),

$$\frac{p}{T} = \frac{R}{v-b} - \frac{ac}{v(uv+a)}, \tag{4.28}$$

or

$$\frac{p}{T} = \frac{R}{v-b} - \frac{c}{v} + \frac{cu}{uv+a}, \tag{4.29}$$

which can be considered an equation of state in the entropy representation, since the right hand side depend on u and v. Integrating these equations of state from a reference state (u_0, v_0), for which the molar entropy is s_0, we have

$$s = s_0 + c \ln \frac{(uv + a)v_0}{(u_0 v_0 + a)v} + R \ln \frac{v - b}{v_0 - b}. \tag{4.30}$$

To obtain the fundamental relation $S(U, V, N)$, it suffices to do the replacements $s = S/N$, $u = U/N$ and $v = V/N$,

$$S = Ns_0 + Nc \ln \frac{(UV + N^2 a)v_0}{(u_0 v_0 + a)NV} + NR \ln \frac{V - Nb}{(v_0 - b)N}. \tag{4.31}$$

4.3 Legendre Transformations

Helmholtz Free Energy

The use of the Legendre transformation allows us to get other representations of the fundamental relation. In other words, it allows the use of other thermodynamic quantities as independent variables. Let us begin by performing the change $(S, V, N) \to (T, V, N)$. In this new representation the fundamental relation is given by the Helmholtz free energy $F(T, V, N)$, defined by the Legendre transformation

$$F(T) = \min_S \{U(S) - TS\}, \tag{4.32}$$

which should be understood as follows. Given a certain value of temperature, we vary the entropy until we reach the minimum value of the expression inside the curls, that is, until we find the least difference between $U(S)$ and TS. Since $U(S)$ is differentiable, the minimum occurs when $U'(S) = T$. Inverting the relation, we get $S(T)$ which, replaced in (4.32), gives

$$F(T) = U(S) - TS. \tag{4.33}$$

From (4.33) we obtain $F'(T) = -S$. Therefore the derivatives $T(S) = \partial U/\partial S$ and $S(T) = -\partial F/\partial T$ constitute a pair of inverse functions, as can be seen in Fig. 4.1. This result provides a way of constructing the Legendre transformation: (a) initially, we determine the derivative $T(S)$ of U with respect to S; (b) we take the inverse of $T(S)$ to obtain $S(T)$ which is the derivative of $-F$ with respect to T; (c) the integral of $S(T)$ gives $F(T)$ with the opposite sign, except for a constant.

From the relations (4.33) and (4.4), we find

$$dF = -SdT - pdV + \mu dN, \tag{4.34}$$

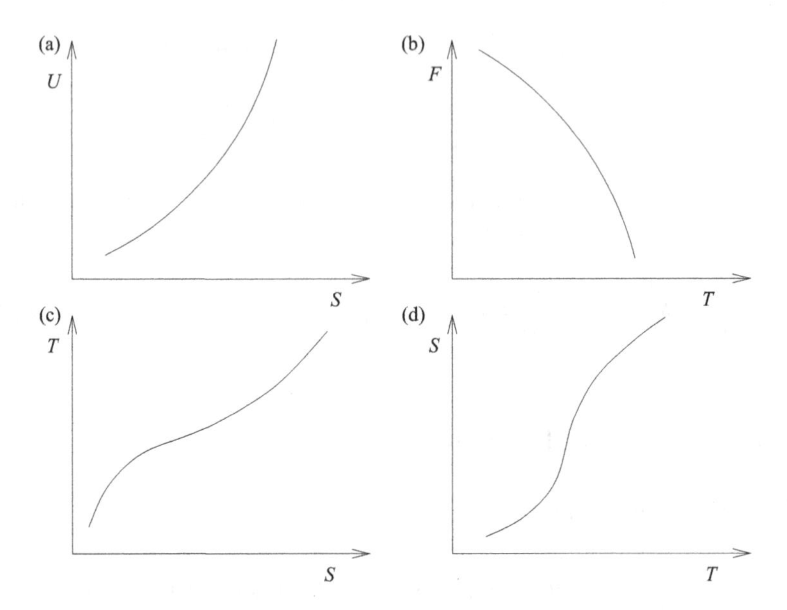

Fig. 4.1 Function $U(S)$ (**a**) and its Lengendre transform $F(T)$ (**b**). Their derivatives (**c**) $T = \partial U/\partial S$ and (**d**) $S = -\partial F/\partial T$ are inverse functions of each other

or equivalently

$$S = -\left(\frac{\partial F}{\partial T}\right)_{VN}, \qquad p = -\left(\frac{\partial F}{\partial V}\right)_{TN}, \qquad \mu = \left(\frac{\partial F}{\partial N}\right)_{TV}, \tag{4.35}$$

which are the equations of state.

Enthalpy

The thermodynamic potential called enthalpy $H(S, p, N)$ is defined by the Legendre transformation

$$H(p) = \min_{V}\{U(V) + pV\}. \tag{4.36}$$

and corresponds to the change of independent variables $(S, V, N) \to (S, p, N)$. Since $U(V)$ is differentiable, the minimum occurs when $U'(V) = -p$. Inverting $p(V)$, we obtain $V(p)$ which, replaced in the expression between curls, gives

$$H(p) = U(V) + pV. \tag{4.37}$$

Deriving with respect to p, we have $H'(p) = V$, so that the derivatives $p(V) = -\partial U/\partial V$ and $V(p) = \partial H/\partial p$ are a pair of inverse functions.

From the relations (4.37) and (4.4), we find

$$dH = TdS + Vdp + \mu dN, \tag{4.38}$$

or equivalently

$$T = \left(\frac{\partial H}{\partial S}\right)_{pN}, \qquad V = \left(\frac{\partial H}{\partial p}\right)_{SN}, \qquad \mu = \left(\frac{\partial H}{\partial N}\right)_{Sp}, \tag{4.39}$$

which are the equations of state.

Gibbs Free Energy

The thermodynamic potential called Gibbs free energy $G(T, p, N)$ is obtained from the Helmholtz free energy by the Legendre transformation

$$G(p) = \min_V \{F(V) + pV\}. \tag{4.40}$$

Using relation (4.34), we get

$$dG = -SdT + Vdp + \mu dN, \tag{4.41}$$

or equivalently

$$S = -\left(\frac{\partial G}{\partial T}\right)_{pN}, \qquad V = \left(\frac{\partial G}{\partial p}\right)_{TN}, \qquad \mu = \left(\frac{\partial G}{\partial N}\right)_{Tp}, \tag{4.42}$$

which are the equations of state. Alternatively, we may define the Gibbs free energy from the enthalpy by the Legendre transformation

$$G(T) = \min_S \{H(S) - TS\}, \tag{4.43}$$

which lead us again to relation (4.41) if we use (4.38).

Grand Thermodynamic Potential

Other thermodynamic potentials can be obtained considering transformations that involve the number of moles. One of them is the grand thermodynamic potential $\Phi(T, V, \mu)$, defined by the Legendre transformation,

$$\Phi(\mu) = \min_N \{F(N) - \mu N\}, \tag{4.44}$$

obtained from the Helmholtz free energy. From relation (4.34), we find

$$d\Phi = -SdT - pdV - Nd\mu, \tag{4.45}$$

which allows to write

$$S = -\left(\frac{\partial\Phi}{\partial T}\right)_{V\mu}, \qquad p = -\left(\frac{\partial\Phi}{\partial V}\right)_{T\mu}, \qquad N = -\left(\frac{\partial\Phi}{\partial\mu}\right)_{TV}. \tag{4.46}$$

Molar Thermodynamic Potentials

The extensivity property of the Helmholtz free energy $F(T, V, N)$ allows the molar Helmholtz free energy $f(T, v)$, which is defined by $f = F/N$, to be a function of temperature T and molar volume v only. From $f(T, v)$, we can recover $F(T, V, N)$ by

$$F(T, V, N) = Nf(T, \frac{V}{N}). \tag{4.47}$$

From this relation, we get

$$s = -\frac{\partial f}{\partial T}, \qquad p = -\frac{\partial f}{\partial v}, \tag{4.48}$$

and

$$df = -sdT - pdv. \tag{4.49}$$

Similarly, from the extensivity of enthalpy, we conclude that the molar enthalpy $h = H/N$ is a function of the molar entropy s and pressure p only. From $h(s, p)$ we recover the enthalpy by

$$H(S, p, N) = Nh(\frac{S}{N}, p). \tag{4.50}$$

From this relation we get

$$T = \frac{\partial h}{\partial s}, \qquad v = \frac{\partial h}{\partial p}, \tag{4.51}$$

and

$$dh = Tds + vdp. \tag{4.52}$$

Due to the extensivity, the Gibbs free energy $G(T, p, N)$ can be written in the form

$$G(T, p, N) = Ng(T, p), \tag{4.53}$$

where g is the molar Gibbs free energy $g = G/N$ and depends only on T and p, because G is a function of just one extensive variable. From this relation, we get

$$s = -\frac{\partial g}{\partial T}, \qquad v = \frac{\partial g}{\partial p}, \tag{4.54}$$

and

$$dg = -sdT + vdp. \tag{4.55}$$

In addition, taking into account that $\mu = \partial G/\partial N$, we conclude that $\mu = g$. That is, for a system constituted by a pure substance, the chemical potential is identified with the molar Gibbs free energy. Therefore, from the last equation we obtain the Gibbs-Duhem equation

$$d\mu = -sdT + vdp. \tag{4.56}$$

The grand thermodynamic potential Φ is also a function of just one extensive variable, so that we may write

$$\Phi(T, V, \mu) = V\phi(T, \mu), \tag{4.57}$$

where ϕ depends only on the temperature T and chemical potential μ. The following relations can be easily obtained:

$$\bar{s} = -\frac{\partial \phi}{\partial T}, \qquad \bar{\rho} = -\frac{\partial \phi}{\partial \mu}, \tag{4.58}$$

where $\bar{s} = S/V$ is the entropy per unit volume and $\bar{\rho} = N/V$ is the number of moles per unit volume, or

$$d\phi = -\bar{s}dT - \bar{\rho}d\mu. \tag{4.59}$$

In addition, since $p = -\partial\Phi/\partial V$, then $p = -\phi$, so that

$$dp = \bar{s}dT + \bar{\rho}d\mu, \tag{4.60}$$

which is equivalent to the Gibbs-Duhem equation. It suffices to recall that $\bar{s} = s/v$ and $\bar{\rho} = 1/v$.

Thermodynamic Potentials of an Ideal Gas

From the internal energy of an ideal gas we can determine the Helmholtz free energy by means of a Legendre transformation. To this end, we replace $U = NcT$ in expression (4.22) to obtain the entropy

$$S = N\{s_0 + c \ln \frac{T}{T_0} + R \ln \frac{V}{Nv_0}\} \tag{4.61}$$

as a function of T and V, where $T_0 = u_0/c$. Substituting these two results in $F = U - TS$, we get

$$F = -TN\{c \ln \frac{T}{T_0} + a + R \ln \frac{V}{Nv_0}\}, \tag{4.62}$$

where $a = s_0 - c$. The Gibbs free energy is obtained from the transformation $G = F + pV$. Using $V = NRT/p$ to eliminate V, we obtain the result

$$G = N\{-bT - \gamma cT \ln \frac{T}{T_0} + RT \ln \frac{p}{p_0}\}, \tag{4.63}$$

where $b = a - R$.

The respective molar quantities are given by

$$s = s_0 + c \ln \frac{T}{T_0} + R \ln \frac{v}{v_0}, \tag{4.64}$$

$$f = -T\{c \ln \frac{T}{T_0} + a + R \ln \frac{v}{v_0}\}, \tag{4.65}$$

and

$$g = -bT - \gamma cT \ln \frac{T}{T_0} + RT \ln \frac{p}{p_0}. \tag{4.66}$$

Heat, Work and Thermodynamic Potentials

The heat received by a system in an isochoric process equals the variation of the internal energy, since the work vanishes, that is,

$$\Delta U = Q_\mathrm{v}. \tag{4.67}$$

If, on the other hand, we consider an isobaric process, then the variation of energy will be

$$\Delta U = Q_\mathrm{p} - p\Delta V, \tag{4.68}$$

because the work done is $p\Delta V$ for an isobaric process. But the variation of enthalpy in an isobaric process is $\Delta H = \Delta U + p\Delta V$ and

$$\Delta H = Q_{\mathrm{p}}, \tag{4.69}$$

that is, the heat received by the system in an isobaric process equals the variation of enthalpy.

The work consumed by a system in a quasi-static adiabatic process equals the variation of the internal energy, since there is no heat exchanged, that is,

$$\Delta U = -W_{\mathrm{ad}}. \tag{4.70}$$

If, on the other hand, we consider an isothermal process, then the heat received is given by $Q_{\mathrm{T}} = T\Delta S$, so that the variation of internal energy will be

$$\Delta U = T\Delta S - W_{\mathrm{T}}, \tag{4.71}$$

where W_{T} is the work done along an isothermal process. Taking into account that in an isothermal process the variation of the Helmholtz free energy is $\Delta F = \Delta U - T\Delta S$, then

$$\Delta F = -W_{\mathrm{T}}, \tag{4.72}$$

that is, the work consumed by a system along an isothermal process equals the variation of the Helmholtz free energy.

The isochoric heat capacity C_v is defined as the limit of the ratio $Q_{\mathrm{v}}/\Delta T$, when $\Delta T \to 0$, between the heat introduced at constant volume and the increment in temperature. Taking into account the equality (4.67), the ratio becomes equal to $\Delta U/\Delta T$ along an isochoric line process and therefore

$$C_v = (\partial U/\partial T)_V. \tag{4.73}$$

The isobaric heat capacity C_p is defined as the limit of the ratio $Q_{\mathrm{p}}/\Delta T$, when $\Delta T \to 0$, between the heat introduced at constant pressure and the increment in temperature. Taking into account the equality (4.69), the ratio becomes equal to $\Delta H/\Delta T$ along an isobaric process and therefore

$$C_p = (\partial H/\partial T)_p. \tag{4.74}$$

The molar heat capacities are related with the molar energy and molar enthalpy by $c_v = (\partial u/\partial T)_v$ and $c_p = (\partial h/\partial T)_p$.

4.4 Convexity

Clausius-Gibbs Principle

We saw in the previous chapter that the principle of minimum energy can be translated by the equation

$$U(S, V) - U_0 \geq T_0(S - S_0) - p_0(V - V_0), \tag{4.75}$$

established for a system in which the number of moles is kept constant, where $U_0 = U(S_0, V_0)$. To include processes in which the number of moles may vary, we generalize this inequality to include the number of moles. To this end, we first divide both sides of (4.75) by the number of moles N to obtain

$$u(s, v) - u_0 \geq T_0(s - s_0) - p_0(v - v_0), \tag{4.76}$$

where $u_0 = u(s_0, v_0)$.

Next, we use the Euler equation in the form $u_0 = T_0 s_0 - p_0 v_0 + \mu_0$ to eliminate u_0 in expression (4.76) and to obtain the inequality

$$u(s, v) \geq T_0 s - p_0 v + \mu_0. \tag{4.77}$$

Multiplying both members of the inequality by the number of moles N, we get

$$U(S, V, N) \geq T_0 S - p_0 V + \mu_0 N, \tag{4.78}$$

which subtracted from the Euler equation, $U_0 = T_0 S_0 - p_0 V_0 + \mu_0 N_0$, gives

$$U - U_0 \geq T_0(S - S_0) - p_0(V - V_0) + \mu_0(N - N_0), \tag{4.79}$$

which is the expression of the principle of minimum energy in all three variables.

The entropy obeys the principle of maximum entropy expressed as

$$S - S_0 \leq \frac{1}{T_0}(U - U_0) + \frac{p_0}{T_0}(V - V_0) - \frac{\mu_0}{T_0}(N - N_0), \tag{4.80}$$

which is obtained directly from the inequality (4.79). The inequalities (4.79) and (4.80) are the expression of the Clausius-Gibbs principle.

Convex and Concave Functions

The principle of minimum energy is equivalent to saying that energy is a convex function of all extensive variables jointly. Similarly, the principle of maximum

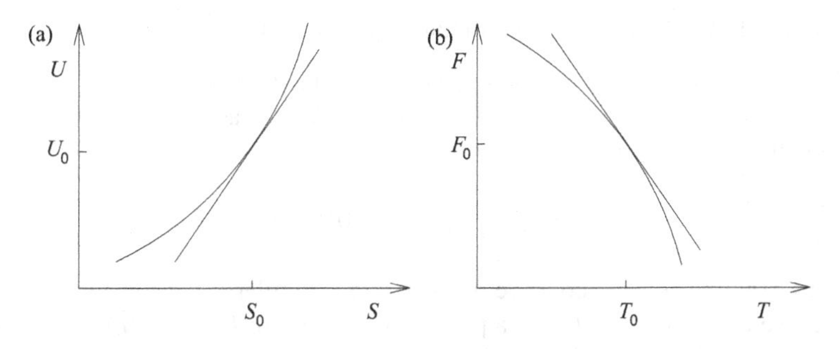

Fig. 4.2 (**a**) Example of a convex function: the internal energy $U(S)$ versus entropy S. (**b**) Example of a concave function: the Helmholtz free energy $F(T)$ versus temperature T

entropy is equivalent to saying that entropy is a concave function of all extensive variables jointly. Furthermore, an important property in respect of convex and concave functions is that a Legendre transformation converts a convex function into a concave function and vice versa. That is, both functions that are linked by a Legendre transformation form a pair of function such that one is convex and the other is concave. It is therefore appropriate to examine the properties of convex and concave functions. We analyze initially a convex function of a single variable, the internal energy $U(S)$ as a function of the entropy S, and a concave function of a single variable, the Helmholtz free energy, $F(T)$ as a function of temperature T. The variables V and N are taken as constant.

The convexity of a convex or concave function manifest itself if we consider any tangent to the curve described by the function. All the curve will be entirely on one side of the tangent line. The convex functions are those curves that lie above any tangent to the curve whereas the concave curves are those that lie below any tangent to the curves, as shown in Fig. 4.2.

Consider the tangent to the convex function $U(S)$ at the point (S_0, U_0), where $U_0 = U(S_0)$. The equation that describes the tangent line is

$$y(S) = U(S_0) + U_1(S_0)(S - S_0), \tag{4.81}$$

where $U_1(S_0)$ is the slope of the tangent line. Taking into account that $U(S)$ is differentiable, then $U_1(S_0)$ may be identified with the derivative of U calculated at $S = S_0$. Since $U(S)$ is a convex function, then $U(S) \geq y(S)$, that is,

$$U(S) \geq U(S_0) + U_1(S_0)(S - S_0), \tag{4.82}$$

valid for any value of S, as can be seen in Fig. 4.2a. This inequality is the same inequality (4.79) for the case in which the volume and the number of moles remains constant. It suffices to recall that $U_1(S_0) = T_0$.

A convex function $U(S)$ has two fundamental properties:

(a) The derivative $U_1(S)$ is monotonic increasing function of S.
(b) If the function $U(S)$ has the second derivative $U_{11}(S)$ at a certain point $S = S_0$, then the convexity condition implies

$$U_{11}(S_0) \geq 0. \tag{4.83}$$

Let us examine now the concave function $F(T)$. Similarly, we consider a tangent to the curve described by $F(T)$ at a point (T_0, F_0), where $F_0 = F(T_0)$. The heights of the points on the tangent line is given by

$$y(T) = F(T_0) + F_1(T_0)(T - T_0), \tag{4.84}$$

where $F_1(T_0)$ is the slope of the line. At the points where $F(T)$ is differentiable, $F_1(T_0)$ is identified with the derivative of $F(T)$ calculated at T_0. Since $F(T)$ is concave then $F(T) \leq y(T)$, that is,

$$F(T) \leq F(T_0) + F_1(T_0)(T - T_0), \tag{4.85}$$

valid for any value of T, as seen in Fig. 4.2b.

The concave function $F(T)$ has the fundamental properties:

(a) The derivative of $F_1(T)$ is a monotonic decreasing function of T.
(b) If the function $F(T)$ has the second derivative $F_{11}(T)$ at a certain point $T = T_0$, then the convexity condition implies

$$F_{11}(T_0) \leq 0. \tag{4.86}$$

Convex and Concave Function in Many Variables

Equation (4.82) may be generalized to the case of a function of more than one variable. Let us consider initially the case of a function of two variable: the internal energy $U(S, V)$ as a function of the entropy S and the volume V, only. We are assuming that the number of moles N remains constant. Taking into account that $U(S, V)$ is differentiable, then its convexity is expressed by the inequality

$$U(S, V) \geq U(S_0, V_0) + U_1(S_0, V_0)(S - S_0) + U_2(S_0, V_0)(V - V_0), \tag{4.87}$$

where $U_1(S_0, V_0)$ and $U_2(S_0, V_0)$ are the derivatives of $U(S, V)$ with respect to S and V, respectively, calculated at the point (S_0, V_0). The right hand site of this inequality describes a tangent plane to the surface described by $U(S, V)$ at the point (S_0, V_0). The surface lies entirely above the plane. This inequality is identified with (4.79). It suffices to recall that $U_1(S_0, V_0) = T_0$ and $U_2(S_0, V_0) = -p_0$.

Since $U(S, V)$ is convex in the two variables jointly, it follows that $U_{11}(S_0, V_0)$, $U_{12}(S_0, V_0)$ and $U_{22}(S_0, V_0)$ satisfy the properties

$$U_{11} \geq 0, \qquad U_{22} \geq 0 \tag{4.88}$$

and

$$(U_{12})^2 - U_{11}U_{22} \leq 0. \tag{4.89}$$

Next we analyze the Helmholtz free energy $F(T, V)$ as a function of temperature T and volume V, holding the number of moles N fixed. The Helmholtz free energy is obtained from the internal energy by means of a Legendre transformation. This makes $F(T, V)$ a concave function of T while remaining a convex function of V. It therefore obeys the inequalities

$$F(T, V) \geq F(T, V_0) + F_2(T, V_0)(V - V_0) \tag{4.90}$$

and

$$F(T, V) \leq F(T_0, V) + F_1(T_0, V)(T - T_0), \tag{4.91}$$

where the coefficients $F_1(T, V_0)$ and $F_2(T_0, V)$ of the tangent lines are identified with the derivatives of $F(T, V)$ with respect to T and V, respectively, calculated at the points where $F(T, V)$ is differentiable. The Helmholtz free energy $F(T, V)$ is an example of a concave-convex function.

Since $F(T, V)$ is concave in T and convex in V, it follows that its second derivatives $F_{11}(T_0, V_0)$ and $F_{22}(T_0, V_0)$ satisfies the properties

$$F_{11} \leq 0, \qquad F_{22} \geq 0. \tag{4.92}$$

The enthalpy is also obtained from the internal energy by a Legendre transformation. For fixed N, the enthalpy $H(S, p)$ is a convex function of S and concave in p and obeys the inequalities

$$H(S, p) \geq H(S_0, p) + H_1(S_0, p)(S - S_0) \tag{4.93}$$

and

$$H(S, p) \leq H(S, p_0) + H_2(S, p_0)(p - p_0), \tag{4.94}$$

where the coefficients $H_1(S_0, p)$ and $H_2(S, p_0)$ of the tangent lines are identified with the derivatives of $H(S, p)$ with respect to S and p, respectively, calculated at the points where $H(S, p)$ is differentiable. The enthalpy is also an example of a concave-convex function.

Since $H(S,p)$ is convex in S and concave in p, it follows that the second derivatives $H_{11}(S_0,p_0)$ and $H_{22}(S_0,p_0)$ satisfies the properties

$$H_{11} \geq 0, \qquad H_{22} \leq 0. \tag{4.95}$$

We now examine the Gibbs free energy $G(T,p)$ as a function of temperature T and pressure p, at constant number of moles N. The Gibbs free energy is obtained from the internal energy $U(S,V)$ by means of two successive Legendre transformations. This makes $G(T,P)$ concave in both variables T and p, jointly. The convexity manifests itself by the inequality

$$G(T,p) \leq G(T_0,p_0) + G_1(T_0,p_0)(T-T_0) + G_2(T_0,p_0)(p-p_0), \tag{4.96}$$

where $G_1(T_0,p_0)$ and $G_2(T_0,p_0)$ are the coefficients that defines the slope of the tangent plane and are identified with the derivatives of $G(T,p)$ with respect to T and p, respectively, calculated in the points where $G(T,p)$ is differentiable.

Since $G(T,p)$ is concave in two variable jointly, it follows that its second derivatives $G_{11}(T_0,p_0)$, $G_{12}(T_0,p_0)$ and $G_{22}(T_0,p_0)$ satisfies the properties

$$G_{11} \leq 0, \qquad G_{22} \leq 0 \tag{4.97}$$

and

$$(G_{12})^2 - G_{11}G_{22} \leq 0. \tag{4.98}$$

Convex Hull

In some situations, the description of a system in thermodynamic equilibrium is done by means of a potential defined as the convex hull of a function $F(V)$, as the one shown in Fig. 4.3a, which is devoid of the convexity property. The convex hull $F_{\rm ch}(V)$ of $F(V)$ is obtained as follows. We start by determining the Legendre transformation

$$G(p) = \min_V \{F(V) + pV\}, \tag{4.99}$$

which is a concave function. Next, we perform the inverse Legendre transformation

$$F_{\rm ch}(V) = \min_p \{G(p) - Vp\}, \tag{4.100}$$

which gives us a convex function, the convex hull of $F(V)$.

The convex hull $F_{\rm ch}(V)$ may be obtained more simply by the construction of a double tangent, as seen in Fig. 4.3a. The double tangent AB gives rise to a horizontal

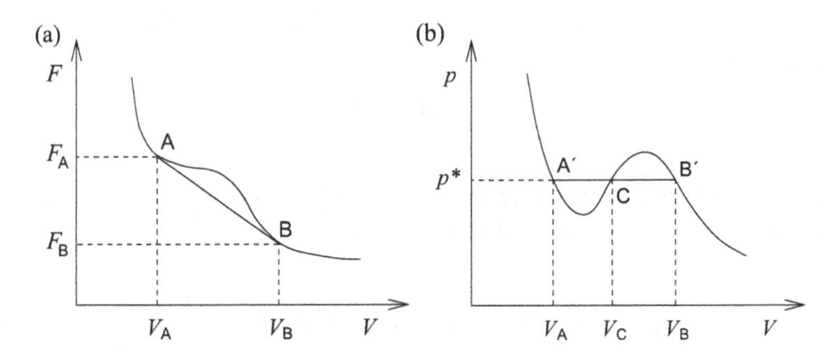

Fig. 4.3 (**a**) Convex hull. (**b**) Maxwell construction

line segment A′B′ in the derivative $p_{ch}(V) = -\partial F_{ch}/\partial V$ of the convex hull, as seen in Fig. 4.3b. This horizontal segment guarantees that $p_{ch}(V)$ is a monotonic function although the derivative $p(V) = -\partial F/\partial V$ of the original function does not have this property.

To set up the convex hull $F_{ch}(V)$ it suffices to determine V_A and V_B, the values of V at the tangential points A and B, respectively. Let $-p^*$ be the slope of the double tangent. According to Fig. 4.3a,

$$F(V_A) - F(V_B) = p^*(V_B - V_A). \tag{4.101}$$

Taking into account that the slopes at the tangential points A and B are equal, then

$$p(V_A) = p^*, \qquad p(V_B) = p^*. \tag{4.102}$$

These three equations determine p^*, V_A and V_B from $F(V)$ and its derivative.

Equations (4.101) and (4.102), that determine the localization of the double tangent AB and therefore the horizontal segment A′B′, are valid even if $F(V)$ has a singular behavior in the interval between the points V_A and V_B, as happens when $F(V)$ is defined by two unconnected branches. When the function $F(V)$ is well behaved, the position of the segment A′B′ is such that the area above A′B′ is equal to the area below A′B′, as seen in Fig. 4.3b. The trace of the segment A′B′ made according to this result is called Maxwell construction and is equivalent to the making of the double tangent. To demonstrate this result we proceed as follows. Integrating $p(V)$ between V_A and V_B, we get

$$\int_{V_A}^{V_B} p(V)dV = F(V_A) - F(V_B). \tag{4.103}$$

Using result (4.101), we see that

$$\int_{V_A}^{V_B} p(V)dV = p^*(V_B - V_A), \tag{4.104}$$

or

$$\int_{V_A}^{V_B} \{p^* - p(V)\} dV = 0. \tag{4.105}$$

If we denote by C the point where the segment A′B′ cuts the curve $p(V)$, then we can separate the integral in two and obtain

$$\int_{V_A}^{V_C} \{p^* - p(V)\} dV = \int_{V_C}^{V_B} \{p(V) - p^*\} dV, \tag{4.106}$$

which is the desired result since the right and left hand sides of these equations are, respectively, equal to the areas above and below the segment that join the two points A′ and B′.

Fundamental Properties of the Thermodynamic Potentials

Below we summarize the properties of the thermodynamic potentials, which are obtained by successive Legendre transformations. We recall initially that the internal energy is a function of extensive variables only. The thermodynamic fields are obtained by differentiation of the internal energy. A given thermodynamic field obtained by differentiation of the internal energy with respect to a certain extensive variable, constitutes with it a pair of conjugate variables.

At each Legendre transformation, a extensive variable is replaced by its conjugate thermodynamic field. Each thermodynamic potential is, therefore, a function of a set of extensive variables and a set of thermodynamic fields, which constitute the thermodynamic space related to this potential. Notice that just one component of a pair of conjugate variables makes part of the thermodynamic space. The other pair component is obtained by differentiation of the thermodynamic potential.

The fundamental properties of a thermodynamic potential associated to a thermodynamic space are the followings:

(a) *Continuity of the potential.*
The potential is a continuous function of all variables of the thermodynamic space.
(b) *Continuity of the thermodynamic field.*
The potential is differentiable with respect to the extensive variables of the thermodynamic space. This property is equivalent to the following. A thermodynamic field obtained by differentiation of the potential is a continuous function of all variables of the thermodynamic space.
(c) *Extensivity of the potential.*
A thermodynamic potential is an extensive function of the extensive variable or, equivalently, it is homogeneous of the first order in the extensive variables.

(d) *Convexity of the potential.*
A thermodynamic potential is a convex function of the set of extensive variables and a concave function of the set of thermodynamic fields.

The thermodynamic potentials related to the thermodynamic space containing just one extensive variable are particularly useful, mainly in the study of phase transitions. The thermodynamic quantity defined as the ratio between one of such potentials and the extensive variable depends only on the thermodynamic fields and is also identified as a thermodynamic field.

Problems

4.1 For each of the fundamental relations below, show that S is extensive. Determine the equations of state for each case. Find the fundamental relation in the energy representation.
(a) $S = A(UVN)^{1/3}$, (b) $S = a(U^2V^3N)^{1/6}$,
(c) $S = B(U^3V)^{1/4}$, (d) $S = b(UV)^{1/2}$.

4.2 For each of the fundamental relations below, show that U is extensive. Determine the equations of state for each case.
(a) $U = aS^3/VN$, (b) $U = AS^3/(V^3N)^{1/2}$,
(c) $U = b(S^4/V)^{1/3}$, (d) $U = BS^2/V$.

4.3 For each of the potentials of the previous problem, determine F and H by means of Legendre transformations. For the potentials of items (a) and (b) determine also G and Φ.

4.4 Determine the three equations of state for the system that obeys the equation $u = av^{-1}s^2e^{s/R}$.

4.5 Find the fundamental equation of a system that obeys the following relations $U = pV$ and $p = BT^2$. Do the same for the system that follows the equations $u = 3pv/2$ and $u^{1/2} = BTv^{1/3}$.

4.6 Determine the molar entropy $s(u, v)$ of a system whose equations of state are $p = aT^4$ and $u = 3pv$.

4.7 Show that the compressibilities are related with the density $\bar{\rho}$ by $\bar{\rho}\kappa_T = (\partial\bar{\rho}/\partial p)_T$ e $\bar{\rho}\kappa_s = (\partial\bar{\rho}/\partial p)_s$.

4.8 Consider the equations of state

$$\frac{1}{T} = \frac{a}{u} + bv, \qquad \frac{p}{T} = \frac{c}{v} + f(u).$$

Determine $f(u)$ and the fundamental equation knowing that $f(0) = 0$.

4.9 From the equations $p = RT/v$ and $c_v = c$, valid for an ideal gas, obtain the fundamental relation in the Helmholtz free energy representation.

4.10 Show the following relations between pressure p, density $\bar{\rho}$ and chemical potential μ of an ideal gas

$$p = a\, e^{\mu/RT}, \qquad \bar{\rho} = b\, e^{\mu/RT},$$

where a and b depend only on the temperature. Determine a and b.

4.11 Obtain the fundamental relation of an ideal gas in the grand thermodynamic potential representation.

4.12 From the van der Waals equation

$$p = \frac{RT}{v-b} - \frac{a}{v^2},$$

and assuming that the molar heat capacity is a constant, $c_v = c$, obtain the fundamental relation of a van der Waals fluid in the Helmholtz free energy representation. Show that $u = cT - a/v$.

4.13 From the fundamental relation of a van der Waals fluid in the entropy representation, obtain the fundamental relation in the energy representation. From this relation determine the fundamental relation in the Helmholtz free energy representation by a Legendre transformation.

4.14 Show that the functions of the problem 4.1 are concave and those of problem 4.2 are convex.

4.15 Determine the conditions that must be imposed on the parameters a, b and c so that $S = r\, U^a V^b N^c$ is extensive and has the convexity property.

4.16 Show that the functions x^2, $-\ln x$, e^x are convex. Determine their Legendre transformations and show that they are concave functions.

4.17 Determine the convex hull of the functions $\exp\{|x| - 1\} - |x|$ and $(|x| - 1)^2$.

4.18 Show that the functions

$$f(x) = \begin{cases} (|x| - 1)^2, & |x| \geq 1, \\ 0, & |x| < 1 \end{cases}$$

and

$$f(x) = \begin{cases} \exp\{|x| - 1\} - |x|, & |x| \geq 1, \\ 0, & |x| < 1 \end{cases}$$

are convex. Determine, for each one of them, the Legendre transform $g(p) = \min_x\{f(x) - px\}$ and show that it is concave. Sketch the plots of $f(x)$, $f'(x)$, $g(p)$, and $g'(p)$. Show that $f'(x)$ and $g'(p)$ are inverse of each other.

4.19 Determine the convex hull of the function

$$f(x) = x^4 - ax^2 + bx,$$

where a and b are constants and $a > 0$. Find the points x_1 and x_2 corresponding to the double tangent as functions of a and b. Sketch the plots of $f(x)$, of the convex hull and of their derivatives.

Chapter 5
Thermodynamic Identities

5.1 Consistency of the Equations of State

Exact Differential

We have seen that the equations of state are equations deduced from the fundamental relations by differentiation of the thermodynamic potential. For this reason, there is an inter-relationship among them which we will reveal. Consider the equations of state

$$\frac{1}{T} = s_1(u, v) \qquad \text{and} \qquad \frac{p}{T} = s_2(u, v) \tag{5.1}$$

in the entropy representation. Since $1/T = \partial s/\partial u$ and $p/T = \partial s/\partial v$, then

$$\frac{\partial s}{\partial u} = s_1(u, v) \qquad \text{and} \qquad \frac{\partial s}{\partial v} = s_2(u, v). \tag{5.2}$$

Taking into account that $\partial^2 s/\partial v \partial u = \partial^2 s/\partial u \partial v$, we conclude that

$$\frac{\partial}{\partial v} s_1(u, v) = \frac{\partial}{\partial u} s_2(u, v), \tag{5.3}$$

that is, the functions $s_1(u, v)$ and $s_2(u, v)$ and, therefore, the equations of state (5.1) cannot be arbitrary. They must obey the consistency relation given by (5.3). This condition is equivalent to saying that

$$ds = \frac{1}{T} du + \frac{p}{T} dv \tag{5.4}$$

is an exact differential.

M.J. de Oliveira, *Equilibrium Thermodynamics*, Graduate Texts in Physics,
DOI 10.1007/978-3-662-53207-2_5

As an example, let us imagine that a given system obeys the equations of state,

$$p = Tw(v) - q(v) \tag{5.5}$$

and

$$u = cT - r(v), \tag{5.6}$$

where $w(v)$, $q(v)$ and $r(v)$ are functions of v only and c is a constant. From (5.6),

$$\frac{1}{T} = \frac{c}{u + r(v)} = s_1(u, v). \tag{5.7}$$

From (5.5) and using (5.7), we get

$$\frac{p}{T} = w(v) - q(v)\frac{c}{u + r(v)} = s_2(u, v). \tag{5.8}$$

Using now the consistency equation (5.3), we reach the result

$$q(v) = -r'(v), \tag{5.9}$$

which says that the functions $q(v)$ and $r(v)$ cannot be arbitrary. For the van der Waals fluid, $w = R/(v - b)$ and $r = a/v$, so that $q = a/v^2$.

Maxwell Relations

Consider the potentials U, F, H and G and their differentials

$$dU = TdS - pdV, \tag{5.10}$$

$$dF = -SdT - pdV, \tag{5.11}$$

$$dH = TdS + Vdp, \tag{5.12}$$

$$dG = -SdT + Vdp. \tag{5.13}$$

Since they are all exact differentials then

$$\left(\frac{\partial T}{\partial V}\right)_S = -\left(\frac{\partial p}{\partial S}\right)_V, \tag{5.14}$$

$$\left(\frac{\partial S}{\partial V}\right)_T = \left(\frac{\partial p}{\partial T}\right)_V, \tag{5.15}$$

$$\left(\frac{\partial T}{\partial p}\right)_S = \left(\frac{\partial V}{\partial S}\right)_p, \tag{5.16}$$

$$-\left(\frac{\partial S}{\partial p}\right)_T = \left(\frac{\partial V}{\partial T}\right)_p, \tag{5.17}$$

which are the Maxwell relations.

The Maxwell relations are useful, for example, when one wishes to obtain information about the entropy from the equation of state $p(T, V)$. The second Maxwell relation above says that the isothermal variation of the entropy with the volume is equal to the isochoric increase of pressure with temperature. These relations are also useful to obtain further thermodynamic identities as we shall see below.

5.2 Identities

Identities Involving Derivatives

In the manipulation of the thermodynamic derivatives it is necessary to use certain mathematical identities, which we now demonstrate. Consider five generic variables X, Y, Z, U, and W, assume that $X(Y, Z)$ is a function of Y and Z and that $Y(U, W)$ and $Z(U, W)$ are functions of U and W. Then by the chain rule,

$$\left(\frac{\partial X}{\partial U}\right)_W = \left(\frac{\partial X}{\partial Y}\right)_Z \left(\frac{\partial Y}{\partial U}\right)_W + \left(\frac{\partial X}{\partial Z}\right)_Y \left(\frac{\partial Z}{\partial U}\right)_W. \tag{5.18}$$

1. For the case $U = Y$,

$$\left(\frac{\partial X}{\partial Y}\right)_W = \left(\frac{\partial X}{\partial Y}\right)_Z + \left(\frac{\partial X}{\partial Z}\right)_Y \left(\frac{\partial Z}{\partial Y}\right)_W. \tag{5.19}$$

2. For the case $W = Z$,

$$\left(\frac{\partial X}{\partial U}\right)_Z = \left(\frac{\partial X}{\partial Y}\right)_Z \left(\frac{\partial Y}{\partial U}\right)_Z. \tag{5.20}$$

3. For $W = X$ and $U = Y$,

$$0 = \left(\frac{\partial X}{\partial Y}\right)_Z + \left(\frac{\partial X}{\partial Z}\right)_Y \left(\frac{\partial Z}{\partial Y}\right)_X \tag{5.21}$$

or

$$\left(\frac{\partial Z}{\partial Y}\right)_X = -\frac{\left(\frac{\partial X}{\partial Y}\right)_Z}{\left(\frac{\partial X}{\partial Z}\right)_Y}. \tag{5.22}$$

4. For $U = X$ and $W = Z$,

$$\left(\frac{\partial X}{\partial Y}\right)_Z \left(\frac{\partial Y}{\partial X}\right)_Z = 1 \tag{5.23}$$

or

$$\left(\frac{\partial Y}{\partial X}\right)_Z = \frac{1}{\left(\frac{\partial X}{\partial Y}\right)_Z}. \tag{5.24}$$

Relation Between the Heat Capacities

As an example of the use of these equalities, we see, from (5.22),

$$\left(\frac{\partial p}{\partial T}\right)_V = -\frac{\left(\frac{\partial V}{\partial T}\right)_p}{\left(\frac{\partial V}{\partial p}\right)_T} = \frac{\alpha}{\kappa_T}. \tag{5.25}$$

Taking into account that $\kappa_T \geq 0$, this equality says that bodies that expand with temperature ($\alpha > 0$), get hotter in an isochoric compression ($(\partial p/\partial T)_V > 0$). Those that contract with temperature ($\alpha < 0$), get colder in an isochoric compression ($(\partial p/\partial T)_V < 0$).

Another example can be obtained from (5.19). From this equation,

$$\left(\frac{\partial S}{\partial T}\right)_p = \left(\frac{\partial S}{\partial T}\right)_V + \left(\frac{\partial S}{\partial V}\right)_T \left(\frac{\partial V}{\partial T}\right)_p, \tag{5.26}$$

or, by Maxwell relation (5.15),

$$\left(\frac{\partial S}{\partial T}\right)_p = \left(\frac{\partial S}{\partial T}\right)_V + \left(\frac{\partial p}{\partial T}\right)_V \left(\frac{\partial V}{\partial T}\right)_p. \tag{5.27}$$

Using the definitions of the thermodynamic coefficients and the result (5.25) for $(\partial p/\partial T)_V$, we get

$$C_p = C_v + TV\frac{\alpha^2}{\kappa_T}, \tag{5.28}$$

which gives the relation between the heat capacities.

This relation, which is one of the best known of thermodynamics, is useful in determining C_v for solids and liquids because in these cases it is very difficult to make measurements of quantities defined at constant volume. It is easier to make measurements of quantities defined at constant pressure, as is the case of the quantities C_p and α. As the second term on the right hand side of (5.28) is never negative, we conclude that the heat capacity at constant pressure is always greater than or equal to the heat capacity at constant volume, $C_p \geq C_v$.

Another relation particularly useful is obtained from (5.22), namely,

$$\left(\frac{\partial V}{\partial p}\right)_S = -\frac{\left(\frac{\partial V}{\partial S}\right)_p}{\left(\frac{\partial p}{\partial S}\right)_V}. \tag{5.29}$$

Using the relation (5.20), we can write the numerator and the denominator in the forms

$$\left(\frac{\partial V}{\partial S}\right)_p = \left(\frac{\partial V}{\partial T}\right)_p \left(\frac{\partial T}{\partial S}\right)_p = \left(\frac{\partial V}{\partial T}\right)_p \frac{T}{C_p} \tag{5.30}$$

and

$$\left(\frac{\partial p}{\partial S}\right)_V = \left(\frac{\partial p}{\partial T}\right)_V \left(\frac{\partial T}{\partial S}\right)_V = \left(\frac{\partial p}{\partial T}\right)_V \frac{T}{C_v}, \tag{5.31}$$

so that

$$\left(\frac{\partial V}{\partial p}\right)_S = -\frac{\left(\frac{\partial V}{\partial T}\right)_p C_v}{\left(\frac{\partial p}{\partial T}\right)_V C_p} = \left(\frac{\partial V}{\partial p}\right)_T \frac{C_v}{C_p}, \tag{5.32}$$

where we have used again the relation (5.22). From the definitions of the compressibilities, we get

$$\frac{\kappa_T}{\kappa_s} = \frac{C_p}{C_v}. \tag{5.33}$$

The ratio κ_T/κ_s between the compressibilities is therefore equal to the ration C_p/C_v between the heat capacities. Replacing (5.28) into (5.33), we find the result

$$\kappa_T = \kappa_s + TV\frac{\alpha^2}{C_p}. \tag{5.34}$$

Since the second term is never negative, then the isothermal compressibility is always greater or equal to the adiabatic compressibility, $\kappa_T \geq \kappa_s$.

Reduction of Derivatives

A thermodynamic system can be described in various ways, according to the representation we have chosen. This variety of descriptions leads us to a large number of equivalent results, which can be revealed by the thermodynamic identities given above. In general, these identities involve the derivatives of thermodynamic variables. To achieve identities in a systematic way, we present a scheme to reduce a given derivative in terms of other previously chosen. In the scheme we present below, we chose the following quantities: the heat capacity at constant pressure, the coefficient of thermal expansion and the isothermal compressibility, given by

$$C_p = T\left(\frac{\partial S}{\partial T}\right)_p, \qquad \alpha = \frac{1}{V}\left(\frac{\partial V}{\partial T}\right)_p, \qquad \kappa_T = -\frac{1}{V}\left(\frac{\partial V}{\partial p}\right)_T, \tag{5.35}$$

respectively.

Notice that the expressions derived earlier for the heat capacities and for the compressibilities are under this scheme. In fact, from (5.28) and (5.34),

$$C_v = C_p - TV\frac{\alpha^2}{\kappa_T}, \qquad \kappa_s = \kappa_T - TV\frac{\alpha^2}{C_p}. \tag{5.36}$$

The rules for the reduction of derivatives are as follows.

1. Bring the thermodynamic potentials U, F, H and G to the numerator, using identities (5.22) or (5.24) and, next, eliminate them, using the identities (5.10)–(5.13).
2. Bring the entropy to the numerator, using identities (5.22) or (5.24) and, then eliminate them, by the use of the Maxwell relations (5.15) and (5.17) or through the definitions of the heat capacities C_p or C_v. If these two alternatives are not appropriate, use the identities

$$\left(\frac{\partial S}{\partial p}\right)_V = \left(\frac{\partial S}{\partial T}\right)_V \left(\frac{\partial T}{\partial p}\right)_V = \frac{C_v}{T}\left(\frac{\partial T}{\partial p}\right)_V \tag{5.37}$$

or

$$\left(\frac{\partial S}{\partial V}\right)_p = \left(\frac{\partial S}{\partial T}\right)_p \left(\frac{\partial T}{\partial V}\right)_p = \frac{C_p}{T}\left(\frac{\partial T}{\partial V}\right)_p. \tag{5.38}$$

3. The remaining derivatives are of the type $(\partial X/\partial Y)_Z$ where X, Y and Z are V, p, and T. Therefore, they can be eliminate by bringing the volume to the numerator and using the definition of α and κ_T.

5.3 Applications

Compression

We examine two types of compression: the isothermal and adiabatic. In an isothermal compression we are interested in determining the heat given way $Q = -T\Delta S$ when the pressure increases by an amount Δp. The appropriate quantity is then

$$-T\left(\frac{\partial S}{\partial p}\right)_T. \tag{5.39}$$

Using the rules above,

$$-T\left(\frac{\partial S}{\partial p}\right)_T = T\left(\frac{\partial V}{\partial T}\right)_p = TV\alpha. \tag{5.40}$$

In an adiabatic compression we are interested in determining the variation of temperature ΔT when the pressure increases by an amount Δp. The appropriate quantity is then

$$\left(\frac{\partial T}{\partial p}\right)_S. \tag{5.41}$$

Using the rules above,

$$\left(\frac{\partial T}{\partial p}\right)_S = -\frac{\left(\frac{\partial S}{\partial p}\right)_T}{\left(\frac{\partial S}{\partial T}\right)_p} = \frac{T\left(\frac{\partial V}{\partial T}\right)_p}{C_p} = \frac{TV\alpha}{C_p}. \tag{5.42}$$

Since $C_p \geq 0$, this equality allows us to say that the bodies that expand with temperature ($\alpha > 0$), get hotter in an adiabatic compression ($(\partial T/\partial p)_S > 0$). Those that contract with temperature ($\alpha < 0$), get colder in an adiabatic compression ($(\partial T/\partial p)_S < 0$).

Free Expansion

In a free expansion the internal energy is constant and the volume of the system increases. If we wish to determine whether there is a variation ΔT in temperature with the change ΔV in volume it is appropriate to calculate the ratio $(\Delta T/\Delta V)$ at constant energy, that is

$$\left(\frac{\partial T}{\partial V}\right)_U. \tag{5.43}$$

Using the scheme for the reduction of derivatives, we start from the identity (5.22) to get

$$\left(\frac{\partial T}{\partial V}\right)_U = -\frac{\left(\frac{\partial U}{\partial V}\right)_T}{\left(\frac{\partial U}{\partial T}\right)_V}. \tag{5.44}$$

Next, we use the chain rule and the identity (5.10) to get

$$\left(\frac{\partial U}{\partial V}\right)_T = T\left(\frac{\partial S}{\partial V}\right)_T - p \tag{5.45}$$

and

$$\left(\frac{\partial U}{\partial T}\right)_V = T\left(\frac{\partial S}{\partial T}\right)_V. \tag{5.46}$$

To eliminate the entropy, we use the Maxwell relation (5.15) in the first and the definition of C_v in the second, obtaining

$$\left(\frac{\partial U}{\partial V}\right)_T = T\left(\frac{\partial p}{\partial T}\right)_V - p \tag{5.47}$$

and

$$\left(\frac{\partial U}{\partial T}\right)_V = C_v. \tag{5.48}$$

Finally, we use the identity (5.22) in the first to get

$$\left(\frac{\partial U}{\partial V}\right)_T = T\frac{\alpha}{\kappa_T} - p. \tag{5.49}$$

Then,

$$\left(\frac{\partial T}{\partial V}\right)_U = \frac{1}{C_v}(p - T\frac{\alpha}{\kappa_T}). \tag{5.50}$$

It is worth noting that, for an ideal gas, $\alpha = 1/T$ and $\kappa_T = 1/p$, leading to the results $(\partial U/\partial V)_T = 0$ and $(\partial T/\partial V)_U = 0$. From these results, we conclude that the internal energy of an ideal gas does not depend on the volume, but only on temperature or, equivalently, that there is no variation in temperature in a free expansion of an ideal gas.

Joule-Thomson Process

In a Joule-Thomson process, also called throttling process, a gas is forced through a porous wall so that the enthalpy of the gas remains unchanged. The pressure difference on both sides of the wall is Δp and we wish to know the temperature variation ΔT. For small pressure differences, the variation of temperature is $\Delta T = \mu_{\rm JT}\Delta p$ where $\mu_{\rm JT}$ is the Joule-Thomson coefficient

$$\mu_{\rm JT} = \left(\frac{\partial T}{\partial p}\right)_H . \tag{5.51}$$

From the identity (5.22),

$$\left(\frac{\partial T}{\partial p}\right)_H = -\frac{\left(\frac{\partial H}{\partial p}\right)_T}{\left(\frac{\partial H}{\partial T}\right)_p} . \tag{5.52}$$

Using the chain rule and the identity (5.12), we get

$$\left(\frac{\partial H}{\partial p}\right)_T = T\left(\frac{\partial S}{\partial p}\right)_T + V \tag{5.53}$$

and

$$\left(\frac{\partial H}{\partial T}\right)_p = T\left(\frac{\partial S}{\partial T}\right)_p . \tag{5.54}$$

Using Maxwell relation (5.17) in the first equation and the definition of C_p in the second, then

$$\left(\frac{\partial H}{\partial p}\right)_T = -T\left(\frac{\partial V}{\partial T}\right)_p + V = -TV\alpha + V \tag{5.55}$$

and

$$\left(\frac{\partial H}{\partial T}\right)_p = C_p . \tag{5.56}$$

Therefore

$$\left(\frac{\partial T}{\partial p}\right)_H = \frac{V}{C_p}(T\alpha - 1) . \tag{5.57}$$

For an ideal gas $\alpha = 1/T$, so that $(\partial H/\partial p)_T = 0$ and $(\partial T/\partial p)_H = 0$. From these results we conclude that the enthalpy of an ideal gas do not depend on pressure, but only on temperature or, equivalently, that there is no variation in temperature of an ideal gas under a Joule-Thomson process.

5.4 Gas Properties

Virial Expansion

For sufficiently low densities, the gases follows the ideal behavior $pV = NRT$ which we write as

$$p = RT\bar{\rho}, \tag{5.58}$$

where $\bar{\rho} = N/V$ is the number of moles per unit volume. The plot represented in Fig. 5.1a shows the pressure of hydrogen as a function of density for several temperatures. For low densities we see that indeed p is proportional to $\bar{\rho}$ and therefore the hydrogen behaves as an ideal gas in this regime. However, deviations from this behavior can be seen in this figure, if the density is not sufficiently small.

To examine the deviation from the ideal behavior it is convenient to plot the compressibility factor $Z = p/RT\bar{\rho} = pv/RT$ as a function of $\bar{\rho}$, where $v = V/N = 1/\bar{\rho}$ is the molar volume. Figure 5.1b shows such a plot for hydrogen. It is seen that Z deviates from the value $Z = 1$, valid for an ideal gas. We may represent the

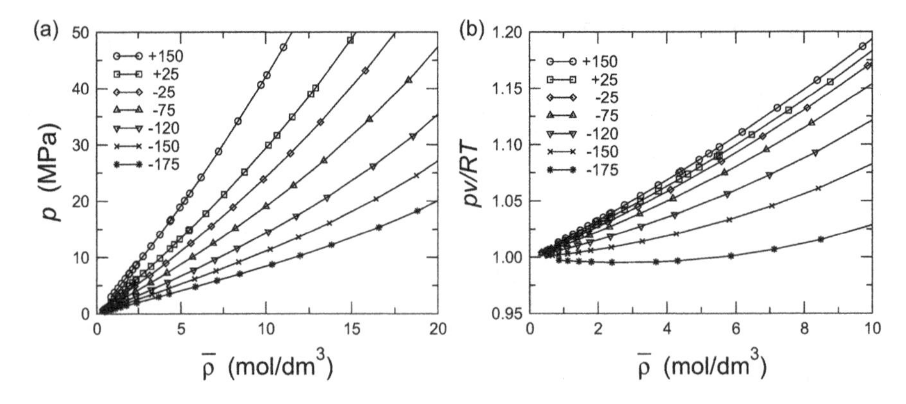

Fig. 5.1 Isotherms of the hydrogen obtained experimentally by Michels et al. [106] at various temperatures between −175 and 150 °C. (**a**) Pressure p versus number of moles per unit volume $\bar{\rho}$. (**b**) Compressibility factor $Z = pv/RT$ versus $\bar{\rho}$

behavior of the compressibility factor Z by the expression

$$\frac{pv}{RT} = 1 + B_2\bar{\rho} + B_3\bar{\rho}^2 + \ldots, \tag{5.59}$$

called virial expansion. The coefficients $B_2, B_3, \ldots$, depend only on temperature.

Equation (5.59) may describe the general behavior of gases, as long as the expansion contains a sufficient number of terms. Alternatively, it is possible to describe the general behavior by means of approximate equations of state such as the van der Waals equation

$$(p + \frac{a}{v^2})(v - b) = RT, \tag{5.60}$$

the Dieterici equation

$$p(v - b) = RTe^{-a/RTv}, \tag{5.61}$$

and the Berthelot equation

$$(p + \frac{a}{v^2 T})(v - b) = RT. \tag{5.62}$$

These equations of state, unlike the virial expansion, are capable of describing the liquid phase, in addition to the gas phase, and the transition between them.

Measurement of Compressibility

An experiment to measure the compressibility of gases is performed as follows. We consider a vessel with adiabatic walls, in the form of a bottle, which encloses the gas whose compressibility we want to measure. In the neck of the bottle, which has a cylindrical shape, there is a small object of mass m, that can move without friction. The weight of the object is such that it is in equilibrium position when the volume of the gas is V_0 and the pressure of the gas is p_0. The object is then displaced from its equilibrium position and starts to oscillate with a certain angular frequency ω. The equation of motion of the object is given by

$$m\frac{d^2x}{dt^2} = (p - p_0)A, \tag{5.63}$$

where A is the cross-sectional area of the bottle neck. The pressure at the equilibrium point p_0 is equal to the pressure due to the weight of the object plus the atmospheric air pressure p_{atm}, that is, $p_0 = p_{\text{atm}} + mg/A$, where g is the acceleration of gravity. We are assuming that the equilibrium point occurs at $x = 0$, so that the volume depends on x as $V = V_0 + Ax$.

Assuming that the oscillations are adiabatic, the pressure and the volume are connected by an adiabatic line, so that, for small oscillations, that is, for small volume variations,

$$\frac{p - p_0}{V - V_0} = \left(\frac{\partial p}{\partial V}\right)_S = -\frac{1}{V_0 \kappa_s}, \tag{5.64}$$

and therefore

$$m\frac{d^2x}{dt^2} = -kx, \tag{5.65}$$

where the constant k is given by $k = A^2/\kappa_s V_0$. But the angular frequency ω of oscillations is related to the constant k by $\omega = \sqrt{k/m}$, from which we get

$$\omega = \sqrt{\frac{A^2}{\kappa_s V_0 m}}. \tag{5.66}$$

Therefore, measuring the angular frequency ω, knowing the dimensions of the bottle, that is, the volume V_0 and the area A, and the mass m of the object, one can determine the adiabatic compressibility.

If we know the isothermal compressibility κ_T independently, then this experiment allows us to determined the ratio between the heat capacities $\gamma = C_p/C_v$ because, according to (5.33), $\kappa_s = \kappa_T/\gamma$. For instance, if the gas can be considered ideal then $\kappa_T = 1/p_0$.

The speed of sound in a gas is given by $v_{\text{som}} = \sqrt{(\partial p/\partial \rho)_s}$ where ρ is the gas density, related to $\bar{\rho}$ by $\rho = M\bar{\rho}$ where M is the gas molar mass. Since $(\partial p/\partial \rho)_s = 1/\rho\kappa_s$, the speed of sound is related to the gas compressibility in accordance with

$$v_{\text{som}} = \sqrt{\frac{1}{\rho\kappa_s}} = \sqrt{\frac{\gamma}{\rho\kappa_T}}. \tag{5.67}$$

The second equality is obtained from the relation $\kappa_s = \kappa_T/\gamma$. If we know the adiabatic compressibility, or if we know the isothermal compressibility and γ, we can determine the speed of sound. If the gas may be considered as ideal, then $\rho\kappa_T = M\bar{\rho}/p = M/RT$ so that

$$v_{\text{som}} = \sqrt{\frac{\gamma RT}{M}}. \tag{5.68}$$

Considering the atmospheric air as an ideal gas, we can use formula (5.68) to compute the speed of sound. For normal conditions of pressure and temperature, that is, pressure of 1 atm and temperature of 0 °C, the experimental values are

$\rho = 1.293\,\text{kg/m}^3$ and $\gamma = 1.403$. With these data we find $v_{\text{som}} = 331.6\,\text{m/s}$. For comparison, the value of the speed of sound in atmospheric air measured by direct means $v_{\text{som}} = 331.5\,\text{m/s}$.

Inversion Curve

One way of achieving lower temperatures is through the Joule-Thomson process. To decrease the temperature of the gas when it passes through the porous wall, it is necessary that the Joule-Thomson coefficient μ_{JT} is positive since the pressure of the gas necessarily decreases. For a given gas, we can draw in the p-T diagram the region for which $\mu_{\text{JT}} > 0$, which is the region useful for cooling through the Joule-Thomson process. This region is bounded by the curve $\mu_{\text{JT}} = 0$, called the inversion curve.

Let us determine the inverse curve for a fluid obeying the van der Waals equation (5.60). According to the result (5.57), the inversion curve is given by $T = 1/\alpha$, that is, by

$$T = v\left(\frac{\partial T}{\partial v}\right)_p. \tag{5.69}$$

Determining the derivative $\partial T/\partial v$ from (5.60) and substituting into the inversion curve equation, we arrive at the result

$$\frac{RTb}{2a} = \left(1 - \frac{b}{v}\right)^2. \tag{5.70}$$

Solving for v and substituting into (5.60), we obtain the inversion curve in the p-T diagram, given by

$$p = \frac{a}{b^2}\left(1 - \sqrt{\frac{bRT}{2a}}\right)\left(3\sqrt{\frac{bRT}{2a}} - 1\right). \tag{5.71}$$

Figure 5.2 shows the inversion curve, according to this equation and the inverse curve of nitrogen, obtained experimentally. It is seen that the inversion curve reaches the liquid-vapor coexistence line. We shall see later that the van der Waals equation in fact provides a liquid-vapor coexistence line and a critical point, shown in Fig. 5.2a. The critical temperature T_c and critical pressure p_c are related to the parameters a and b by $T_c = 8a/27bR$ and $p_c = a/27b^2$.

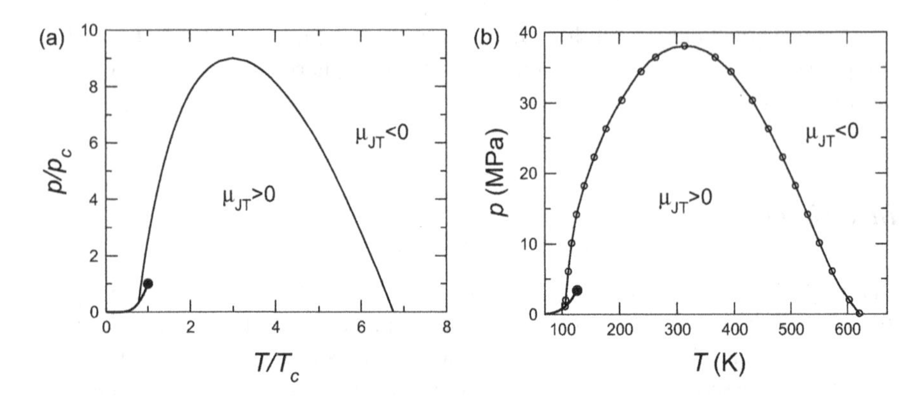

Fig. 5.2 (**a**) Inversion curve for a van der Waals fluid according to (5.71). (**b**) Inversion curve for nitrogen according to Roebuck and Osterberg [107]. In both plots, the *thick curve* represents the liquid-vapor coexistence and the *full circle*, the critical point

Molar Heat Capacity

The molar heat capacity of the ideal gases, that is, those that obey the equation of state $pv = RT$, depends only on temperature. In fact, from (5.47) it is seen that for an ideal gas $(\partial u/\partial v)_T = 0$ and, therefore, the molar internal energy u is independent of v. Since the molar enthalpy $h = u + pv = u + RT$, then h is also independent of v. Therefore, the molar heat capacities $c_v = (\partial u/\partial T)_v$ and $c_p = (\partial h/\partial T)_p$, depend only on T. Moreover, taking into account that $h = u + RT$ then $c_p = c_v + R$. Figure 5.3a shows c_p of hydrogen as a function of temperature, for several values of pressure. Notice that, at high temperatures, c_p becomes independent of pressure and approaches the value $(7/2)R$.

Figure 5.3b shows c_p of several gases that have behavior close to the ideal behavior. We assume that for these gases c_v is a sum of the terms

$$c_v = c_{\text{trans}} + c_{\text{rot}} + c_{\text{vib}} \tag{5.72}$$

corresponding to the translation motion, rotation and vibration of the molecules. According to the equipartition of energy $c_{\text{trans}} = (3/2)R$. The rotational part is $c_{\text{rot}} = R$ for diatomic gases and $c_{\text{rot}} = (3/2)R$ for polyatomic gases. Therefore, for monoatomic gases ideal gases that have only translational degrees of freedom, $c_v = (3/2)R$ and $c_p = (5/2)R$ as seen in Fig. 5.3b for argon.

For diatomic ideal gases, $c_v = (5/2)R$ and $c_p = (7/2)R$, as seen in Fig. 5.3b for nitrogen and approximately for carbon monoxide. Hydrogen has the same behavior as long as the temperature is not so low. In the low temperature regime, hydrogen behave as a monoatomic gas. The behavior of c_{vib} with temperature is more complicated and will be object of study in the next chapter.

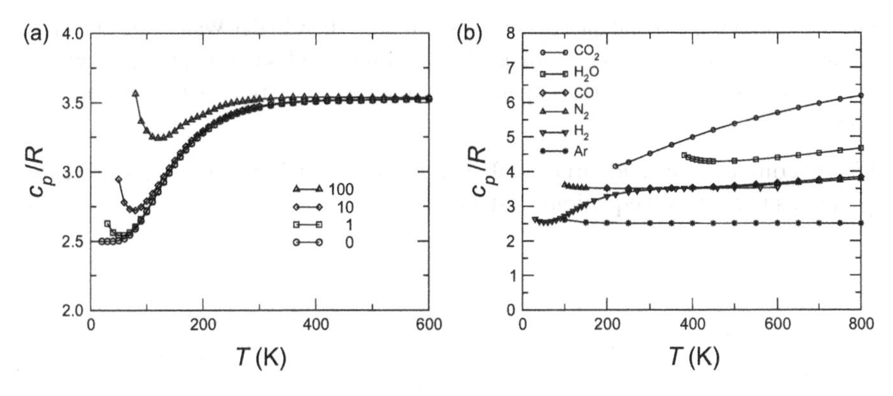

Fig. 5.3 Molar isobaric heat capacity as a function of temperature of (**a**) hydrogen for the pressures indicated in atmospheres and (**b**) several gases at the pressure of 1 atm. *Source*: TT

Problems

5.1 Show that

$$\left(\frac{\partial u}{\partial v}\right)_T = T\left(\frac{\partial p}{\partial T}\right)_v - p.$$

Use this identity to show that, for a fluid that obeys the equations of state of the type

$$p(T, v) = Tw(v) - q(v) \qquad \text{and} \qquad u(T, v) = cT - r(v).$$

where $w(v)$, $q(v)$ and $r(v)$ depend only on v, the functions $q(v)$ and $r(v)$ are related by $r'(v) = -q(v)$. Determine $r(v)$ for the van der Waals gas for which $w = R/(v - b)$ and $q(v) = a/v^2$.

5.2 Reduce the following derivatives to expressions containing α, κ_T and C_p.

$$\left(\frac{\partial T}{\partial p}\right)_V, \left(\frac{\partial V}{\partial S}\right)_p, \left(\frac{\partial T}{\partial V}\right)_U, \left(\frac{\partial T}{\partial p}\right)_H, \left(\frac{\partial T}{\partial p}\right)_U, \left(\frac{\partial V}{\partial p}\right)_S.$$

5.3 Show the following expressions for the heat capacity $C_{VN} = T(\partial S/\partial T)_{VN} = (\partial U/\partial T)_{VN}$ at constant volume and number of moles,

$$C_{VN} = C_{V\mu} - T\frac{(\partial N/\partial T)^2_{V\mu}}{(\partial N/\partial \mu)_{VT}} = \left(\frac{\partial U}{\partial T}\right)_{V\mu} - \left(\frac{\partial U}{\partial \mu}\right)_{VT} \frac{(\partial N/\partial T)_{V\mu}}{(\partial N/\partial \mu)_{VT}},$$

where $C_{V\mu} = T(\partial S/\partial T)_{V\mu}$ is the heat capacity at constant volume but at constant chemical potential. These expressions allows the determination of C_{VN} from S, U and N as functions of T, V and μ.

5.4 Show that the internal energy of gases that obey the van der Waals, the Dieterici and the Berthelot equations must depend on volume, besides temperature. Show that this is also valid for the virial expansion as long as the virial coefficients are not independent of temperature.

5.5 Determine the second virial coefficient B_2 corresponding to the van der Waals, Dieterici and Berthelot equations of state.

Chapter 6
Nernst-Planck Principle

6.1 Nernst Postulate

Entropy

At sufficiently low temperatures, it is experimentally shown that the thermal capacity of the solids decreases appreciably and vanishes in the limit of zero absolute temperature. This behavior however does not necessarily follow from the laws of thermodynamics seen so far. It is one of the consequences of the Nernst postulate according to which the entropy of a system does not decrease without limits, but approaches a constant in the limit of zero absolute temperature, that is,

$$S(T) \to S_0 \qquad \text{when} \qquad T \to 0. \tag{6.1}$$

Although the existence of a lower bound for the entropy seems to be quite evident, we must remember that this does not occur in a system whose heat capacity is constant, like an ideal gas whose internal energy is linear with temperature. The entropy of a system with constant heat capacity behaves as $\ln T$ and therefore decreases without bounds as $T \to 0$. Therefore, the heat capacity can not remain constant at low enough temperatures if S has a lower bound.

If the temperature of a system is decreased along an isobaric line, the entropy $S(T)$ can be determined by

$$S(T) - S_0 = \int_0^T \frac{C_p}{T} dT, \tag{6.2}$$

where the integral is performed along and isobaric line. Since S and S_0 are finite, the integral must also be finite. The integral will be finite if the heat capacities at constant pressure C_p vanish when $T \to 0$. Similarly, we conclude that the heat capacity at constant volume vanishes when $T \to 0$.

M.J. de Oliveira, *Equilibrium Thermodynamics*, Graduate Texts in Physics,
DOI 10.1007/978-3-662-53207-2_6

The Nernst postulate has a second part: the entropy S_0 is the same for any state of a system at temperature zero. This means to say that the isotherm corresponding to the absolute zero coincides with an isentropic line. It should be noted that only in this case an isotherm coincides with an isentropic line. For nonzero temperatures, an isotherm never coincides with an isentropic line. We also note that S_0 is the smallest value of the entropy of a thermodynamic system because the entropy is a monotonic increasing function of temperature

Being S_0 a constant, the derivative $(\partial S/\partial p)_T$ vanishes at $T = 0$. Since by the Maxwell relation $(\partial S/\partial p)_T = -(\partial V/\partial T)_p$, then the coefficient of thermal expansion $\alpha = (1/V)(\partial V/\partial T)_p$ also vanishes at the absolute zero of temperature. Similarly, the derivative $(\partial S/\partial V)_T$ vanishes when $T = 0$. Using the Maxwell relation $(\partial S/\partial V)_T = (\partial p/\partial T)_v$, we conclude that the coefficient $\beta = (\partial p/\partial T)_v$ also vanishes at zero absolute temperature. On the other hand, the compressibilities $\kappa_T = -(1/V)(\partial V/\partial p)_T$ and $\kappa_S = -(1/V)(\partial V/\partial p)_S$ do not necessarily vanish. In general, these two quantities remain nonzero. We remark that they become identical at absolute zero, since the isotherm $T = 0$ coincides with the isentropic line $S = S_0$.

As an experimental illustration of the Nernst postulate we consider the entropy of sulfur. This substance can be found in two crystalline forms known as rhombic and monoclinic. Below $T_0 = 368.6$ K, the stable form is the rhombic. Above this temperature, the stable form is the monoclinic. At the transition temperature T_0, rhombic sulfur consumes $q = 402 \pm 2$ J/mol, where q is the latent heat of transition from rhombic to monoclinic. In this conversion, carried out at constant temperature, the entropy of sulfur increases by the value $q/T_0 = 1.091 \pm 0.005$ J/mol K.

From the measurement of heat capacity of rhombic sulfur, it is possible to determine the variation of molar entropy $s^{\text{rhomb}} - s_0$ between $T = 0$ and $T = T_0$ from the formula (6.2). Using the same formula, we can also determine the change in entropy $s^{\text{mon}} - s_0$ in the same temperature range for the monoclinic sulfur. Although the monoclinic form is not stable below T_0, the transformation speed to the rhombic form is very slow, enough for the experimental determination of its heat capacity. Therefore, the entropy difference $s^{\text{mon}} - s^{\text{rhomb}}$ can be determined by

$$s^{\text{mon}} - s^{\text{rhomb}} = \int_0^{T_0} \frac{\Delta c_p}{T} dT \tag{6.3}$$

where $\Delta c_p = c_p^{\text{mon}} - c_p^{\text{rhomb}}$ is the difference between the molar heat capacities of the two forms of sulfur. We are assuming according to the Nernst postulate that S_0 is the same for both allotropic forms. Figure 6.1 shows the thermal capacities and the difference between them as a function of temperature. The integral in (6.3), performed numerically from the experimental data presented in Fig. 6.1, provides the value $s^{\text{mon}} - s^{\text{rhomb}} = 0.90 \pm 0.17$ J/mol K, close to the value obtained above.

Another example is the entropy of tin. This substance has two allotropic forms known as white tin, which is metallic and has a tetragonal structure, and gray tin, which is semiconductor and has a cubic structure. The gray tin is stable below 13 °C and white tin is stable above this temperature. Under 13 °C, white tin is metastable

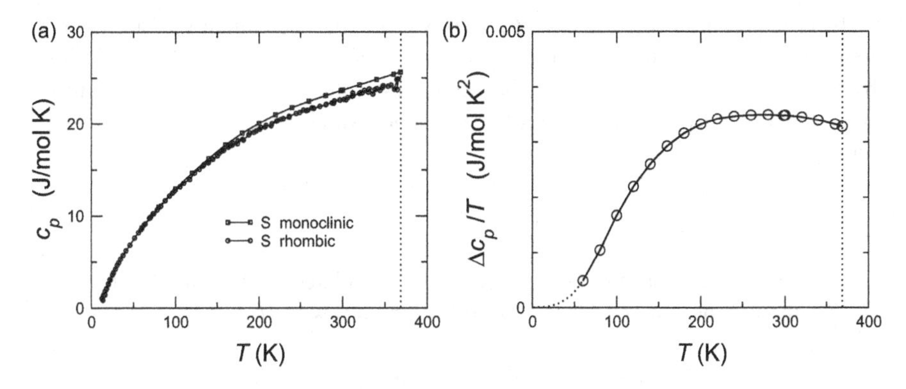

Fig. 6.1 Experimental data of the molar heat capacity of sulfur obtained by Eastman and McGavock [111]. (**a**) Molar heat capacity of sulfur in the rhombic and monoclinic form. (**b**) Difference Δc_p between the molar heat capacities of sulfur in the rhombic and monoclinic form. The *vertical line* indicates the transition temperature from the rhombic to the monoclinic form

and transforms itself into gray tin, a phenomenon known as tin pest. However, the transformation is slow enough so that they heat capacity measurements can be made. From the results of the thermal capacity of the two forms, one obtains the value 8.1 J/mol K for the change in entropy. On the other hand, from the latent heat, which is 2.28 kJ/mol, we obtain the value 8.0 J/mol K very close to the previous value

Thermodynamic Potentials

Decreasing the temperature of a system at constant volume, the internal energy also diminishes. In fact, along an isochoric line, the energy is a monotonic increasing function of temperature because $(\partial U/\partial T)_v = C_v \geq 0$. In the limit of absolute zero, the internal energy approaches its lowest value, which we postulate to be finite. That is, the internal energy has a lower bound which is reached when $T \to 0$, that is,

$$U \to U_0 \qquad \text{when} \qquad T \to 0. \tag{6.4}$$

The corresponding state is called ground state.

Taking into account that the Helmholtz free energy is related to the internal energy through $F = U - TS$, then in the limit of absolute zero F coincides with U, that is,

$$F \to F_0 = U_0 \qquad \text{when} \qquad T \to 0, \tag{6.5}$$

because S_0 is finite.

Using the relation $(\partial H/\partial T)_p = C_p \geq 0$, we see that the enthalpy of a system is monotonic increasing function of temperature, at constant pressure. In the limit of absolute zero, the enthalpy approaches its lowest value, which we postulate to be finite. The enthalpy has therefore a lower bound, which is reached when $T \to 0$, that is,

$$H \to H_0 \qquad \text{when} \qquad T \to 0. \tag{6.6}$$

Taking into account that the Gibbs free energy is related to the enthalpy by $G = H - TS$, then in the limit of absolute zero G coincides with H, that is,

$$G \to G_0 = H_0 \qquad \text{when} \qquad T \to 0. \tag{6.7}$$

We should note that, unlike S_0, the thermodynamic potentials at zero temperature U_0, F_0, H_0 and G_0 may depend on other thermodynamic variables.

6.2 Heat Capacity of Solids

Dulong-Petit Law

The heat capacity of the solids is usually measured at constant pressure and not at constant volume, since the low compressibility of the solid makes it difficult to maintain the volume constant. However, from c_p we get c_v using the identity

$$c_p = c_v + \frac{Tv\alpha^2}{\kappa_T}, \tag{6.8}$$

provided the coefficient of thermal expansion α and the isothermal compressibility κ_T are known. In any event, the second term is small but not negligible for solids. Table 6.1 shows the isobaric heat capacity of several solids at room temperature.

According to the Dulong-Petit law, the molar heat capacity of a solid is constant and has the same value $3R$ for any solid. We assume that the Dulong-Petit law is valid for the isochoric molar heat capacity,

$$c_v = 3R. \tag{6.9}$$

From Table 6.1 and taking into account that c_v is close to c_p, we see that the Dulong-Petit Law is reasonably obeyed at room temperature, but with some significant deviations. However, the greatest deviation from the Dulong-Petit law is not apparent in Table 6.1, because it occurs at low temperatures. In this regime the heat capacity becomes small and vanishes when $T \to 0$.

Table 6.1 Isobaric heat capacity of solids at temperature of 25 °C and pressure of 1 kPa. The table shows the specific heat $\tilde{c}_p$ and the molar heat capacity c_p, in addition to the molar mass M and the Debye temperature Θ_D. *Source*: CRC, AIP

Solid	$\tilde{c}_p$ (J/gK)	M (g/mol)	$c_p/3R$	Θ_D (K)
Lithium	3.582	6.941	0.997	344
Graphite	0.709	12.01	0.341	420
Diamond	0.509	12.01	0.245	2230
Aluminium	0.903	26.98	0.976	428
Silicon	0.705	28.09	0.793	640
Sodium	1.228	22.99	1.132	158
Magnesium	1.023	24.31	0.997	400
Potassium	0.757	39.10	1.187	91
Calcium	0.647	40.08	1.040	230
Titanium	0.523	47.87	1.005	420
Chromium	0.449	52.00	0.936	630
Manganese	0.479	54.94	1.055	410
Iron	0.449	55.84	1.006	467
Nickel	0.444	58.69	1.045	450
Cobalt	0.421	58.93	0.995	445
Copper	0.385	63.55	0.980	343
Zinc	0.388	65.39	1.018	327
Germanium	0.320	72.61	0.931	370
Niobium	0.265	92.91	0.986	275
Silver	0.235	107.9	1.016	225
Cadmium	0.232	112.4	1.043	209
Tin	0.228	118.7	1.087	199
Tellurium	0.202	127.6	1.032	153
Cesium	0.242	132.9	1.291	38
Gadolinium	0.236	157.2	1.485	195
Tungsten	0.132	183.4	0.973	400
Platinum	0.133	195.1	1.037	240
Gold	0.129	197.0	1.019	165
Lead	0.129	207.2	1.068	105
Bismuth	0.122	209.0	1.023	119

From the microscopic point of view, the Dulong-Petit law can be understood if we consider a harmonic solid, that is, a solid whose interatomic potential energy contains terms up to second order in the deviations of atoms from their equilibrium positions. It can be shown that such a system is equivalent to a set of independent one-dimensional oscillators whose frequencies of oscillation are the frequencies of the normal modes of vibration of the solid. Appealing to the theorem of equipartition of energy, valid for high temperatures, we assume that the kinetic and potential energies of each oscillator contributes each with a term equal to $k_B T/2$ to the total energy of the solid, where k_B is the Boltzmann constant. Since the number of oscillators is equal to the number of degrees of freedom $3n$, where n is the total number of atoms of the solid, then the contribution of the total energy of oscillators is $3nk_BT = 3NRT$. Taking into account the potential energy of atoms in its equilibrium position, which we denote by U_0, then the molar energy u of the solid is

$$u = u_0 + 3RT \tag{6.10}$$

where $u_0 = U_0/N$ depends only on the molar volume v. From this result we get $c_v = (\partial u/\partial T)_v = 3R$, which is the Dulong-Petit law.

Debye Theory

At low temperatures, the heat capacity of solids deviates appreciably from the Dulong-Petit law. According to Debye theory, the heat capacity of the solids behaves, in this regime, according to the law

$$c_v = aT^3, \tag{6.11}$$

and vanishes as $T \to 0$, in accordance with the Nernst postulate. The same behavior applies to c_p, as we shall see, which can be seen in Fig. 6.2 where the experimental data of c_p/T for KCl are plotted as a function of T^2.

The Debye theory also provides an interpolation between this law, valid for low temperatures, and the law of Dulong-Petit. According to Debye, the isochoric molar heat capacity of a solid harmonic behaves as

$$c_v = 3R\,\mathrm{C}(\frac{\Theta_\mathrm{D}}{T}), \tag{6.12}$$

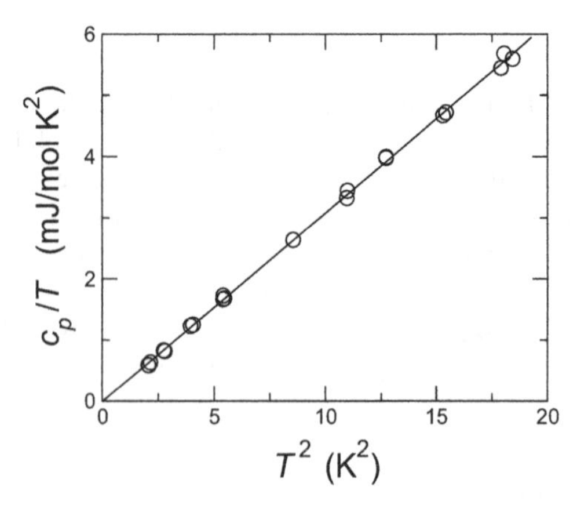

Fig. 6.2 Molar heat capacity of KCl at low temperatures obtained experimentally by Keesom and Pearlman [115]

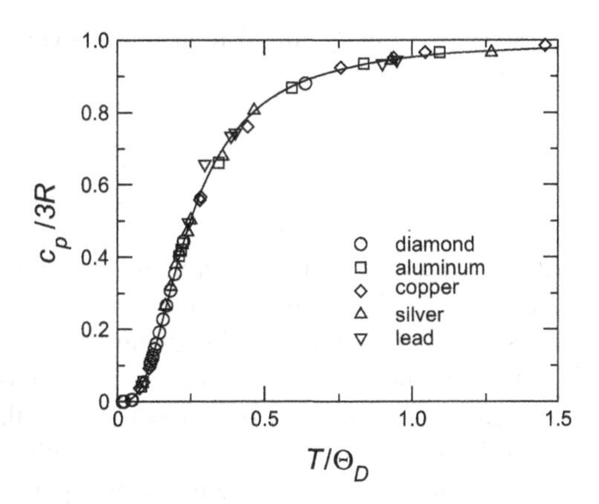

Fig. 6.3 Molar heat capacity of several solids obtained experimentally, according to Debye [110]. The *continuous line* corresponds to the Debye function given by (6.13)

where $\mathsf{C}(x)$ is the Debye function defined by (Fig. 6.3)

$$\mathsf{C}(x) = \frac{3}{2x^3} \int_0^x \frac{\xi^4}{\cosh \xi - 1} d\xi \tag{6.13}$$

and has the following properties: (a) $\mathsf{C}(x) = 1 - x^2/20$ for $x << 1$ and (b) $\mathsf{C}(x) = 4\pi^4/5x^3$ for $x >> 1$. At high temperatures, $T >> \Theta_\text{D}$, we use the first property to obtain the Dulong-Petit law. At low temperatures, $T << \Theta_\text{D}$, we use the second property to obtain the result

$$c_v = \frac{12\pi^4 R T^3}{5\Theta_\text{D}^3}, \tag{6.14}$$

which is the law (6.11) with

$$a = \frac{12\pi^4 R}{5\Theta_\text{D}^3}. \tag{6.15}$$

The parameter Θ_D, called the Debye temperature, is a characteristic of the solid and is related to the normal frequencies of vibration of the solid. Usually, these frequencies are affected by variations of the distances R_{ij} between equilibrium positions of atoms. This can be understood if we examine the coefficients of the quadratic terms of the expansion of the potential energy in the deviations of atoms from their equilibrium positions. These coefficients generally depend on R_{ij}. As the normal frequencies are obtained from the coefficients of second order, then they depend on the distances R_{ij}. On the other hand, the distances R_{ij} are proportional to $v^{1/3}$, from which follows that the normal frequencies and therefore Θ_D, depend on the molar volume v. Only for an ideal harmonic solid, for which the second order coefficients are independent of the volume, the normal frequencies and Θ_D are independent of the volume.

The Debye temperature is related to the velocity of propagation of elastic waves v_{som} by

$$\Theta_{\text{D}} = \frac{\hbar v_{\text{som}}}{k_B} \left(\frac{6\pi^2}{v} N_A \right)^{1/3} . \tag{6.16}$$

From the experimental measurement of c_v at low temperatures, we obtain the coefficient a and from this coefficient we may obtain Θ_{D} by (6.15). Experimental values of Θ_{D} for several solids obtained in this manner are shown in Table 6.1. The Debye temperature obtained by this method can then be compared with that obtained from the velocity of elastic waves by the use of formula (6.16).

The Debye theory is valid for solids with harmonic interactions. At low temperatures one expects that the harmonic approximation is reasonable, considering that the atoms perform small oscillations in this regime, so that the law (6.11) is in fact obeyed at low temperatures. At high temperatures, anharmonic terms of the interatomic interactions become important and deviations from the Debye theory, in this regime, and from the Dulong-Petit law, begin to be observed.

It is important to note also that some solids have very high Debye temperatures such that the regime of the Dulong-Petit law is not reached at room temperature. This is the case of diamond whose Debye temperature is 2230 K, well above room temperature. We also note that some solids melt before the regime of the Dulong-Petit law is reached.

Thermal Expansion

The thermal expansion α of solids is related, as we will see, to the Grüneisen coefficient γ_{G}, defined by

$$\gamma_{\text{G}} = -\frac{d \ln \Theta_{\text{D}}}{d \ln v} . \tag{6.17}$$

To determine α, we use the relation $\alpha/\kappa_T = (\partial p/\partial T)_v$ and the Maxwell relation $(\partial p/\partial T)_v = (\partial s/\partial v)_T$ to write

$$\alpha = \kappa_T \left(\frac{\partial s}{\partial v} \right)_T . \tag{6.18}$$

The entropy is obtained by integrating the expression $(\partial s/\partial T)_v = c_v/T$. Using (6.12), we get

$$s = s_0 + 3R \int_0^T \mathsf{C}(\frac{\Theta_{\text{D}}}{T'}) \frac{dT'}{T'} , \tag{6.19}$$

where s_0 is the zero temperature entropy. Then we derive this expression with respect to v, recalling that Θ_D depends on v, to obtain the relation

$$\left(\frac{\partial s}{\partial v}\right)_T = -3R\mathrm{C}(\frac{\Theta_D}{T})\frac{1}{\Theta_D}\frac{d\Theta_D}{dv} = -c_v\frac{d\ln\Theta_D}{dv}, \tag{6.20}$$

where we have taken into account, by the Nernst postulate, that s_0 is independent of v. Using the definition of the Grüneisen coefficient, the following expression follows

$$\alpha = \frac{c_v}{v}\kappa_T\gamma_G. \tag{6.21}$$

It is worth noting that for an ideal harmonic solid, $\gamma_G = 0$ because Θ_D is independent of the volume, so that $\alpha = 0$ and therefore $c_p = c_v$.

Assuming that the compressibility κ_T is weakly dependent on the temperature and remembering that γ_G is independent of T, we may conclude that the ratio α/c_v is independent of the temperature from which follows that α has the same behavior of c_v. Therefore, in the regime where the Dulong-Petit law is valid, α is constant. In this regime, the relation (6.8) implies that the molar isobaric heat capacity c_p has a correction proportional to temperature. In the regime of low temperatures, α must have the behavior

$$\alpha = bT^3. \tag{6.22}$$

This result together with the relation (6.8) implies that the difference between c_p and c_v at low temperatures is of the order T^7, what justifies to say that c_p has the same behavior as c_v in this regime.

Metals

Metals are characterized by electrical conduction, resulting from the mobility of valence electrons of atoms. These electrons contribute to the heat capacity of solid, adding to the heat capacity of the lattice described by the Debye theory. Assuming that the free electrons behave like an ideal gas, we would expect an electronic contribution equal to $3R/2$ for the molar heat capacity at least for high temperatures, a value that is half the contribution of the crystal lattice. But, the metals at room temperature show no discrepancy in the heat capacity, when compared with an insulator, which could be attributed to the electrons.

According to Sommerfeld, the electronic heat capacity behaves linearly with the temperature in accordance with

$$c_v = \frac{\pi^2}{2}R\frac{T}{\Theta_F}, \tag{6.23}$$

where $\Theta_{\rm F}$ is the Fermi temperature given by

$$\Theta_{\rm F} = \frac{\hbar^2}{2mk_B}\left(\frac{3\pi^2}{v}n_e\right)^{2/3} \tag{6.24}$$

where m is the mass of the electron and n_e is the number of electrons per mole.

The result (6.23) is valid at low temperatures, more precisely at temperatures much smaller than the Fermi temperature. However, given that the Fermi temperature of metals is greater than 10^4 K, we can use the result (6.23) to room temperature and concluded that the electronic contribution to the heat is negligible when compared to the contribution of the crystal lattice.

To observe experimentally the electronic contribution we should consider temperatures much lower than the Debye temperature, in which case the two contributions have the same order of magnitude. Thus, in this regime, the behavior of the heat capacity of metals is given by

$$c_v = \gamma T + aT^3, \tag{6.25}$$

where the constant γ is

$$\gamma = \frac{\pi^2 R}{2\Theta_{\rm F}}. \tag{6.26}$$

It is convenient thus to present the experimental data of the molar heat capacity in a plot of c_v/T versus T^2, as that shown in Fig. 6.4 for copper at low temperatures.

To determine the coefficient of thermal expansion of metals, we start from the molar Helmholtz free energy f of free electrons, given, according to Sommerfeld, by

$$f = \frac{3}{5}R\Theta_{\rm F} - \frac{\pi^2 RT^2}{4\Theta_{\rm F}}. \tag{6.27}$$

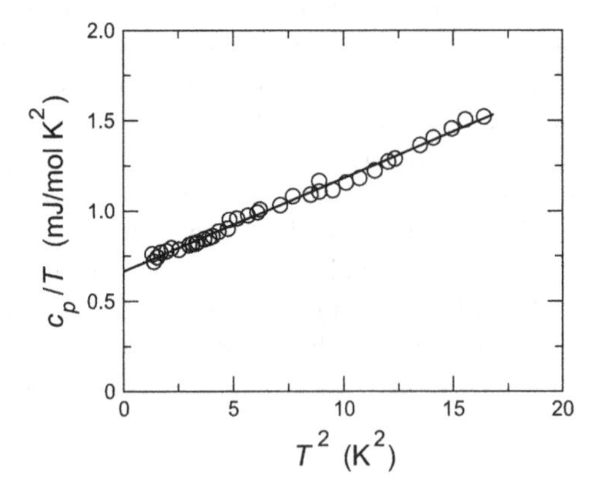

Fig. 6.4 Molar heat capacity of copper at low temperatures obtained experimentally by Corak et al. [109]. The *straight line* fitted to the data points is described by $c/T = 0.665 + 0.0516T^2$

The electronic pressure $p = -\partial f/\partial v$ is obtained from f recalling that Θ_F depends on v. Using the result (6.24) and taking into account that n_e is constant, then $d\Theta_\mathrm{F}/dv = (-2/3)(\Theta_\mathrm{F}/v)$ and we get

$$p = \frac{2R\Theta_\mathrm{F}}{5v} + \frac{\pi^2 RT^2}{6v\Theta_\mathrm{F}}. \tag{6.28}$$

The isothermal compressibility κ_T is obtained from this equation by means of $1/\kappa_T = -v(\partial p/\partial v)_T$. The dominant term in (6.28) gives the result

$$\kappa_T = \frac{3v}{2R\Theta_\mathrm{F}}. \tag{6.29}$$

Using the identity $\alpha/\kappa_T = (\partial p/\partial T)_v$ and the result (6.29), we obtain from (6.28) the following result for the coefficient of thermal expansion of metals

$$\alpha = \frac{\pi^2 T}{2\Theta_\mathrm{F}^2}, \tag{6.30}$$

that is, α depends linearly on temperature. Therefore, the electronic coefficient of thermal expansion has a behavior similar to the electronic heat capacity and moreover $\alpha/c_v = 1/R\Theta_\mathrm{F}$.

The coefficient of thermal expansion of a metal at low temperatures has also a contribution corresponding to the lattice, which is proportional to T^3. Therefore, α has a behavior similar to the heat capacity, namely,

$$\alpha = \delta T + bT^3, \tag{6.31}$$

where the first and second terms are the electronic and the lattice contribution, respectively. Figure 6.5 shows this behavior for copper at low temperatures.

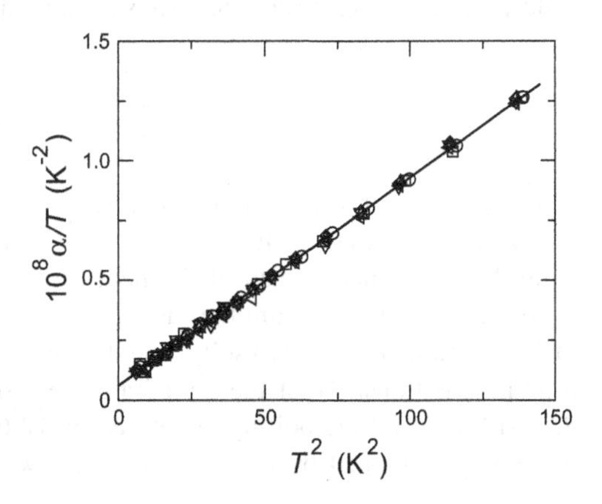

Fig. 6.5 Coefficients of thermal expansion α of cooper a low temperatures obtained experimentally by White and Collins [123]. The *straight line* fitted to the data points is described by $10^{10}\alpha/T = 6.0 + 0.87T^2$

6.3 Planck Postulate

Entropy

According to the Nernst postulate, the molar entropy s of a system approaches a finite value s_0 in the limit $T \to 0$ and that, moreover, s_0 is independent of thermodynamic variables, that is, a constant. One could imagine this constant as a characteristic of the substance and, therefore, different for each substance. However, Planck postulated that s_0 has the same value for any substance in thermodynamic equilibrium then assigning the value zero to the entropy at absolute zero. Therefore, the Nernst postulate together with the Planck postulate, which we call the Nernst-Planck principle or the third law of thermodynamics, tells us that

$$s(T) \to 0 \qquad \text{when} \qquad T \to 0, \tag{6.32}$$

for any substance in thermodynamic equilibrium.

To verify the Planck postulate we must define a reference state whose entropy s_R is known. Next, we determine the entropy difference Δs between the reference state and the ground state. Since the entropy of the ground state is zero then we must have $s_R = \Delta s$. In general, the reference state is considered to be a gas at high temperatures and low densities, so that it can be considered an ideal gas. Let us first examine how to determine Δs. Subsequently, we examine the entropy of the reference state.

Consider for simplicity an isobaric process that takes a system composed by a pure substance from one state at sufficiently low temperature to the reference state. In this process the system undergoes the transition from solid to liquid, from liquid to gas, and possible transitions between different crystalline states. In each transition, occurring at a well defined temperature, the system gains entropy which is equal to the latent heat divided by the transition temperature. From the experimental data of the heat capacity c_p and latent heats q_i of each transition, the entropy difference Δs between the reference state and the ground state is determined by

$$\Delta s = \int_0^T \frac{c_p}{T} dT + \sum_i \frac{q_i}{T_i}, \tag{6.33}$$

where T_i are the transition temperatures.

Table 6.2 shows the values of the several terms of the entropy of xenon, obtained from the experimental data of the heat capacity of solid and liquid xenon at the pressure of 1 atm, and from the fusion and boiling latent heats. Under the pressure of 1 atm xenon undergoes a solid-liquid transition at the temperature of 161.3 K and a liquid-gas transition at the temperature of 165.13 K. The melting latent heat is 2.293 kJ/mol and the boiling latent heat is 12.62 kJ/mol. Between 0 and 10 K the entropy is estimated by using the Debye law assuming $\Theta_D = 55$ K. According

Table 6.2 Entropy of xenon according to the experimental data obtained by Clusius and Riccoboni [108] at the pressure of 1 atm

State	Interval in T (K)	Δs (J/mol K)
Solid	$0 \rightarrow 161.3$	62.4
Fusion	161.3	14.2
Liquid	$161.3 \rightarrow 165.13$	3.8
Boiling	165.13	76.4
Correction		0.6
Solid $\rightarrow$ vapor	$0 \rightarrow 165.13$	157.4
Sackur-Tetrode	165.13	157.3

to Table 6.2, the difference in entropy between xenon in the vapor state, under the pressure of 1 atm at the boiling temperature, and the solid xenon, at zero temperature, is $\Delta s = 157.4$ J/mol K.

Ideal Gases

In the regime of low densities, the average distance between gas molecules is large enough so that the interaction between them can be neglected. Thus, in this regime, the gas obeys the equation of state of ideal gas $pv = RT$ and the internal energy u depends only on temperature. As a result, $c_v = (\partial u/\partial T)_v$ depends only on temperature and the relation between c_p and c_v becomes simply $c_p = c_v + R$ since $c_p = (\partial h/\partial T)_p$ and $h = u + pv = u + RT$.

If the interaction between molecules is negligible, it suffices to consider the motion of molecules in isolation. The movement of a molecule is composed of translation, rotation and vibration and other movements that we will not consider here. The translation motion is always independent of the other types of motion. The rotation and vibration of the molecules are in general interdependent movements. However, the separation into independent motions constitutes a good approximation. Thus, we assume that the thermodynamic potential of an ideal gas is equal to the sum of three terms corresponding to the three types of motions. The isochoric heat capacity is therefore a sum of three terms

$$c_v = c_{\text{trans}} + c_{\text{rot}} + c_{\text{vib}}. \tag{6.34}$$

Similarly, the entropy will be the sum of three terms

$$s = s_{\text{trans}} + s_{\text{rot}} + s_{\text{vib}}. \tag{6.35}$$

We assume that each type of motion is characterized by a certain energy value below which the movement is frozen and the corresponding degrees of freedom suppressed. Thus, we introduce the characteristic energy of each type of motion,

Table 6.3 Characteristic temperatures of rotation and vibration of molecules of several gases, in K. The indices ^{d} and ^{t} indicate the double and triple degeneracies, respectively. *Source*: LB

Gas	$\Theta_{r,1}$	$\Theta_{r,2}$	$\Theta_{r,3}$	$\Theta_{v,1}$	$\Theta_{v,2}$	$\Theta_{v,3}$	$\Theta_{v,4}$
H_2	85.3			5995			
D_2	43.0			4300			
O_2	2.11			2238			
F_2	1.27			1283			
Cl_2	0.36			801			
N_2	2.92			3352			
HCl	15.3			4160			
CO	2.82			3081			
NO	2.45			2701			
CO_2	0.57			961^d	1924	3379	
CS_2	0.15			571^d	943	2190	
HCN	2.16			1024^d	3002	4765	
N_2O	0.62			847^d	1851	3198	
H_2S	6.84	10.23	15.05	1777	3753	3861	
SO_2	3.04	0.47	0.41	755	1647	1918	
H_2O	39.4	21.0	13.7	2293	5259	5401	
NO_2	11.6	0.63	0.60	1079	1901	2324	
NH_3	14.5	14.5	9.34	1365	2344^d	4794	4911^d
CH_4	7.41	7.41	7.41	1877^t	2186^d	4190	4343^t

which we denote by ϵ_{trans}, ϵ_{rot} and ϵ_{vib}. It is convenient to define the characteristic temperatures $\Theta_{\text{t}} = \epsilon_{\text{trans}}/k_B$, $\Theta_{\text{r}} = \epsilon_{\text{rot}}/k_B$ and $\Theta_{\text{v}} = \epsilon_{\text{vib}}/k_B$. Extending the Nernst-Planck principle to each type of motion, we conclude that the heat capacity and the entropy associated to a particular type of movement vanish at temperatures well below the corresponding characteristic temperature.

In general, the characteristic temperatures are such that

$$\Theta_{\text{t}} << \Theta_{\text{r}} << \Theta_{\text{v}}. \tag{6.36}$$

Table 6.3 shows values of the characteristic temperatures of rotation and vibration of some gases. Therefore, as the temperature of the gas is decreasing, the vibrational, rotational and translational modes will be successively removed. This can be seen in Fig. 5.3, where we present the molar heat capacity at constant pressure of several gases. We must remark however that in general the gases condense long before the regime where $T < \Theta_{\text{t}}$ could be reached. Equivalently, we can say that the gases are usually found in the regime $T >> \Theta_{\text{t}}$.

To determine the characteristic energy of translation Θ_{t}, we use the relation $p = h/\lambda$ between the moment of a molecule and its de Broglie wavelength λ, where h is the Planck constant. The corresponding kinetic energy is $p^2/2m = h^2/2m\lambda^2$. The lowest energy, identified as ϵ_{trans}, is given by the largest value of λ, that we consider to be equal to the distance between gas molecules, that is, $\lambda = (\bar{v})^{1/3}$, where $\bar{v}$ is the volume per molecule. That is, $\epsilon_{\text{trans}} = h^2/2m(\bar{v})^{2/3}$ so that

$$\Theta_{\text{t}} = \frac{h^2}{2m(\bar{v})^{2/3}k_B}. \tag{6.37}$$

This temperature can be estimated by assuming an intermolecular distance of the order of 10^{-8} m. For helium, the lightest of gases, we obtain the result $\Theta_{\text{t}} = 2.4$ mK. For other gases the characteristic temperature is even smaller.

Sackur-Tetrode Formula

The translational part of the molar heat capacity of an ideal gas is $c_v = (3/2)R$ because each molecule has three translational degrees of freedom. If the ideal gas is monoatomic this is the only contribution to the heat capacity. Using the relation $(\partial s/\partial T)_v = c_v/T = 3R/2T$ and taking into account that $(\partial s/\partial v)_T = (\partial p/\partial T)_v = R/v$, we reach the following expression for the molar entropy of an ideal gas

$$s = R\{\frac{3}{2}\ln T + \ln \bar{v} + a\}, \tag{6.38}$$

where $\bar{v} = v/N_A$ is the volume per molecule, N_A is the Avogadro constant and a is a constant, independent of T and v.

The constant a must depend only on the characteristic of the gas with respect to the translational kinetic energy. In this sense, the constant a depends solely on the mass of the molecules, which is the parameter associated to the translational kinetic energy. Sackur assumed that a depends on the mass m of the gas molecules in accordance with

$$a = \frac{3}{2}\ln m + b, \tag{6.39}$$

where b is a universal constant. This constant, determined by Tetrode by means of the semi-classical theory is given by

$$b = \frac{3}{2}\ln\frac{2\pi k_B}{h^2} + \frac{5}{2}, \tag{6.40}$$

where h is the Planck constant.

Replacing these results into (6.38) gives the Sackur-Tetrode formula

$$s = R\{\ln \bar{v} + \frac{3}{2}\ln\frac{2\pi m k_B T}{h^2} + \frac{5}{2}\}, \tag{6.41}$$

which allows us the determination of the translation absolute entropy of the ideal gases. Sackur-Tetrode formula can also be written as

$$s = R\{\frac{3}{2}\ln\frac{T}{\Theta_t} + \frac{3}{2}\ln\pi + \frac{5}{2}\}, \tag{6.42}$$

recalling that Θ_t depends on $\bar{v}$.

The determination of the entropy through the Sackur-Tetrode formula enables us to verify the Planck postulate for substances that turn into monoatomic ideal gas at high temperatures. To this end it suffices to compare s, obtained by Sackur-Tetrode formula with Δs obtained from (6.33). As an example, we determine the entropy of vapor xenon in coexistence with its liquid at the pressure of 1 atm. Formula (6.41) provides the following value for the absolute molar entropy $s = 157.3$ J/mol K, which should be compared with that obtained from (6.33), shown in Table 6.2.

We remark that an ideal gas of diatomic or polyatomic molecules can behave as a monoatomic gas, that is, as a gas having only translational degrees of freedom if it can be found in the regime where $T << \Theta_r$. However, the diatomic and polyatomic gases in general are found in the opposite regime, where $T >> \Theta_r$. An exception is hydrogen, which remains in the gaseous state even at temperatures below the characteristic temperature of rotation $\Theta_r = 85.3$ K. In this regime, the molar heat capacity c_v of H_2 approaches the value $(3/2)R$, corresponding to a monoatomic gas, as shown in Fig. 5.3.

Rotational Modes

For temperatures well above Θ_r, the molar heat capacity of gases composed by linear diatomic or polyatomic molecules is given by

$$c_{rot} = R, \tag{6.43}$$

because such molecules have only two rotational degrees of freedom. The rotational molar entropy has, therefore, the form

$$s_{rot} = R\{\frac{1}{2}\ln T + a'\}, \tag{6.44}$$

where a' is a constant. This constant must depend on the characteristic of the gas with respect to the rotation motion. In this sense, the constant a' must depend on the moment of inertia I of the linear molecule, which is the parameter associated to rotation.

A semi-classical calculation leads us to the following expression

$$s_{rot} = R\{\ln\frac{T}{\Theta_r} - \ln\sigma + 1\}, \tag{6.45}$$

where Θ_r is the characteristic temperature of rotation and is given by

$$\Theta_r = \frac{\hbar^2}{2Ik_B}. \tag{6.46}$$

The parameter σ is related to the symmetries of molecules. For diatomic molecules consisting of distinct type of atoms $\sigma = 1$ and for the same type of atoms $\sigma = 2$. For asymmetric polyatomic molecules, like N_2O (NNO), $\sigma = 1$ and for symmetric polyatomic molecules, like CO_2 (OCO), $\sigma = 2$.

For noncolinear polyatomic molecules, the number of rotational degrees of freedom is three, so that the rotational contribution to the molar heat capacity is

$$c_{\text{rot}} = \frac{3}{2}R. \tag{6.47}$$

The corresponding molar entropy is

$$s_{\text{rot}} = R\{\frac{3}{2}\ln\frac{T}{\Theta_r} - \ln\sigma + \frac{1}{2}\ln\pi + \frac{3}{2}\}, \tag{6.48}$$

where

$$\Theta_r = (\Theta_{r,1}\Theta_{r,2}\Theta_{r,3})^{1/3}. \tag{6.49}$$

The characteristic temperatures $\Theta_{r,i}$ are related with the principal moments of inertia I_1, I_2 and I_3 of the molecules by

$$\Theta_{r,i} = \frac{\hbar^2}{2I_ik_B}. \tag{6.50}$$

Characteristic temperatures of rotational modes of several molecules are presented in Table 6.3. The parameter σ is related to the symmetries of the molecule. For H_2O (isosceles triangle), $\sigma = 2$; for NH_3 (pyramid with triangular base), $\sigma = 3$; and for CH_4 (tetrahedron), $\sigma = 12$. We remark that the formulas for entropy and the heat capacity are valid for temperatures well above the characteristic temperatures of rotation.

Vibrational Modes

The vibrational part of entropy and of the heat capacity can be determined assuming that the oscillations are harmonic. For a gas of diatomic molecules, the vibrational molar entropy is given by

$$s_{\text{vib}} = R\{\frac{\Theta_v}{T}\frac{1}{e^{\Theta_v/T}-1} - \ln(1-e^{-\Theta_v/T})\}, \tag{6.51}$$

and the molar heat capacity by

$$c_{\text{vib}} = R\frac{\Theta_{\text{v}}^2}{T^2}\frac{e^{\Theta_{\text{v}}/T}}{(e^{\Theta_{\text{v}}/T}-1)^2}. \tag{6.52}$$

where the characteristic temperature Θ_{v} is related to the frequency ν of the sole mode of vibration of the diatomic molecule by

$$\Theta_{\text{v}} = \frac{h\nu}{k_B}. \tag{6.53}$$

Result (6.52) was obtained originally by Einstein for the heat capacity of solids assuming that they oscillate with the same frequency.

Polyatomic molecules in general have various modes of vibration, each one associated to a natural frequency. The vibrational entropy and heat capacity of polyatomic gases are determined in a manner similar to that of the diatomic gases. For each normal mode of oscillation of the molecule, we add to the molar entropy a term equal to (6.51) and, to the molar heat capacity, a term equal to (6.52). The term corresponding to the normal mode of frequency ν_i must contain the characteristic temperature

$$\Theta_{\text{v},i} = \frac{h\nu_i}{k_B}. \tag{6.54}$$

Characteristic temperatures of the vibrational modes of various molecules are shown in Table 6.3.

In the regime where $T >> \Theta_{\text{v}}$, the terms of the entropy and of the heat capacity corresponding to each normal mode are given by

$$s_{\text{vib}} = R\{\ln\frac{T}{\Theta_{\text{v}}} + 1\} \tag{6.55}$$

and

$$c_{\text{vib}} = R, \tag{6.56}$$

respectively. We remark, however, that this regime is never reached because in general the polyatomic molecules decompose well before this happens.

Residual Entropy of Ice

At the pressure of 1 atm, water freezes at 0 °C becoming a crystalline solid, that we call ice. A way of determining the absolute entropy s_0 of ice at zero temperature

Table 6.4 Entropy of water from experimental data for the heat capacity presented in Fig. 6.6 and from the latent heats of fusion and boiling

State	Interval in T (K)	Δs (J/mol K)
Ice	$0 \to 273.15$	38.05
Ice melting	273.15	22.00
Water	$273.15 \to 373.15$	23.49
Boiling water	373.15	108.96
Ice → vapor	$0 \to 373.15$	192.50

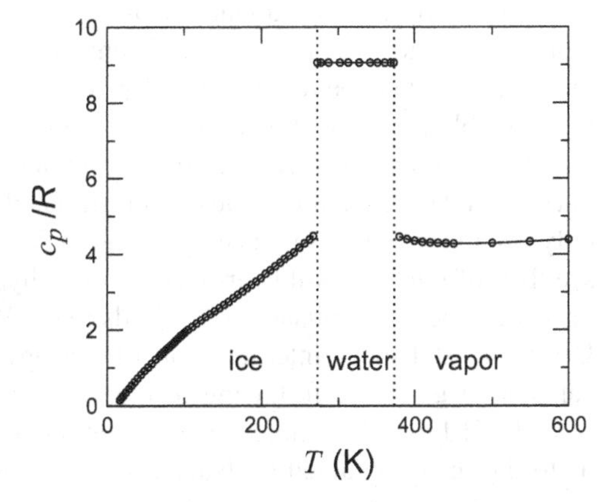

Fig. 6.6 Molar isobaric heat capacity c_p of water in the three states at the pressure of 1 atm. The *dashed lines* indicate the ice-water transition at 273.15 K and the water-vapor transition at 373.15 K. The results for ice were obtained experimentally by Giauque and Stout [114]

consists in calculating the difference in entropy Δs between the state at zero temperature and a reference state, which we choose as the vapor in coexistence with water at the pressure of 1 atm and temperature of 373.15 K. Table 6.4 shows the variations of entropy of ice and water, obtained from the experimental data for the heat capacity, presented in Fig. 6.6.

The integral of c_p/T for ice gives 38.05 J/mol K and for the liquid, 23.49 J/mol K. The melting latent heat of ice is 6.006 kJ/mol, so that the entropy at the ice-water transition increases by 22.00 J/mol K. The boiling latent heat of water is 40.660 kJ/mol, so that the entropy at the water-vapor transition increases by 108.96 J/mol K. Therefore, the difference between the molar ice entropy at 0 K and the vapor at 1 atm is $\Delta s = 192.50$ J/mol K, as long as no other transition takes places in addition to the ones mentioned.

The molar entropy s_R of the reference state, which we assumed to be that of an ideal gas, is determined from the formulas for the translational, rotational and vibrational. The translational entropy gives the result $s_{\text{trans}} = 149.33$. To determine the rotational entropy we use the rotational characteristic temperatures of H_2O presented in Table 6.3. We obtain the following result $s_{\text{rot}} = 46.50$ J/mol K. The vibrational contribution to the entropy is small. From the data of Table 6.3 we obtain the value $s_{\text{vib}} = 0.12$ J/mol K. Therefore, the entropy of water vapor is $s_R = 195.95$ J/mol K.

From the values obtained above, we see that ice has a residual entropy $s_0 = s_R - \Delta s = 3.45$ J/mol K, in apparent contradiction with the Nernst-Planck principle. The contradiction disappears if we assume, for example, that at low temperatures ice finds itself in a metastable state. In fact, in this regime one observes experimentally another more stable solid phase, called ice XI, whose crystalline structure is distinct from ordinary ice. The transition between the ordinary ice, called ice I, and ice XI occurs at 72 K, with a decrease of entropy.

Pauling considers that the hydrogen bonds existing in the ice structure have a fundamental role in the explanation of the residual entropy. A hydrogen bond binds atoms of oxygen belonging to different molecules and the hydrogen, actually a hydrogen ion, may be found next to one or the other oxygen atom, that is, in one of two possible positions. Since the number of hydrogen bonds is equal to twice the number n of molecules then the number of microscopic state $W = w^n$ would be such that $w = 4$. However, this value is overestimated because each oxygen ion must have only two hydrogen ions in its neighborhood. Taking into account this restriction, we see that, of the 16 possible arrangements of hydrogen ions around an oxygen ion, only 6 can be accomplished, what leads to the Pauling result $w = 4(6/16) = 3/2$. Using the statistical interpretation of entropy, $S = k_B \ln W$, this approximate calculation gives the following residual molar entropy $s_0/R = \ln(3/2)$, that is, $s_0 = 3.371$ J/mol K. A more correct counting of the microscopic arrangements leads us to the result obtained by Nagle, $w = 1.5069$, that is, $s_0 = 3.409$ J/mol K, in accordance with the observed value, $s_0 = 3.45 \pm 0.20$ J/mol K, taking into account the experimental errors.

We remark finally that ice XI, in contrast to the ordinary ice, has a crystalline structure such that the hydrogen ions are ordered resulting in zero entropy at $T = 0$.

Problems

6.1 At low temperatures, the molar Helmholtz free energy $f(T, v)$ of a solid is given by

$$f = -aT^4 v + bv(\ln v + c).$$

Determine s, p, u, c_v, c_p, κ_T, κ_s and α. Show that these quantities behave according to the Nernst-Planck principle.

6.2 At low temperatures, the molar heat capacity at constant volume of a solid behaves as $c_v = avT^3$. Determine the molar entropy and the quantity $\gamma = (\partial p/\partial T)_v$. Show that the ratio c_v/γ is independent on temperature. Solve the same problem for the case $c_v = av^{2/3}T$.

Table 6.5 Entropy of nitrogen according to the experimental data obtained by Giauque and Clayton [113] at the pressure of 1 atm

State	Interval in T (K)	Δs (J/mol K)
Solid II	$0 \to 35.61$	27.1
Transio II→I	35.61	6.4
Solid I	$35.61 \to 63.14$	23.4
Fuso	63.14	11.4
Liquid	$63.14 \to 77.32$	11.4
Boiling	77.32	72.1
Solid II → vapor	$0 \to 77.32$	151.8

6.3 According to the Debye theory, the molar free energy f of a solid is given by

$$f = u_0 + 3RT\ln(1 - e^{-\Theta_D/T}) - RT\mathsf{D}(\frac{\Theta_D}{T}),$$

where $\mathsf{D}(x)$ is the function defined by

$$\mathsf{D}(x) = \frac{3}{x^3}\int_0^x \frac{\xi^3}{e^\xi - 1}d\xi.$$

Show that the molar energy u is given by $u = u_0 + 3RT\mathsf{D}(\Theta_D/T)$. Obtain, from this result, the expression (6.12) for the molar heat capacity c_v. Determine the behavior of f, u and c_v at low and high temperatures.

6.4 Demonstrate the Slater relation

$$\gamma_G = -\frac{1}{6} - \frac{v}{2B_S}\frac{\partial B_S}{\partial v} = -\frac{1}{6} - \frac{1}{2}\frac{\partial \ln B_S}{\partial \ln v}$$

from the relation between $v_{\rm som}$ and the bulk modulus B_S.

6.5 Determine the absolute molar entropy s of the nitrogen vapor at the pressure of 1 atm and boiling temperature T_e. Compare it with the variation of the molar entropy Δs between the solid nitrogen at $T = 0$ K and the nitrogen vapor at $T = T_e$, obtained by calorimetric measurements shown in Table 6.5. Notice that the solid nitrogen undergoes a structural transition in addition to the two solid-liquid and liquid-vapor transitions.

Chapter 7
Phase Transition

7.1 Pure Substance

State of Matter

A substance may occur in different states of aggregation. At high temperatures and low pressures it occurs in gaseous form. At low temperatures and high pressures, it occurs in solid form. At intermediate temperatures and pressures, it may be a liquid. A gas is characterized by its high compressibility and low density. A solid and a liquid, in contrast, are less compressible, and have moderate or high densities. A solid, contrary to a liquid, offers resistance to deformation.

At the microscopic level the states of matter are characterized by the form of aggregation of the molecules that make up the substance. Either as a liquid or a gas the molecules are very mobile and therefore are spatially disordered. However, the mobility of the molecules and the average distance between molecules in a gas are much larger. In a solid, on the other hand, the mobility is greatly reduced and the movement of the molecules is only local, enabling the regular spatial structures of molecules, which characterize the crystalline solids. A solid material may have one or more crystalline structures.

Besides these three states of aggregation of matter, other states may arise in substances composed of more complex molecules. For example, liquid crystals consist of long molecules favoring the emergence of a state, called nematic, which has orientational order but is devoid of spatial order. We must also mention the glass, traditionally made by the rapid cooling of fused silica. Although a solid material, glass and other amorphous materials do not have the regular structural found in crystalline solids. But, the glassy state is not a state of thermodynamic equilibrium and therefore will not be the subject of our study.

M.J. de Oliveira, *Equilibrium Thermodynamics*, Graduate Texts in Physics,
DOI 10.1007/978-3-662-53207-2_7

Thermodynamic Phases

A homogeneous system, that is, completely uniform with regard to specific properties, constitutes a thermodynamic phase. From the microscopic point of view, the molecules that comprise a thermodynamic phase are in the same type of aggregation. The gaseous state, the liquid state and the various crystalline states are possible thermodynamic phases of a system. A system may also be heterogeneous, that is, display simultaneously two or more thermodynamic phases. In this case, if the system is in equilibrium, we are facing a coexistence of phases.

The phenomenon of phase coexistence is particularly useful in determining the purity of a substance. Suppose that a system consisting of two phases in coexistence receive heat at constant pressure. In this process, one phase grows at the expenses of the other which decreases. If the system consists of a pure substance, the temperature remains unchanged as long as there are phases in coexistence. Thus, for a given pressure, the transition between two phases of a pure substance occurs at a well defined temperature.

From the microscopic point of view, a pure substance consists of a single type of atom (simple substance) or a single type of molecule (composite substance). As an example, distilled water, formed by molecules of H_2O. Many minerals consist of pure or nearly pure substances such as diamond and graphite, which are different crystalline forms of carbon (C), and quartz, consisting almost exclusively of silica (SiO_2).

Phase Diagram

The various thermodynamic phases of a pure substance are represented in a pressure-temperature diagram, called phase diagram. The different phases are represented by regions. The coexistence of two thermodynamic phases is represented by a line in the diagram since, for a given pressure, the coexistence of two phases must occur at a well defined temperature and not in a range of temperatures. Two lines of coexistence may meet. The meeting point of coexistence lines is a triple point, that is, the point of coexistence of three phases. The liquid-vapor coexisting line always ends in a point, called critical point, where the liquid and vapor become indistinct.

Figure 7.1 shows the phase diagram of pure water (H_2O). At a pressure of 1 atm (101,325 Pa), ice and water coexist at a temperature of 0 °C (273.15 K). At the same pressure, water and vapor coexist at a temperature of 100 °C (373.15 K). Heating water, at the pressure of 1 atm, from the room temperature, its temperature will increase to the point where it starts to boil when reaching 100 °C. While boiling, its temperature remains unchanged. The temperature will rise again only when all water has been evaporated. Similarly, if we remove heat of a certain amount of water at a pressure of 1 atm, starting from the room temperature, its temperature will decrease until reaching 0 °C when it starts to freeze. While water is being transformed into ice, the temperature remains unchanged and will decrease only when all water has been transformed into ice.

The line of coexistence of water and its vapor has a positive slope which means that at pressures lower than 1 atm boiling occurs at lower temperatures. In places of high altitudes, where the atmospheric air pressure is lower, the water boils at lower temperatures. At pressures higher than 1 atm, water boils at temperatures higher than 100 °C, which is the working principle of pressure cookers. At still higher pressures, the line of coexistence of water-vapor eventually ends at a critical point, where water and steam become indistinct, which happens at the pressure of 22.06 MPa (217.7 atm) and at temperature of 374.00 °C (647.15 K), as shown if Fig. 7.1. The line of coexistence of ice-water has a negative slope, in contrast to most substances, whose lines of coexistence solid-liquid have positive slope. Decreasing the pressure, the melting temperature slightly increases. The two lines of transition, ice-water and water-vapor, meet at the triple point that occurs at the pressure of 611.7 Pa (0.006037 atm) and at the temperature of 0.01 °C (273.16 K).

Figure 7.2 shows the phase diagram of carbon dioxide (CO_2). The triple point occurs at a temperature of −56.57 °C (216.58 K) and at the pressure of 0.518 MPa (5.11 atm). As the pressure of the triple point is greater than 1 atm, the carbon dioxide does not exist in the liquid state at a pressure of 1 atm. At this pressure, the solid carbon dioxide passes directly to the gaseous state at a temperature of −78.45 °C (194.7 K), a phenomenon known as sublimation. However, it is possible to obtain liquid carbon dioxide, for example, at the temperature of 20 °C, increasing the pressure to 5.73 MPa (56.5 atm). The critical point of CO_2 occurs at the temperature of 30.99 °C (304.14 K) and at the pressure of 7.375 MPa (72.79 atm).

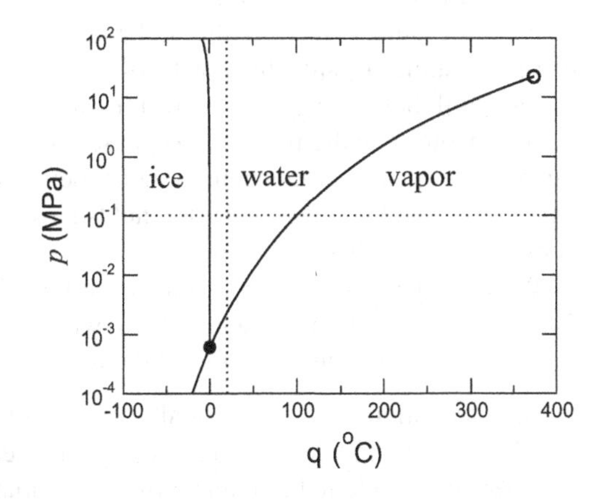

Fig. 7.1 Phase diagram of water. The *dotted horizontal line* indicates the pressure of 1 atm and the *vertical dotted line*, the temperature of 20 °C. The triple point is represented by a *full circle* and the critical point by an *empty circle*. Source: LB

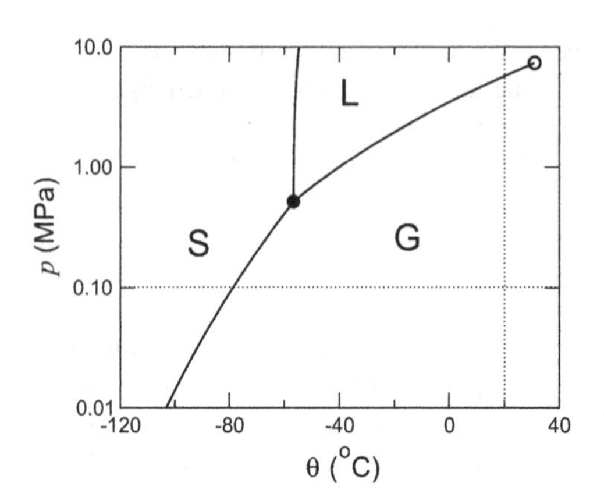

Fig. 7.2 Phase diagram of carbon monoxide (CO_2) showing the gas (G), the liquid (L) and solid (S) phases. The *horizontal dotted line* indicates the pressure of 1 atm and the *vertical dotted line*, the temperature of 20 °C. The triple point is represented by a *full circle* and the critical point by an *empty circle*. Source: LB

7.2 Discontinuous Transition

Densities and Fields

For a pure substance, the fundamental relation in the energy representation is constituted by the internal energy $U(S, V, N)$ as function of entropy S, volume V and number of moles N of the substance. As the internal energy is extensive, it is possible to describe the thermodynamic properties of a pure substance by the molar internal energy of $u(s, v)$ as a function of the molar entropy s and the molar volume v. The fundamental relation is recovered by means of $U(S, V, N) = Nu(S/N, V/N)$. Performing the Legendre transformations starting from $u(s, v)$, we obtain the molar potentials $f(T, v)$, $h(s, p)$ and $g(T, p)$ which describe equivalently the thermodynamic properties of a pure substance.

The variables such as temperature and pressure are known as thermodynamic fields while others, such as the molar entropy and molar volume, are known as thermodynamic density. In a state of thermodynamic equilibrium all parts of a system will have the same value for each one of the fields. If the system is homogeneous (one thermodynamic phase), not only the fields but densities have the same values. In a heterogeneous system (coexistence of one or more thermodynamic phase) on the other hand, while the fields have the same values for each phase, the density will have different values.

We state below the fundamental properties of the molar thermodynamic potentials $u(s, v)$, $f(T, v)$, $h(s, p)$ and $g(T, p)$, with respect to their thermodynamic space. Let ϕ one of these potentials and Ω the respective thermodynamic space. Then:

(a) ϕ is a continuous function of all variables of Ω;
(b) ϕ is differentiable with respect to the densities of Ω; or, equivalently, a thermodynamic field obtained by differentiation of ϕ is a continuous function of all variable of Ω;

(c) ϕ is a convex function of the densities of Ω and a concave function of the fields of Ω;
(d) ϕ is analytic function inside a one-phase region of Ω.

The following formulas allows us to obtain the equation of state from each one of the molar thermodynamic potential:

$$du = Tds - pdv, \tag{7.1}$$

$$df = -sdT - pdv, \tag{7.2}$$

$$dh = Tds + vdp, \tag{7.3}$$

$$dg = -sdT + vdp, \tag{7.4}$$

For a pure substance, the molar Gibbs free energy identifies itself with the chemical potential, that is, $g = \mu$, so that

$$d\mu = -sdT + vdp. \tag{7.5}$$

Equivalently

$$dp = -\bar{s}dT + \bar{\rho}d\mu, \tag{7.6}$$

where $\bar{\rho} = 1/v$ is the number of moles per unit volume, and $\bar{s} = s/v$ is the entropy per unit volume.

Volume Change

Suppose that a pure substance in the liquid state is in coexistence with its vapor whose pressure is kept constant while receiving a certain amount of heat. The heat introduced transforms part of the liquid into vapor thus increasing the total volume of the liquid-vapor system. In this process the temperature remains unchanged as long as the phases are in coexistence. Moreover, considering that each phase is homogeneous, the densities of the liquid and of the vapor are well defined and remain invariant. That is, the molar volumes of the liquid v_{L} and the molar volume of vapor v_{V} remain invariant. The volume change is caused by the transformation of the liquid, which has smaller molar volume, into vapor, which has greater molar volume.

If a certain number of mole N of the liquid is transformed into vapor, the variation in volume will be $\Delta V = Nv_{\mathrm{G}} - Nv_{\mathrm{L}}$, where $v_{\mathrm{G}} = 1/\bar{\rho}_{\mathrm{G}}$ and $v_{\mathrm{L}} = 1/\bar{\rho}_{\mathrm{L}}$ are the molar volumes of the vapor and liquid, respectively. Therefore, the variation $\Delta v = \Delta V/N$, corresponding to the transformation of one mole of the substance in the liquid state into vapor, is equal to the difference between the molar volumes of vapor and liquid,

Table 7.1 Liquid-vapor transition at the pressure of 1 atm of several pure substances. The table presents the boiling temperature T_e, the specific boiling latent heat ℓ_e, the specific volume of the liquid $\tilde{v}_L$ and of the vapor $\tilde{v}_V$. Source: ICT, LB, EG, CRC

Substances		T_e K	ℓ_e J/g	$\tilde{v}_L$ cm^3/g	$\tilde{v}_V$ dm^3/g
Helium	He	4.222	20.7	8.006	0.059
Neon	Ne	27.07	84.8	0.831	0.1051
Argon	Ar	87.293	161.0	0.716	0.173
Krypton	Kr	119.92	108.4	0.413	0.112
Xenon	Xe	165.10	96.1	0.327	0.102
Hydrogen	H_2	20.28	445	14.12	0.7468
Oxygen	O_2	90.188	213.1	0.871	0.2239
Nitrogen	N_2	77.35	198.8	1.239	0.2164
Fluorine	F_2	85.03	174	0.666	0.178
Chlorine	Cl_2	239.11	288	0.640	0.270
Bromine	Br_2	331.9	187.5	0.335	0.17
Carb. monox.	CO	81.6	215.6	1.267	0.230
Ammonia	NH_3	239.82	1370	1.466	1.16
Hydr. chlor.	HCl	188.05	443.1	0.840	0.48
Water	H_2O	373.15	2257	1.04346	1.6731
Methane	CH_4	111.67	510.8	2.367	0.551
Ethylene	C_2H_4	169.38	482.3	1.761	0.480
Ethane	C_2H_6	184.5	488.5	1.830	0.487
Propane	C_3H_8	231.0	431	1.718	0.413
Butane	C_4H_{10}	272.6	386.1	1.663	0.370
Benzene	C_6H_6	353.24	393.3	1.23	0.36
Methanol	CH_4O	337.7	1099	1.332	0.82
Ethanol	C_2H_6O	351.44	837	1.358	0.606
Acetone	C_3H_6O	329.20	501	1.333	0.5
Diethyl ether	$C_4H_{10}O$	307.6	358.8	1.436	0.316

that is, $\Delta v = v_G - v_L$. Table 7.1 presents the specific volumes of liquid and vapor at the transition, at the pressure of 1 atm, of various pure substances.

Suppose that a certain quantity of gas is compressed isothermally. Figure 7.3c shows the representation of this process. The volume of the gas decreases and the pressure increases to the point that the vapor begins to condense. From this point, the decrease in volume does not change the pressure, known as the vapor pressure, until the condensation is complete when the pressure then increases again.

The transition from one phase to another along an isotherm is represented by a horizontal straight line in the p-v diagram because the pressure remains unchanged. The end points of the segment have coordinates (v_L, p^*) and (v_G, p^*), where p^* is the vapor pressure. These two end points represent the thresholds of the liquid-vapor coexistence and the line segment that joins them is called the tie line.

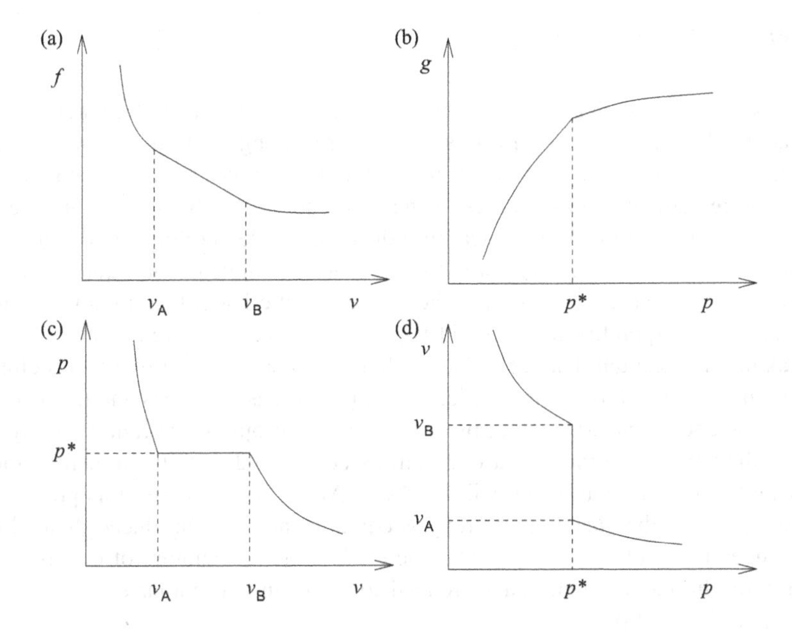

Fig. 7.3 The figures show an isotherm corresponding to a discontinuous phase transition. (**a**) Molar Helmholtz free energy f versus molar volume v. (**b**) Molar Gibbs free energy g versus pressure p. (**c**) Pressure $p = -\partial f/\partial v$ versus v. (**d**) Molar volume $v = \partial g/\partial p$ versus p. Notice that the plots (**a**) and (**b**) form a pair of functions related by a Legendre transformation whereas (**c**) and (**d**) form a pair of inverse functions

Given any point (v, p^*) on the tie line we can determine the fraction in number of moles of each phase. The volume occupied by the liquid is $V_{\rm L} = v_{\rm L} N_{\rm L}$ and by the vapor is $V_{\rm G} = v_{\rm G} N_{\rm G}$, where $N_{\rm L}$ and $N_{\rm G}$ are the number of moles of the liquid and vapor, respectively. The fractions in number of mole are $r_{\rm L} = N_{\rm L}/N$ and $r_{\rm G} = N_{\rm G}/N$, where $N = N_{\rm L} + N_{\rm G}$ is the total number of moles. Therefore, $V = V_{\rm L} + V_{\rm G} = v_{\rm L} r_{\rm L} N + v_{\rm G} r_{\rm G} N$, so that $v = V/N$ is given by

$$v = v_{\rm L} r_{\rm L} + v_{\rm G} r_{\rm G}, \tag{7.7}$$

which is the lever rule. Since $r_{\rm L} + r_{\rm G} = 1$, we can then obtain from it the fractions in number of moles

$$r_{\rm L} = \frac{v_{\rm G} - v}{v_{\rm G} - v_{\rm L}} \quad \text{and} \quad r_{\rm G} = \frac{v - v_{\rm L}}{v_{\rm G} - v_{\rm L}}. \tag{7.8}$$

Therefore, the fraction in number of moles of one phase is proportional do the distance between the point considered and the end point of the tie line corresponding to the other phase.

Latent Heat

Let us suppose again that a pure substance in the liquid state is in coexistence with its vapor, whose pressure is kept constant while receiving a certain amount of heat. The heat introduced transforms part of the liquid into vapor without changing the temperature. The number of moles of the substance in the liquid state that turns into vapor is proportional to the heat introduced L_e, called boiling latent heat. The amount of heat $\ell_e = L_e/N$ required to evaporate one mole of the liquid is called molar boiling latent heat. Table 7.1 shows the specific latent heat of several pure substances corresponding to the liquid-vapor transition at pressure of 1 atm.

Taking into account that each phase is homogeneous, each have a well-defined molar entropy, which remains invariant if the pressure is kept constant and as long as there is coexistence. Denoting by s_L the molar entropy of the liquid and by s_G that of the vapor, then the increase in entropy corresponding to the transformation of N moles of liquid into vapor will be $\Delta S = Ns_G - Ns_L$. Because this process is isothermal, then the change in entropy is equal to the heat introduced divided by the temperature so that $L_e = T^* \Delta S$, where T^* is the temperature of coexistence. Therefore, the molar latent heat is related to the change in molar entropy $\Delta s = s_G - s_L$ by $\ell_e = T^* \Delta s$.

In the T-s diagram, the phase transition along an isobaric line is also represented by a tie line between the points (s_L, T^*) and (s_G, T^*), which indicate the threshold of liquid-vapor coexistence, as shown in Fig. 7.4. For a point (s, T^*), the lever rule also holds

$$s = s_L r_L + s_G r_G. \tag{7.9}$$

The fractions in number of moles of each phase are given by

$$r_L = \frac{s_G - s}{s_G - s_L} \quad \text{e} \quad r_G = \frac{s - s_L}{s_G - s_L}. \tag{7.10}$$

Phase Transition

The phase transition, in the diagram of Helmholtz free energy versus volume, is represented by an inclined line segment whose slope is equal to $-p^*$ because $p = -\partial f/\partial v$, as shown in Fig. 7.3a. In the range $v_L \leq v \leq v_G$ the free energy $f = (N_L f_L + N_G f_G)/N$ is the weighted average of f_L and f_G, that is,

$$f(T, v) = \frac{v_G - v}{v_G - v_L} f(T, v_L) + \frac{v - v_L}{v_G - v_L} f(T, v_G). \tag{7.11}$$

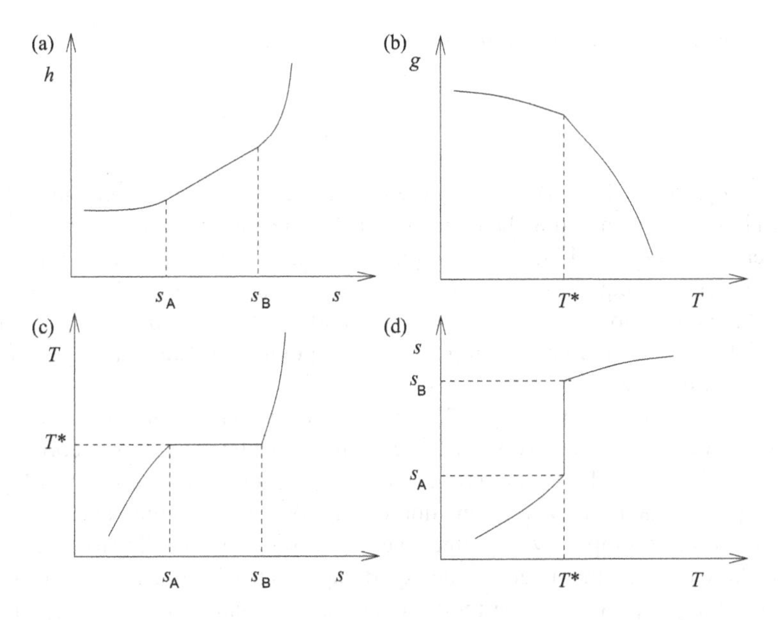

Fig. 7.4 The figures show an isobaric line corresponding to a discontinuous phase transition. (**a**) Molar enthalpy h versus molar entropy s. (**b**) Molar Gibbs free energy g versus temperature T. (**c**) Temperature $T = \partial h/\partial s$ versus s. (**d**) Molar entropy $s = -\partial g/\partial T$ versus T. Notice that the plots (**a**) and (**b**) form a pair of functions related by a Legendre transformation whereas (**c**) and (**d**) form a pair of inverse functions

In this same range, the pressure is

$$p^*(T) = \frac{1}{v_G - v_L} f(T, v_L) - \frac{1}{v_G - v_L} f(T, v_G). \tag{7.12}$$

Inverting the plot of p versus v, we get the plot of v versus p, shown in Fig. 7.3d. In this case the transition is represented by a jump in the molar volume, equal to $\Delta v = v_G - v_L$, at the vapor pressure $p = p^*$. Taking into account that $v = \partial g/\partial p$, the phase transition is represented, in the diagram of Gibbs free energy versus pressure, by a point whose derivatives at left and at right are distinct, as shown in Fig. 7.3b.

Similarly, the phase transition, in the diagram of enthalpy versus entropy, is represented by an inclined line segment because $T = \partial h/\partial s$. In the range $s_L \le s \le s_G$ the enthalpy $h = (N_L h_L + N_G h_G)/N$ is the weighted average of h_L and h_G, that is,

$$h(s, p) = \frac{s_G - s}{s_G - s_L} h(s_L, p) + \frac{s - s_L}{s_G - s_L} h(s_G, p). \tag{7.13}$$

In the same range the temperature is

$$T^*(p) = \frac{1}{s_G - s_L} h(s_G, p) - \frac{1}{s_G - s_L} h(s_L, p). \tag{7.14}$$

The slope is equal to the temperature of coexistence T^* at the pressure p^*, as shown in Fig. 7.4a. Inverting the plot T versus s, shown in Fig. 7.4c, we get the plot of s versus T, shown if Fig. 7.4d. The phase transition is represented by a jump in the molar entropy equal to $\Delta s = s_G - s_L = \ell_e/T^*$ that occurs at the temperature $T = T^*$. Taking into account that $s = -\partial g/\partial T$, the phase transition is represented, in the Gibbs free energy versus temperature, by a point whose derivatives at left and right are different, as seen in Fig. 7.4b.

The graphs presented in Figs. 7.3 and 7.4 correspond to the isothermal and isobaric processes, respectively. In the T-p diagram, these processes correspond to straight lines parallel to the axis T and p, that intersect at the point (T^*, p^*). At this point, where the phase transition occurs, the molar volume and the molar entropy present a jump. If we imagine other processes corresponding to trajectories in the phase diagram that intersect the liquid-vapor transition line at the same point (T^*, p^*), the jump in v will be the same because v_G and v_L depend only on the point on the coexistence line. Similarly, the jump in s as well as the molar boiling latent heat ℓ_e will be the same because s_G, s_L and ℓ_e depend only on the point on the coexistence line.

The phase transition just described, where the molar volume and molar entropy present a jump when one crosses the coexistence line in the T-p diagram is called discontinuous or first-order phase transition.

Clausius-Clapeyron Equation

Consider two nearby points A' and B' in the region corresponding to the liquid phase in the T-p diagram. The difference in the molar Gibbs free energy $\Delta g'$ between these two points is given by

$$\Delta g' = -s_L \Delta T + v_L \Delta p, \tag{7.15}$$

since they are close together, where s_L and v_L are the molar entropy and the molar volume of the liquid phase and ΔT and Δp are the difference in temperature and pressure, respectively, between the two points. Similarly, consider another pair of nearby points A'' and B'' in the region corresponding to the gas phase such that the differences in temperature and pressure of these two points are also ΔT and Δp. The difference in the molar Gibbs free energy $\Delta g''$ between these two points are

$$\Delta g'' = -s_G \Delta T + v_G \Delta p, \tag{7.16}$$

since the two points are close together, where s_G and v_G are the molar entropy and molar volume of the gas phase.

Imagine now two points A and B on the coexistence line such that the difference in temperature and pressure between these points are also ΔT and Δp. Suppose next that the points A' and A'' approach the point A. As a consequence, B' and B'' will approach the point B. Since the Gibbs free energy is a continuous function of T and p, then $\Delta g'' = \Delta g'$ so that

$$-s_G \Delta T + v_G \Delta p = -s_L \Delta T + v_L \Delta p, \tag{7.17}$$

or

$$\frac{\Delta p}{\Delta T} = \frac{s_G - s_L}{v_G - v_L}. \tag{7.18}$$

Taking the limit in which the points A and B approach each other, then

$$\frac{dp}{dT} = \frac{s_G - s_L}{v_G - v_L}, \tag{7.19}$$

which is the Clausius-Clapeyron equation, which can be written in the form

$$\frac{dp}{dT} = \frac{\ell_e}{T(v_G - v_L)}, \tag{7.20}$$

where $\ell_e = T(s_G - s_L)$ is the molar boiling latent heat and T is the transition temperature.

The Clausius-Clapeyron equation relates the slope of the coexistence in the p-T diagram with the discontinuities of entropy and volume when passing from one phase to another. For the liquid-vapor transition, $s_G > s_L$ because the gas phase occur always at a temperature higher than that of the liquid phase and $v_G > v_L$ because the gas always has a molar volume greater than that of the liquid. Therefore $dp/dT > 0$ from which we conclude that the liquid-vapor coexistence curve has a positive slope in the p-T diagram.

The Clausius-Clapeyron equation may also be used to demonstrate that the tie lines in the volume-entropy diagram are perpendicular to the coexistence line as illustrated in Fig. 7.5b. Consider a portion of the coexistence line in the pressure-temperature diagram as shown in Fig. 7.5a. Consider also the volume-entropy diagram constructed so that the axis s is parallel to the axis T and the axis v is antiparallel to the axis p. In this diagram we locate the point L=$(s_L, -v_L)$ and the point G=$(s_G, -v_G)$, corresponding to the liquid and vapor coexisting at the point O of the p-T diagram. Next we perform a translation of the coexistence line from the pressure-temperature diagram into the volume-entropy diagram. The slope of the segment LG is $\lambda = (v_G - v_L)/(s_L - s_G)$ and the slope of the coexistence line is given for dp/dT. But according to the Clausius-Clapeyron equation $dp/dT =$

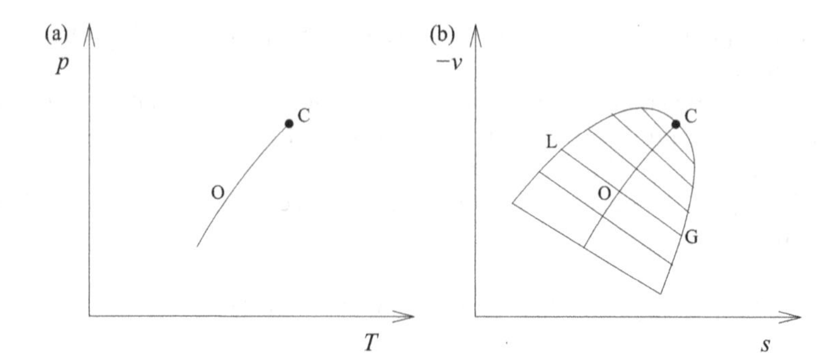

Fig. 7.5 (**a**) Portion of the liquid-vapor coexistence ending on the critical point C, in the pressure-temperature diagram. (**b**) Tie lines in the volume entropy diagram. The tie lines are perpendicular to the line of coexistence obtained by a translation of the (**a**) diagram. The tie line LG is perpendicular to the coexistence line at the point O. Notice the change in sign of v in the vertical axis

$-1/\lambda$, which is the condition of perpendicularity. Therefore, the tie line LG is perpendicular to the coexistence line at the point O.

The Clausius-Clapeyron equation is valid not only for the transition between the liquid and vapor but for the transition between any two phases. For the solid-liquid transition,

$$\frac{dp}{dT} = \frac{s_L - s_S}{v_L - v_S} = \frac{\ell_f}{T(v_L - v_S)}, \tag{7.21}$$

where $\ell_f = T(s_L - s_S)$ is the molar melting latent heat and T is the transition temperature. Table 7.2 presents the specific melting latent heat and the densities of the solid and liquid, corresponding to the solid-liquid transition, at the pressure of 1 atm, of various pure substances. In general, the liquid phase occurs at a temperature higher than that of the solid phase, which implies that the molar entropy of the liquid is greater than the molar entropy of the solid, $s_L > s_S$.

For most pure substances, the density of the liquid is smaller than that of the solid, which means to say that the liquid molar volume is greater than the solid molar volume, $v_L > v_S$. In these case, therefore, $dp/dT > 0$ and the solid-liquid coexistence line has a positive slope in the p-T diagram, as can be seen in Fig. 7.2. An important exception to this general case is the water. As is well known, at the pressure of 1 atm, the ice floats over the water, which means that the density of ice is smaller than the density of water, that is, $\rho_S < \rho_L$ or in an equivalent way $v_S > v_L$. Therefor $dp/dT < 0$ and the water-ice coexistence line has a negative slope in the p-T diagram, as seen in Fig. 7.1. However, there are other crystalline forms of ice that are denser that water, which turns the slope of the coexistence line positive. The substances whose solid phase is less dense than the liquid phase include, besides water, bismuth, silicon, germanium and gallium.

Table 7.2 Solid-liquid transition at the pressure of 1 atm. The table presents the melting temperature θ_f and T_f, the specific melting latent heat ℓ_f and the densities of the solid ρ_S and of the liquid ρ_L. Sources: LB, CRC

Substances	θ_f °C	T_f K	ℓ_f J/g	ρ_S g/cm^3	ρ_L g/cm^3
S	115.21	388.36	53.64	1.919	1.819
P	44.15	317.30	21.3	1.80	1.74
Sb	630.63	903.78	163.2	6.59	6.53
Bi	271.40	544.55	54.07	9.71	10.05
Si	1414	1687	1788	2.31	2.57
Ge	938.25	1211.40	508.7	5.31	5.60
Li	180.5	453.6	432	0.521	0.512
Na	97.80	370.95	113	0.951	0.927
K	63.38	336.53	59.6	0.847	0.828
Mg	650	923	349	1.649	1.584
Fe	1538	1811	247.3	7.23	6.98
Co	1495	1768	275	8.19	7.75
Ni	1455	1728	297.8	8.31	7.81
Pt	1768.4	2041.5	113.6	21.06	19.77
Cu	1084.62	1357.77	208.7	8.36	8.02
Ag	961.78	1234.93	104.8	9.75	9.320
Au	1064.18	1337.33	63.72	18.23	17.31
Zn	419.53	692.68	112	7.06	6.57
Hg	−38.83	234.32	11.4	14.191	13.69
Al	660.32	933.47	397	2.534	2.375
Ga	29.76	302.91	80.2	5.89	6.08
Sn	231.93	505.08	59.2	7.19	6.99
Pb	327.46	600.61	23.0	11.05	10.66
H_2O	0.00	273.15	333.4	0.9173	0.99984
NaCl	800.7	1073.8	482.1	1.91	1.556
KCl	771	1044	356.0	1.83	1.527
KBr	734	1007	214	2.49	2.127
AgCl	455	728	92	5.26	4.83
AgBr	432	705	48.6	6.03	5.577
$NaNO_3$	307	580	170	2.11	1.90

The Clausius-Clapeyron equation can also describe the solid-vapor coexistence. For this transition,

$$\frac{dp}{dT} = \frac{s_G - s_S}{v_G - v_S} = \frac{\ell_s}{T(v_G - v_S)}, \tag{7.22}$$

where $\ell_s = T(s_G - s_S)$ is the molar sublimation latent heat and T is the transition temperature. In sublimation, $s_G > s_S$ because the vapor always occurs at a

Table 7.3 Solid-vapor transition at the pressure of 1 atm. The table presents the sublimation temperature θ_f and T_f, the specific sublimation latent heat ℓ_s and the specific volumes of the solid $\tilde{v}_S$ and of the vapor $\tilde{v}_V$. Sources: LB, EG

		θ_s	T_s	ℓ_s	$\tilde{v}_S$	$\tilde{v}_V$
Substance		°C	K	J/g	cm^3/g	dm^3/g
Carb. diox.	CO_2	−78.50	194.65	570.7	0.640	0.355
Acetylene	C_2H_2	−83.80	189.35	801.4	1.37	0.578
Arsenic	As	615	883	425		
Graphite	C	3600	3873			

temperature higher than that of the solid. Moreover, since $v_G > v_S$, then $dp/dT > 0$ from which we conclude that the sublimation line has a positive slope. At the pressure of 1 atm, some substance pass directly from solid to vapor. Table 7.3 presents some example, that include the carbon dioxide and graphite.

Triple Point

The triple point corresponds to the coexistence of the three phases solid, liquid and vapor. Such a point, for a pure substance, occurs at a unique temperature and pressure since the triple point is the meeting of three lines of coexistence, liquid-vapor, solid-liquid and solid-vapor. Table 7.4 shows the temperature and pressure corresponding to the triple point of several pure substances. At the triple point the latent heat are related by

$$\ell_s = \ell_f + \ell_e \tag{7.23}$$

because $(s_G - s_S) = (s_G - s_L) + (s_L - s_S)$.

At the triple point, the three lines of coexistence meet forming angles always smaller than 180°. In an equivalent manner, we may say that the extension of the line beyond the triple point will be found inside the region between the other two lines of coexistence. To show this rule we proceed as follows. In a system of coordinates consisting by the temperature and pressure, we locate the triple point (T_t, p_t) and the three lines of coexistence, as shown in Fig. 7.6a. In another system of coordinates, set up in such a way that the molar entropy axis is parallel to the temperature and the molar volume axis is antiparallel to the pressure, we locate the points $\mathrm{S} = (-v_S, s_S)$, $\mathrm{L} = (-v_L, s_L)$ and $\mathrm{G} = (-v_G, s_G)$, determined by the molar volumes and molar entropies of the three phases in coexistence. The segments SL, LG and SG constitute tie lines related to the three transitions. Next we perform a translation of the three lines of coexistence and of the triple point to the volume-entropy diagram, so that the triple point is place inside the triangle SLG, as seen in Fig. 7.6b.

Table 7.4 Triple point of some pure substances. The table presents the temperature of the triple point T_t and the pressure of the triple point p_t. Sources: LB, EG, CRC

Substance		T_t K	p_t kPa
Neon	Ne	24.55	43.3
Argon	Ar	83.78	68.7
Krypton	Kr	115.95	73.1
Xenon	Xe	161.35	81.6
Hydrogen	H_2	13.947	7.2
Oxygen	O_2	54.351	0.152
Nitrogen	N_2	63.148	12.53
Fluorine	F_2	53.48	0.252
Chlorine	Cl_2	172.15	1.4
Bromine	Br_2	265.85	6.1
Iodine	I_2	386.85	12.21
Carb. monox.	CO	68.14	15.35
Carb. diox.	CO_2	216.58	518.5
Ammonia	NH_3	195.41	6.08
Hydr. chlor.	HCl	158.91	14.0
Water	H_2O	273.16	0.61166
Methane	CH_4	90.68	11.70
Acetylene	C_2H_2	192.60	128.2
Ethylene	C_2H_4	103.97	0.12
Ethane	C_2H_6	89.28	0.0011
Benzene	C_6H_6	278.69	4.78
Naphthalene	$C_{10}H_8$	353.43	1.00

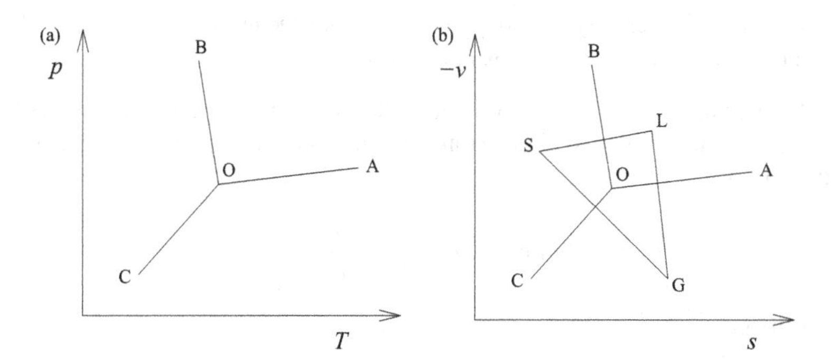

Fig. 7.6 (**a**) The three coexistence lines, liquid-vapor OA, solid-liquid OB, and solid-vapor OC, meet at the triple point O. (**b**) The sides of the triangle SLG, formed by the tie lines, are perpendicular do the three coexistence lines

Each line of coexistence, starting from inside the triangle SLG intercepts one of the sides of the triangle. The lines of coexistence solid-liquid OB, liquid-vapor OA and solid-vapor OC intercept the respective tie lines SL, LG and SG. The interceptions occur at right angles because, as we have seen, the tie line is perpendicular to the coexistence line. Since the angle formed by two lines of coexistence, which comprises one of the phases, and the internal angle of the triangle with respect to this phase are supplementary angles, then, it should be smaller than 180°.

Problems

7.1 Determine the liquid-vapor transition line of a substance whose molar boiling latent heat ℓ_e is independent of temperature. Assume that the vapor behaves as an ideal gas and that the vapor molar volume is much greater than the molar liquid volume. Solve the same problem for the case in which the latent heat depends on temperature in accordance with $\ell_e = a + (b/T)$.

7.2 In the neighborhood of the triple point, the liquid-vapor and the solid-vapor transition lines are described by $\ln(p/a_1) = b_1/T$ and $\ln(p/a_2) = b_2/T$, respectively. Determine the pressure and temperature of the triple point and the boiling, melting and sublimation latent heats around this point.

7.3 In the neighborhood of the solid-liquid transition the molar Helmholtz free energies of the solid and liquid are given by

$$f_S = -RT\ln(v - b_1) + c_1, \qquad f_L = -RT\ln(v - b_2) + c_2,$$

where b_1 and b_2 are constant and c_1 and c_2 depend only of the temperature. Set up the molar Gibbs free energy g_S and g_L of each phase. Determine the pressure p^* and the volumes v_S and v_L of the phases in coexistence.

7.4 At low temperatures, the molar Helmholtz free energy of a van der Waals fluid can be approximated by two expressions that describe the liquid and vapor phases. For the liquid

$$f_L = -RT\ln(v - b) - \frac{a}{b} + \frac{a}{b^2}(v - b) + K,$$

and for the gas

$$f_G = -RT\ln(v - b) + K, \tag{7.24}$$

where K only depends on temperature. Make a Legendre transformation to obtain the molar Gibbs free energy g_L of the liquid and g_G of the gas. Determine the line of coexistence equating the two Gibbs free energy. Determine also v_L, v_G and the latent heat ℓ as functions of temperature.

Chapter 8
Criticality

8.1 Critical Point

Liquid-Vapor Critical Point

Walking along the liquid-vapor coexistence line of a pure substance into direction of high pressures and high temperatures, the vapor density increases and the liquid density decreases to the point where the densities of the two phases become equal. This point which determines the end of the coexistence line and corresponds to the state in which the two phases become the identical, is called the critical point.

The first experimental observations of liquid-vapor critical point were made in alcohol, benzene and ether by Cagniard de La Tour. The critical point of these substances and of others, such as water, which are liquid at room temperature and at the pressure of 1 atm, can be experimentally achieved in two steps. At first, the coexistence line is reached by heating the liquid at constant pressure to the boiling point. Once on the transition line, the substance may be brought to the critical point by confining the liquid and its vapor in sealed vessel, followed by an increase in temperature.

Increasing the temperature, the vapor-liquid system will necessarily be found on the coexistence line, unless one of the two phases disappear, which may happen before the critical point is reached. However, if the liquid-vapor system is prepared so that the number of moles divided by the total volume of the vessel equals the critical density, then the critical point is necessarily reached.

If we use this method to reach the critical point, we see that initially the meniscus between liquid and vapor is well defined. As we approach the critical point, the meniscus becomes less defined and disappears at the critical point. The region where the meniscus was located becomes quite white, meaning that the light scattering is very intense. This phenomenon, known as critical opalescence, is caused by large fluctuations in density that occur around the critical point.

M.J. de Oliveira, *Equilibrium Thermodynamics*, Graduate Texts in Physics,
DOI 10.1007/978-3-662-53207-2_8

In the phase diagram, the critical point (T_c, p_c) lies therefore in the terminal point of the liquid-vapor coexistence line. Near the critical point the coexistence line is represented by the semi-straight line

$$p - p_c = A(T - T_c), \qquad T < T_c, \tag{8.1}$$

where A is a constant strictly positive which is identified as $(\partial p/\partial T)_v$ calculated at the critical point. Using the Clausius-Clapeyron equation, we see that the differences in molar entropy and molar volume in both phases in coexistence are related by

$$(s_\mathrm{G} - s_\mathrm{L}) = A(v_\mathrm{G} - v_\mathrm{L}). \tag{8.2}$$

Therefore, at the critical point, not only the molar volumes v_G and v_L become identical, but the molar entropies s_G and s_L become identical as well and, consequently, the boiling latent heat ℓ_e vanishes.

Liquefaction

The pure substances that are gases under normal conditions of temperature and pressure can be liquefied by compression alone, that is, can pass into the liquid state when subjected isothermally to sufficiently high pressures. This occurs with carbon dioxide, ammonia, ethane, propane, butane and other gases. These substances have a critical temperature higher than the room temperature, so that the coexistence line can be reached by isothermal compression. Carbon dioxide, for example, can be liquefied at a temperature of 20 °C under the pressure of 5.73 MPa. Other gases, on the other hand, such as helium, neon, argon, krypton, xenon, hydrogen, oxygen, nitrogen, carbon monoxide and methane cannot be liquefy at room temperature, no matter what the applied pressure. These substances have critical temperatures below the room temperature. A compression alone at room temperature is insufficient to meet the liquid-vapor coexisting line.

Figure 8.1 shows the carbon dioxide isotherms at various temperatures above and below the critical temperature, which occurs at 31.04 °C. It is seen that, below the critical temperature, the vapor-liquid coexistence may occur and therefore liquefaction may happen by compression alone. Along a subcritical isotherm, the volume decreases by compression and presents a jump at the transition. Along a supercritical isotherm, the volume increases continuously with pressure, without any discontinuity. The critical point is reached by compression alone along the critical isotherm when pressures reaches 72.85 atm. The density of the carbon dioxide at the critical point, where the liquid and vapor become identical, is 0.468 g/cm ^{3}. The first experimental measurements around a critical point were made in carbon dioxide by Andrews, who proved to be possible to continuously convert steam into liquid and vice-versa, bypassing the critical point through an appropriate path.

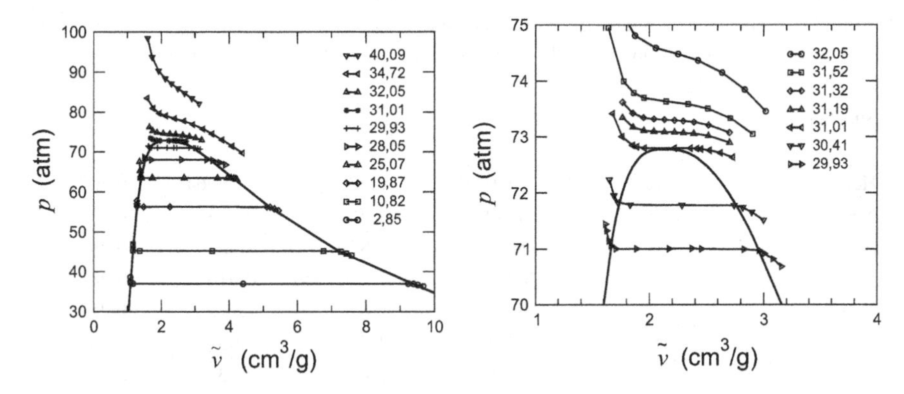

Fig. 8.1 Isotherms of carbon dioxide (CO_2) in the diagram pressure p versus specific volume $\tilde{v}$, for various values of temperature (in °C), obtained experimentally by Michels et al. [127]

Table 8.1 shows experimental data corresponding to the critical point of several pure substances. In addition to critical temperature, pressure and density, the table presents also the compressibility factor $Z_c = p_c v_c / RT_c$ determined at the critical point.

8.2 van der Waals Theory

van der Waals Equation

van der Waals theory provides a quantitative description of the liquid-vapor phase transition and of the corresponding critical point. Although it does not describe accurately the behavior of the thermodynamic properties near the critical point it provides an appropriate description of the phase coexistence and its relation with the critical point.

The van der Waals equation is given by

$$p = \frac{RT}{v - b} - \frac{a}{v^2}, \tag{8.3}$$

where a and b are constants. It describes approximately the behavior of gases and of the liquid-vapor transition. For convenience we call a system obeying this equation a van der Waals fluid or van der Waals gas, although it also describe the liquid phase. From the microscopic point of view, a fluid that undergoes a liquid-vapor transition must consist of molecules that attract each other over long distances and repel at short distances. A van der Waals fluid is to be understood as composed of hard spherical attractive molecules. The repulsion is the result of the rigidity of the

Table 8.1 Critical temperature T_c, critical pressure p_c and critical density ρ_c of various pure substances. The last column shows the corresponding value of the compressibility factor $Z_c = p_c v_c / RT_c$. Source: CRC

Substance		T_c	p_c	ρ_c	Z_c
		K	MPa	g/cm^3	
Helium	He	5.1953	0.22746	0.06964	0.303
Neon	Ne	44.40	2.760	0.484	0.312
Argon	Ar	150.663	4.860	0.531	0.292
Krypton	Kr	209.40	5.500	0.919	0.288
Xenon	Xe	289.73	5.840	1.110	0.287
Hydrogen	H_2	32.98	1.293	0.0310	0.306
Oxygen	O_2	154.581	5.043	0.436	0.288
Nitrogen	N_2	126.20	3.390	0.313	0.289
Fluorine	F_2	144.1	5.172	0.58	0.285
Chlorine	Cl_2	416.9	7.99	0.58	0.284
Bromine	Br_2	588	10.34	1.26	0.269
Carb. monox.	CO	132.91	3.499	0.301	0.294
Carb. diox.	CO_2	304.14	7.375	0.468	0.274
Ammonia	NH_3	405.5	11.35	0.237	0.242
Hydrog. chloride	HCl	324.7	8.31	0.45	0.249
Water	H_2O	647.14	22.06	0.322	0.230
Methane	CH_4	190.56	4.592	0.1627	0.286
Ethane	C_2H_6	305.32	4.872	0.207	0.279
Ethylene	C_2H_4	282.34	5.041	0.214	0.281
Acetylene	C_2H_2	308.33	6.14	0.230	0.271
Propane	C_3H_8	369.83	4.248	0.220	0.276
Butane	C_4H_{10}	425.12	3.796	0.228	0.274
Benzene	C_6H_6	562.0	4.89	0.305	0.268
Naphthalene	$C_{10}H_8$	748	4.10	0.315	0.265
Methanol	CH_4O	512.5	8.1	0.274	0.222
Ethanol	C_2H_6O	514	6.1	0.274	0.241
Diethyl ether	$C_4H_{10}O$	466	3.6	0.265	0.260
Acetone	C_3H_6O	508	4.7	0.278	0.233
Chloroform	$CHCl_3$	536	5.47	0.499	0.294

molecules and is related to the parameter b. The attraction is related to the parameter a, which must be understood as a measure the force of attraction between molecules.

The fundamental relation of a van der Waals fluid, in the Helmholtz free energy representation, is obtained by integrating $p = -(\partial f / \partial v)_T$. Using (8.3), we get

$$f = -RT \ln(v - b) - \frac{a}{v} - K, \tag{8.4}$$

where K depends only on temperature. Therefore, the entropy $s = -(\partial f/\partial T)_v$ is given by

$$s = R\ln(v-b) + K'. \tag{8.5}$$

To determine $K(T)$, we assume that the isochoric molar heat capacity c_v of the van der Waals is constant and equal to c. Thus $T(\partial s/\partial T)_v = c$, that is, $K'' = c/T$, so that

$$K' = c\ln T + c_1 \tag{8.6}$$

and

$$K = c(T\ln T - T) + c_1 T, \tag{8.7}$$

where c_1 is a constant. To determine the molar energy u, we use the relation $u = f + Ts$, from which we get

$$u = cT - \frac{a}{v}. \tag{8.8}$$

Maxwell Construction

At high temperatures, the van der Waals isotherms, in the p-v diagram, are monotonically decreasing and the free energy, given by (8.4) is a convex function. At low temperatures, however, the isotherms are no longer monotonic and free energy, accordingly, loses convexity. Therefore, in this temperature regime, we adopt as the free energy of the van der Waals fluid the convex hull of the function given by (8.4), which is obtained by construction of a double tangential, as shown in Fig. 4.3a.

The convex hull corresponds to perform a Maxwell construction on the van der Waals isotherms, which consists in tracing a straight line parallel to the axis of the volumes, so that the areas between the segment and the isotherm are equal as shown in Fig. 4.3b. The equivalence between the double tangent construction and the construction of Maxwell is shown as follows. Let A and B be the two points where the segment is tangent to the curve f versus v. The coordinates of these points are (v_L,f_G) and (v_G,f_G). Notice that the derivatives $(\partial f/\partial v)$ at the points A and B are the same which means that the pressures corresponding to these two points are equal. This common pressure we denote by p^* and, from the graph Fig. 4.3a, we get

$$f(v_L) - f(v_G) = p^*(v_G - v_L). \tag{8.9}$$

On the other hand

$$f(v_{\mathrm{L}}) - f(v_{\mathrm{G}}) = -\int_{v_{\mathrm{L}}}^{v_{\mathrm{G}}} \left(\frac{\partial f}{\partial v}\right)_T dv = \int_{v_{\mathrm{L}}}^{v_{\mathrm{G}}} p(v)dv, \tag{8.10}$$

and therefore

$$\int_{v_{\mathrm{L}}}^{v_{\mathrm{G}}} p(v)dv = p^*(v_{\mathrm{G}} - v_{\mathrm{L}}), \tag{8.11}$$

which is the algebraic expression of the Maxwell construction. For a better understanding, denote by C the point where the segment cuts the curve in the p-v diagram. Then the expression above is equivalent to

$$p^*(v_{\mathrm{C}} - v_{\mathrm{L}}) - \int_{v_{\mathrm{L}}}^{v_{\mathrm{C}}} p(v)dv = \int_{v_{\mathrm{C}}}^{v_{\mathrm{G}}} p(v)dv - p^*(v_{\mathrm{G}} - v_{\mathrm{C}}). \tag{8.12}$$

The right and left hand sides of this equation are identified as the two areas mentioned above.

To determine the coexistence line, we use (8.11) or (8.9) and the equation

$$p^* = p(v_{\mathrm{L}}) = p(v_{\mathrm{G}}). \tag{8.13}$$

Eliminating v_{L} and v_{G} from these equations, we find $p^*(T)$ as a function of T. Using expressions (8.3) and (8.4), for pressure and free energy, into (8.9) and (8.13), we get equations whose solutions give, for each isotherm, the values of v_{L}, v_{G} and p^*, that are shown in Fig. 8.2.

We should give a justification for the Maxwell construction since it involves an integral along a path that passes necessarily through unstable states, which would

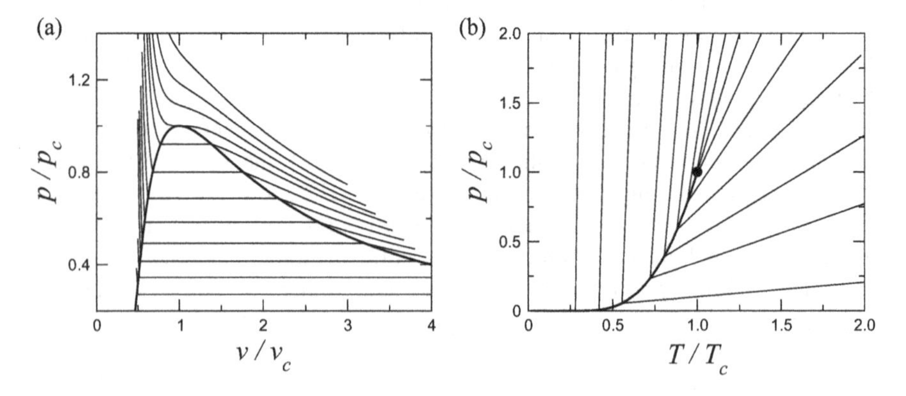

Fig. 8.2 van der Waals fluid. (**a**) Isotherms in the Clapeyron diagram. (**b**) Isochoric lines in the phase diagram pressure versus temperature. The *thick line* is the line of phase coexistence

be illegitimate. The justification would be the construction of the double tangent which, as we have seen, leads us to the Maxwell construction but does not involve unstable states. However, we must remember that, to obtain the free energy from the equation of state, we have performed an integral that necessarily went through the same unstable states, and again illegitimate. The unstable states, on the other hand, can be avoided, as argued in the following if we assume the Griffiths postulate according to which the free energy must be analytical within a single-phase region. With this assumption the Maxwell construction is fully justified.

Suppose that we use only the monotonic part of $p(T, v)$ to determine $f(Tv)$ by integration, thus avoiding unstable states. Below the critical temperature, we must necessarily use two monotonic branches of the same isotherm. The free energy obtained by integration along an isotherm will also have two branches, both convex. These two branches connected by a double tangent, constitute with the double tangent the free energy, which is entirely convex. The two branches, however, are associated with different integration constants $K_1(T)$ and $K_2(T)$, which depend on T. For $f(T, v)$ to be analytical within the single phase region it is required that $K_1(T)$ and $K_2(T)$ are analytical in T within the same region. As a consequence, the analytical continuation of $K_1(T)$ and $K_2(T)$, if they exist, should be the same. But in this case the procedure just presented becomes equivalent to Maxwell construction.

In summary, the placement of the horizontal segment of the isotherm in a position different from that given by the Maxwell construction would make $f(T, v)$ a nonalytic function inside a single-phase region.

Critical Point

Increasing the temperature along the coexistence line, the difference $\Delta v = v_G - v_L$ between the molar volumes decreases and vanishes at a certain temperature that we call critical temperature T_c. At this temperature, $\Delta v = 0$ and the two phases become identical. The corresponding point (T_c, p_c) in the phase diagram marks the end of the coexistence line. Above the critical temperature the van der Waals fluid exhibits a single phase.

The critical point can be determined by

$$\frac{\partial p}{\partial v} = 0 \qquad \text{and} \qquad \frac{\partial^2 p}{\partial v^2} = 0, \tag{8.14}$$

because the critical point is both a stationary point and an inflexion point, as seen in Figs. 8.1 and 8.2. From the van der Waals equation, we get

$$\frac{\partial p}{\partial v} = -\frac{RT}{(v-b)^2} + \frac{2a}{v^3}, \tag{8.15}$$

and

$$\frac{\partial^2 p}{\partial v^2} = \frac{2RT}{(v-b)^3} - \frac{6a}{v^4}, \tag{8.16}$$

from which we obtain the critical molar volume

$$v_c = 3b, \tag{8.17}$$

and the critical temperature

$$T_c = \frac{8a}{27bR}, \tag{8.18}$$

which substituted in van der Waals equation gives the critical pressure

$$p_c = \frac{a}{27b^2}. \tag{8.19}$$

Eliminating a and b from these three equations we reach the following relation

$$\frac{p_c v_c}{RT_c} = \frac{3}{8}. \tag{8.20}$$

The compressibility factor Z of a fluid is defined by $Z = pv/RT$. Therefore, from this equation we conclude that at the critical point $Z_c = 3/8$ for a fluid that satisfies the van der Waals equation. In Table 8.1 we present the values of Z_c obtained from the critical properties of several pure substances. Although the experimental values for $Z_c = p_c v_c/RT_c$ are not equal to $3/8 = 0.375$, they are close to each other, mainly those related to the noble gases.

Expansion Around the Inflexion Point

For temperatures near the critical temperature, the molar volumes of the liquid and vapor phases, v_{L} and v_{G}, are close to each other and close to the inflexion point v_0 of the van der Waals isotherm. Therefore, if we want to determine v_{L} and v_{G} near the critical temperature, it is reasonable to approximate the van der Waals isotherm and the free energy by an expansion around $v = v_0$.

The expansion of $p(v)$ up to cubic terms gives

$$p = p_0 + A(v - v_0) + B(v - v_0)^3, \tag{8.21}$$

where v_0 is given by $p''(v_0) = 0$, $p_0 = p(v_0)$, $A = p'(v_0)$ and $B = p'''(v_0)/6$. The corresponding free energy $f(v)$ is given by

$$f = f_0 - p_0(v - v_0) - \frac{A}{2}(v - v_0)^2 - \frac{B}{4}(v - v_0)^4. \tag{8.22}$$

Notice that f_0, v_0, p_0, A and B depend only on temperature. The expansions above are valid for small values of $|v - v_0|$. For temperatures near T_c, the constants can be obtained as explicit functions of temperature. The expansion of these constants around T_c gives

$$v_0 = v_c(1 + 2\frac{T - T_c}{T_c}), \tag{8.23}$$

$$p_0 = p_c(1 + 4\frac{T - T_c}{T_c}), \tag{8.24}$$

$$A = -6\frac{p_c}{v_c}\frac{T - T_c}{T_c} \tag{8.25}$$

and

$$B = -\frac{3p_c}{2v_c^3}. \tag{8.26}$$

To determine the coexistence line $p^*(T)$ and the values of v_L and v_G, we use the equation that expresses the Maxwell construction

$$f(v_L) - f(v_G) = p^*(v_G - v_L) \tag{8.27}$$

and the equations

$$p^* = p(v_L) = p(v_G). \tag{8.28}$$

The solution of these equations tell us that $p^* = p_0$, that is, the coexistence line (and its extension) is given by

$$p^* = p_c(1 + 4\frac{T - T_c}{T_c}) \tag{8.29}$$

and that v_L and v_G are roots of the equation

$$A(v - v_0) + B(v - v_0)^3 = 0, \tag{8.30}$$

that are distinct from. Solving this equation, we get

$$v_{\mathrm{G}} = v_0 + \sqrt{\frac{-A}{B}} \qquad \text{e} \qquad v_{\mathrm{L}} = v_0 - \sqrt{\frac{-A}{B}}, \tag{8.31}$$

so that

$$v_{\mathrm{G}} - v_{\mathrm{L}} = 2\sqrt{\frac{-A}{B}}. \tag{8.32}$$

If we are very near the critical temperature T_c, then we can use the expressions (8.25) and (8.26) for A and B and write

$$\frac{v_{\mathrm{G}} - v_{\mathrm{L}}}{2} = 2v_c\sqrt{\frac{T_c - T}{T_c}}. \tag{8.33}$$

Notice that

$$\frac{v_{\mathrm{G}} + v_{\mathrm{L}}}{2} = v_0 = v_c(1 + 2\frac{T - T_c}{T_c}). \tag{8.34}$$

Equivalently we may write

$$\frac{v_{\mathrm{G}} - v_{\mathrm{L}}}{2} = v_c\sqrt{\frac{p_c - p^*}{p_c}} \tag{8.35}$$

and

$$\frac{v_{\mathrm{G}} + v_{\mathrm{L}}}{2} = v_0 = v_c(1 + \frac{p^* - p_c}{2p_c}). \tag{8.36}$$

Compressibility

At temperatures above the critical temperature, the isotherms are strictly monotonic decreasing so that the isothermal compressibility $\kappa_T = -(1/v)(\partial v/\partial p)$ is positive (Fig. 8.3). The largest value of κ_T along an isotherm occurs at the inflexion point. At the critical temperature it diverges because the derivative $\partial p/\partial v$ vanishes at this point. According to (8.21), the pressure varies with the molar volume, along the critical isotherm, and around the critical point, according to the equation

$$p = p_c - \frac{3p_c}{2v_c^3}(v - v_c)^3. \tag{8.37}$$

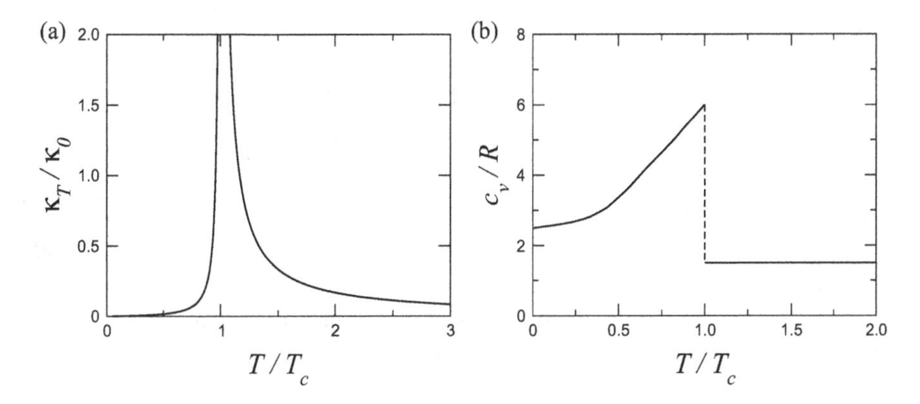

Fig. 8.3 van der Waals fluid. (**a**) Isothermal compressibility κ_T of the fluid along the critical isochoric ($T > T_c$) and of the liquid along the coexistence line ($T < T_c$), where $\kappa_0 = 1/p_c$. (**b**) Isochoric molar heat capacity c_v along the critical isochoric line. The jump of c_v at $T = T_c$ is equal to $9R/2$

Let us determine next the behavior of κ_T around the critical point, along the coexistence line and its extension, defined by (8.29), which coincides with the inflexions points of the isotherms. From (8.21), we get

$$\frac{1}{\kappa_T} = -v\frac{\partial p}{\partial v} = -v[A + 3B(v - v_0)^2]. \tag{8.38}$$

For temperatures above T_c, the molar volume along the extension of the coexistence line is given by $v = v_0$, where v_0 depends on T according to (8.23), so that

$$\frac{1}{\kappa_T} = -v_c A = 6p_c(\frac{T - T_c}{T_c}), \tag{8.39}$$

that is,

$$\kappa_T = \frac{1}{6p_c}(\frac{T_c}{T - T_c}). \tag{8.40}$$

If we wish to determine κ_T along the critical isochoric we substitute $v = v_c$ into (8.38). However, the second term inside the square brackets is of higher order compared to the first and can be neglected. Therefore, the result (8.40) is also valid for the critical isochoric near the critical point.

For temperatures below T_c, along the coexistence line, we should specify whether the compressibility is being determined for the liquid ($v = v_L$) or for the vapor ($v = v_G$). In both cases, however, according to (8.31), we have

$$(v - v_0)^2 = \frac{-A}{B} = 4(\frac{T_c - T}{T_c})v_c^2, \tag{8.41}$$

so that

$$\frac{1}{\kappa_T} = 2v_c A = 12p_c(\frac{T_c - T}{T_c}), \tag{8.42}$$

that is,

$$\kappa_T = \frac{1}{12p_c}(\frac{T_c}{T_c - T}), \tag{8.43}$$

for both phases. The results (8.40) and (8.43) show that the isothermal compressibility diverges at the critical point according to

$$\kappa_T = A_\pm |T_c - T|^{-1}, \tag{8.44}$$

where the coefficients A_+ and A_-, related to the behaviors of the compressibility above and below the critical temperature, differ by a factor of 2, that is, $A_+/A_- = 2$.

Molar Heat Capacity

Let us determine the isochoric molar heat capacity along the critical isochoric near the critical point. Above the critical temperature, there is just a single phase whose internal energy is given by (8.8). Since $c_v = (\partial u/\partial T)_v$, then

$$c_v = c \qquad \text{for} \qquad T > T_c. \tag{8.45}$$

Below the critical temperature, the two phases coexist and along the critical isochoric the internal energy is given by

$$u = x(cT - \frac{a}{v_L}) + (1 - x)(cT - \frac{a}{v_G}), \tag{8.46}$$

where x is the mole fraction of the liquid given by

$$x = \frac{v_G - v_c}{v_G - v_L}. \tag{8.47}$$

Deriving the expression for u with respect to T, we obtain

$$c_v = c + \frac{9}{2}R + c^*(T - T_c) \qquad \text{for} \qquad T < T_c, \tag{8.48}$$

where c^* is a positive constant. Therefore, the molar heat capacity c_v of the van der Waals fluid displays, along the critical isochoric, a jump at $T = T_c$ equal to $9R/2$.

8.3 Critical Behavior

Critical Exponents

Around the critical point, some thermodynamic properties have singular behavior. The compressibility and the molar heat capacity, for example, grow without bounds and diverge at the critical point. Figure 8.4 shows the divergence of molar heat capacity of argon and carbon dioxide at the critical point. It is established that the singularities are well represented by power laws when the quantities are placed as functions of the deviations $(T - T_c)$ and $(p - p_c)$ of temperature and pressure from their critical values. Since the critical point marks the end of the coexistence line, it is natural to examine firstly how the densities of the liquid and vapor approach each other and become identical at the critical point. We assume that the difference between them vanishes at the critical point in accordance with the power law

$$\rho_{\rm L} - \rho_{\rm G} \sim (T_c - T)^{\beta} \qquad \text{or} \qquad v_{\rm G} - v_{\rm L} \sim (T_c - T)^{\beta}. \tag{8.49}$$

Another important quantity in the characterization of the critical behavior in fluids is the isothermal compressibility, which diverges at the critical point. We assume that along the critical isochoric ($v = v_c$ or $\rho = \rho_c$) and for $T > T_c$, it behaves according to

$$\kappa_T \sim |T - T_c|^{-\gamma}. \tag{8.50}$$

Below the critical temperature, $T < T_c$, we observe that the compressibility along the coexistence line. In this case, it can be determined either for the liquid or for the vapor. However, we assume that the compressibility of both phases have the same behavior around the critical point and that it is given by (8.50).

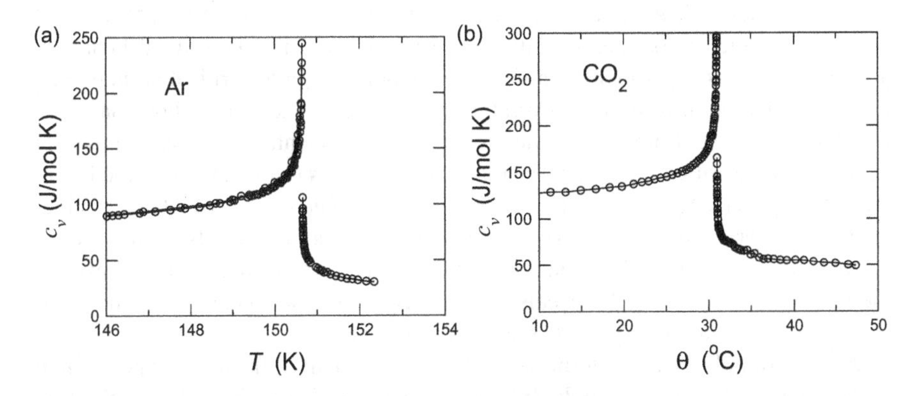

Fig. 8.4 Molar heat capacity c_v along the critical isochoric line as a function of temperature, (**a**) for the argon, with experimental data obtained by Voronel et al. [128], and (**b**) for carbon dioxide, with experimental data obtained by Beck et al. [124]

It is also interesting to determine how the density and pressure approach their critical values. We assume that along the critical isotherm ($T = T_c$) they have the same behavior

$$|p - p_c| \sim |\rho - \rho_c|^\delta \qquad \text{or} \qquad |p - p_c| \sim |v - v_c|^\delta. \tag{8.51}$$

We consider finally the isochoric molar heat capacity c_v. We assume that this quantity behaves along the critical isochoric according to

$$c_v \sim |T - T_c|^{-\alpha}, \tag{8.52}$$

both above and below the critical temperature. In general we expect a weak divergence for the isochoric molar heat capacity. It is possible that this divergence is of the logarithmic type

$$c_v \sim \ln |T - T_c|. \tag{8.53}$$

The notation $a(x) \sim b(x)$ used to characterize the critical behavior, where both a and b diverge or both vanish when $x \to 0$, means that the ratio a/b approaches a finite constant or, equivalently, that $\ln a / \ln b \to 1$ when $x \to 0$.

If the critical behavior corresponds in fact to a power law, the critical exponent is defined by the power law. It is convenient, however, to define the critical exponent more broadly to include other behaviors that are not strict power laws. Thus, we define the critical exponent θ, associated with a quantity $a(x)$, by

$$\theta = \lim_{x \to 0} \frac{\ln a(x)}{\ln x}, \tag{8.54}$$

or equivalently by $a \sim x^\theta$. With this definition we see that a logarithm divergence is associated to an exponent zero.

The exponents β, γ, δ and α contained in the power laws (8.49), (8.50), (8.51) and (8.52), called critical exponents, concisely characterize the critical behavior. Table 8.2 presents experimental values of the exponents for the critical point related to the liquid-vapor transition of several pure substances. The results show that there is a good agreement among the values for each exponent, indicating a universal character of the critical behavior. The results (8.33), (8.44) and (8.37), obtained from the van der Waals equation, give us $\beta = 1/2$, $\gamma = 1$ and $\delta = 3$, which are distinct from the experimental values, which means that the van der Waals equation does not adequately describe the behavior of the fluid near the critical point. However, the theory of van der Waals reveals that the critical exponents are independent of the parameters a and b and in this sense are universal.

As an example of the determination of the exponent β from the experimental measurements done in carbon dioxide (CO_2), we show in Fig. 8.5a the graph of ρ_L and ρ_G versus temperature. To estimate the exponent β and the critical temperature from the data of Fig. 8.5, we plot $\ln[(\rho_L - \rho_G)/\rho_c]$ versus $\ln[|T_c - T|/T_c]$. The value

Table 8.2 Critical exponents α, β, γ and δ related to the liquid-vapor critical point of several pure substances

Substance		α	β	γ	δ
Helium-3	^3He	0.11	0.36	1.19	4.1
Helium-4	^4He	0.13	0.36	1.18	
Neon	Ne		0.33	1.25	
Argon	Ar	0.13	0.34	1.21	
Krypton	Kr		0.36	1.18	
Xenon	Xe	0.11	0.33	1.23	
Hydrogen	H_2		0.33	1.19	
Oxygen	O_2	0.12	0.35	1.25	
Nitrogen	N_2		0.33	1.23	
Carb. diox.	CO_2	0.11	0.32	1.24	
Sulf. hexafl.	SF_6	0.11	0.32	1.28	
Ethylene	C_2H_4		0.33	1.18	4.4
Ethane	C_2H_6	0.12	0.34		

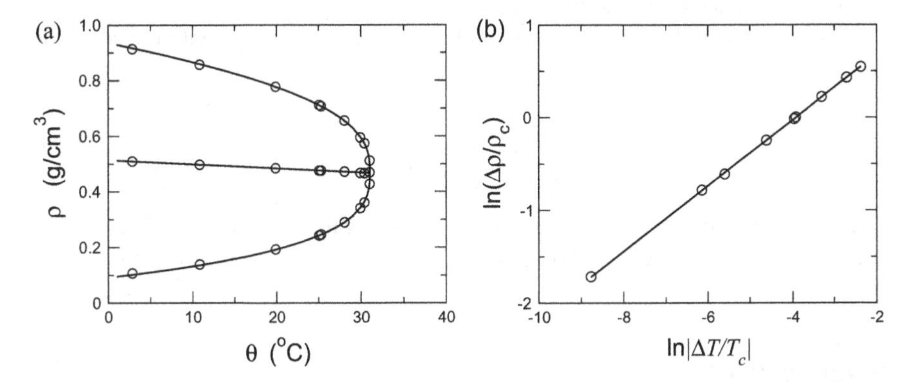

Fig. 8.5 Experimental data for liquid carbon dioxide in coexistence with its vapor, obtained by Michels et al. [127]. In (**a**) we show the density of the liquid ρ_L (upper branch), the density of the vapor ρ_G (lower branch) and the mean $(\rho_L + \rho_G)/2$ versus temperature. In (**b**) we show a log-log plot of $\Delta\rho/\rho_c$ versus $\Delta T/T_c$, where $\Delta\rho = \rho_L - \rho_G$ and $\Delta T = T - T_c$. The slope of the straight line fitted to the data gives the result $\beta = 0.35$

used for the critical temperature is the best fitting of a straight line: $T_c = 31.06$. According to (8.49) the slope of this straight line is identified as the exponent β. From the fitting one obtains $\beta = 0.35$. From this value one obtains also the critical pressure $p_c = 72.85$ atm.

The critical density is obtained in an approximate way by using the Cailletet-Mathias rule. According to this rule, around the critical point, the arithmetic mean $(\rho_L + \rho_G)/2$ is linear with temperature, as seen in Fig. 8.5a The value obtained in this manner gives $\rho_c = 0.467$ g/cm^3. However, as we will see later this rule is not valid very near the critical point.

Other thermodynamic quantities can be analyzed as to the behavior around the critical point. Using the following thermodynamic identity

$$c_p = c_v + Tv \left(\frac{\partial p}{\partial T} \right)_v^2 \kappa_T \tag{8.55}$$

and taking into account that $(\partial p/\partial T)_v$ is bounded and nonzero, we conclude that the isobaric molar heat capacity c_p diverges as κ_T because the divergence of c_v is weak when compared to the divergence of κ_T. From these result and using the identity

$$\kappa_s = \frac{c_v}{c_p} \kappa_T, \tag{8.56}$$

we conclude that the adiabatic compressibility κ_s diverges as c_v. Therefore, along the critical isochoric line, the critical behavior of c_p is related to the exponent γ and that of κ_s is related to the exponent α.

If we recall that the speed of sound in fluids is directly related with the adiabatic compressibility by

$$v_{\text{som}} = \frac{1}{\sqrt{\rho \kappa_s}}, \tag{8.57}$$

then the measurements of the velocity of sound along the critical isochoric line, near the critical point, provides an experimental method to determine the exponent α.

Singular Part

The thermodynamic properties of a pure substance, including the singular behavior around the critical point, are contained in the fundamental relation. In this sense it is desirable to establish a form for the thermodynamic potential, valid around the critical point, from which we power law introduced above could be derived. The thermodynamic potential more convenient for this purpose is the one for which all independent variables are thermodynamic field. We therefore choose the molar Gibbs free energy $g(T, p)$ which is a natural function of T and p.

To proceed with the analysis of the critical behavior, it is convenient to separate the molar Gibbs free energy $g(T, p)$ into one regular part $g_{\text{r}}(T, p)$ and one singular part $g_{\text{s}}(T, p)$ containing the anomalies related to the critical point and the discontinuities along the coexistence line. Thus, we write

$$g(T, p) = g_{\text{r}}(T, p) + g_{\text{s}}(T, p). \tag{8.58}$$

According to this separation, any thermodynamic quantity will also have a regular and a singular part. The singular part of the quantities that are finite at the critical point must be chosen such that it vanishes at this point, which can always be made by the addition of appropriate terms in $g_r(T,p)$. In this way, we guarantee that the singular part of any nondiverging quantity will vanish at the critical point.

The regular part of the Gibbs free energy is assumed to have an expansion around the critical point (T_c,p_c) which, up to linear terms in $(T-T_c)$ and $(p-p_c)$, is given by

$$g_r(T,p) = g_c + v_c(p-p_c) - s_c(T-T_c), \tag{8.59}$$

where $g_c = g(T_c,p_c)$, v_c and s_c are the molar volume and molar entropy at the critical point.

To explore in a systematic way the consequences of a privileged direction in the phase diagram, determined by the transition line, it is convenient to make a change of variables. Near the critical point, the line of coexistence is represented by the semi straight line defined by (8.1). The axes corresponding tho the new variables have origin at the critical point and one of them coincides with the coexistence line (8.1). We adopt the following transformation

$$\varepsilon = (T-T_c) + B(p-p_c) \tag{8.60}$$

and

$$\zeta = -A(T-T_c) + (p-p_c), \tag{8.61}$$

where B is a constant to be defined later. We remember that A is a constant strictly positive. The line of coexistence (8.1) in the new variables is given by $\zeta = 0$ and $\varepsilon < 0$.

In this new reference system we write the singular part of the molar Gibbs free energy as

$$g_s(T,p) = \mathscr{G}(\varepsilon,\zeta), \tag{8.62}$$

so that

$$v = \frac{\partial g}{\partial p} = v_r + B\mathscr{G}_1(\varepsilon,\zeta) + \mathscr{G}_2(\varepsilon,\zeta) \tag{8.63}$$

and

$$s = -\frac{\partial g}{\partial T} = s_r - \mathscr{G}_1(\varepsilon,\zeta) + A\mathscr{G}_2(\varepsilon,\zeta) \tag{8.64}$$

where $\mathscr{G}_1 = \partial\mathscr{G}/\partial\varepsilon$ and $\mathscr{G}_2 = \partial\mathscr{G}/\partial\zeta$ and $v_r = \partial g_r/\partial p$ and $s_r = \partial g_r/\partial T$ are the regular parts of the volume and entropy.

Deriving the expression (8.63) and (8.64) with respect to the pressure and temperature, respectively, we get

$$v\kappa_T = -\left(\frac{\partial v}{\partial p}\right)_T = -B^2\mathscr{G}_{11}(\varepsilon,\zeta) - 2B\mathscr{G}_{12}(\varepsilon,\zeta) - \mathscr{G}_{22}(\varepsilon,\zeta), \tag{8.65}$$

and

$$\frac{1}{T}c_p = \left(\frac{\partial s}{\partial T}\right)_p = -\mathscr{G}_{11}(\varepsilon,\zeta) + 2A\mathscr{G}_{12}(\varepsilon,\zeta) - A^2\mathscr{G}_{22}(\varepsilon,\zeta), \tag{8.66}$$

where $\mathscr{G}_{11} = \partial^2\mathscr{G}/\partial\zeta^2$, $\mathscr{G}_{12} = \partial^2\mathscr{G}/\partial\varepsilon\partial\zeta$ and $\mathscr{G}_{22} = \partial^2\mathscr{G}/\partial\varepsilon^2$. Near the critical point we use the relation $c_v = c_p - TvA^2\kappa_T$, obtained from the identity (8.55), to write

$$\frac{1}{T}c_v = (AB+1)\{(AB-1)\mathscr{G}_{11}(\varepsilon,\zeta) + 2A\mathscr{G}_{12}(\varepsilon,\zeta)\}. \tag{8.67}$$

In these three expressions above we have omitted the regular part by taking into account that these quantities diverges at the critical point and are therefore dominated by the singular part.

Up to now the constant B contained in the transformation of variables (8.60) and (8.61) has not been specified. Next, we choose this constant in such a way that the singular part of the Gibbs free energy is symmetric with respect to the transition line, that is, so that $\mathscr{G}(\varepsilon,\zeta)$ is an even function with respect to the variable ζ,

$$\mathscr{G}(\varepsilon,\zeta) = \mathscr{G}(\varepsilon,-\zeta). \tag{8.68}$$

In addition to this symmetry, the function $\mathscr{G}(\varepsilon,\zeta)$ has the usual properties of the thermodynamic potentials; it is continuous and concave. The derivative of $\mathscr{G}_1(\varepsilon,\zeta)$ is an even function of ζ and is continuous. The derivative $\mathscr{G}_2(\varepsilon,\zeta)$ is an odd function of ζ and continuous except over the line of coexistence where it has a discontinuity. These properties are shown in Fig. 8.6.

Along the coexistence line, the vapor molar volume v_{G} and the liquid molar volume v_{L} are obtained by taking the limits $\zeta \to 0^-$ and $\zeta \to 0^+$, respectively, in expression (8.63):

$$v_{\mathrm{G}} = v_{\mathrm{r}} + B\mathscr{G}_1(\varepsilon,0) + \mathscr{G}_2(\varepsilon,0^-) \tag{8.69}$$

and

$$v_{\mathrm{L}} = v_{\mathrm{r}} + B\mathscr{G}_1(\varepsilon,0) + \mathscr{G}_2(\varepsilon,0^+). \tag{8.70}$$

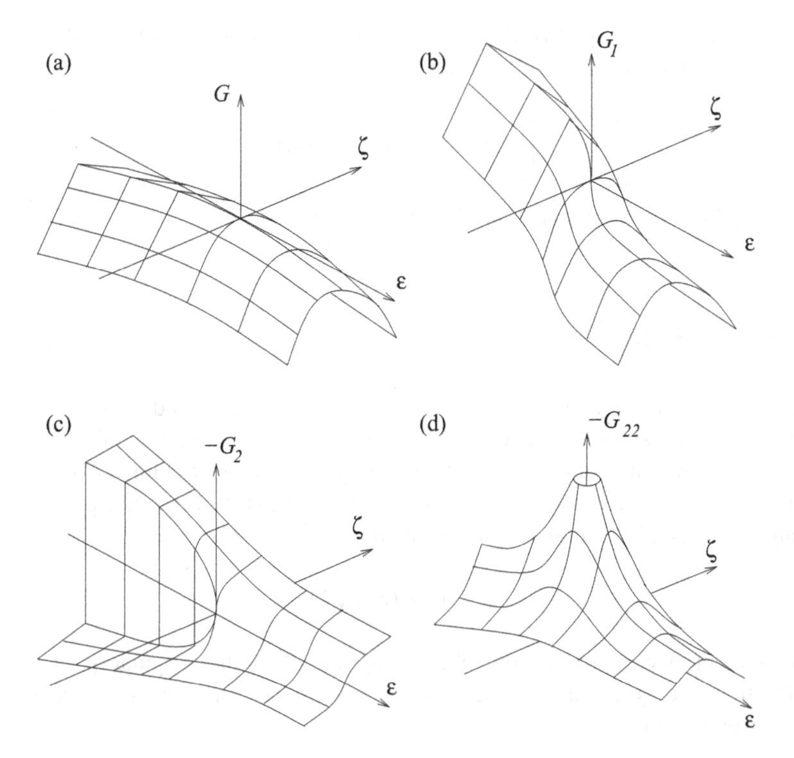

Fig. 8.6 Singular part of the Gibbs free energy $\mathscr{G}$ around the critical point and its derivatives $\mathscr{G}_1$, $\mathscr{G}_2$ as $\mathscr{G}_{22}$ functions of ε and ζ

The difference between the liquid and vapor molar volumes is, therefore, related with the discontinuity of $\mathscr{G}_2(\varepsilon, \zeta)$ by

$$v_{\mathrm{G}} - v_{\mathrm{L}} = \mathscr{G}_2(\varepsilon, 0^-) - \mathscr{G}_2(\varepsilon, 0^+). \tag{8.71}$$

The mean of the liquid and vapor molar volumes is

$$\frac{1}{2}(v_{\mathrm{G}} + v_{\mathrm{L}}) = v_{\mathrm{r}} + B\mathscr{G}_1(\varepsilon, 0). \tag{8.72}$$

Along $\varepsilon = 0$,

$$v = v_{\mathrm{r}} + B\mathscr{G}_1(0, \zeta) + \mathscr{G}_2(0, \zeta) \tag{8.73}$$

From the equality (8.65), we obtain the behavior of the isotherm compressibility along the coexistence line and its extension

$$v\kappa_T = -B^2\mathscr{G}_{11}(\varepsilon, 0) - \mathscr{G}_{22}(\varepsilon, 0). \tag{8.74}$$

Similarly, from the equality (8.67), we obtain the isochoric molar heat capacity along the coexistence line and its extension

$$\frac{1}{T}c_v = (A^2B^2 - 1)\mathcal{G}_{11}(\varepsilon, 0), \tag{8.75}$$

where we have taken into account that $\mathcal{G}_{12}(\varepsilon, 0) = 0$.

Widom Scaling Theory

The Widom scaling theory concerns the behavior of the thermodynamic quantities around the critical point. According to this theory, the value of a certain thermodynamic quantity in a certain point of he phase diagram can be obtained from another point by a change is scale. The critical point is taken as the reference point for the changes of scale. If we consider the following scale transformation

$$(\varepsilon, \zeta) \rightarrow (\lambda\varepsilon, \lambda'\zeta), \tag{8.76}$$

that connect two points of the phase diagram, then the singular parts of the Gibbs free energy corresponding to these two points are connected by a scale transformation, that is,

$$\mathcal{G}(\lambda\varepsilon, \lambda'\zeta) = \lambda''\mathcal{G}(\varepsilon, \zeta). \tag{8.77}$$

The parameters λ, λ' and λ'' are the enlargement factors related to ε, ζ and $\mathcal{G}$, respectively. Without losing generality we may write $\lambda' = \lambda^b$ and $\lambda'' = \lambda^a$, so that

$$\mathcal{G}(\varepsilon, \zeta) = \lambda^{-a}\mathcal{G}(\lambda\varepsilon, \lambda^b\zeta), \tag{8.78}$$

valid for any positive values of λ and that expresses the homogeneity of $\mathcal{G}(\varepsilon, \zeta)$ with respect to the variables ε and ζ. The exponents a and b are connected to the critical exponents introduced above, as we will se later.

The homogeneity property of $\mathcal{G}(\varepsilon, \zeta)$ implies that its derivatives are also homogeneous functions. Indeed, deriving (8.78) with respect to ε and ζ, we get

$$\mathcal{G}_1(\varepsilon, \zeta) = \lambda^{-a+1}\mathcal{G}_1(\lambda\varepsilon, \lambda^b\zeta) \tag{8.79}$$

and

$$\mathcal{G}_2(\varepsilon, \zeta) = \lambda^{-a+b}\mathcal{G}_2(\lambda\varepsilon, \lambda^b\zeta). \tag{8.80}$$

Deriving these expressions with respect to ε and ζ, respectively, we get in addition

$$\mathscr{G}_{11}(\varepsilon,\zeta) = \lambda^{-a+2}\mathscr{G}_{11}(\lambda\varepsilon,\lambda^b\zeta), \tag{8.81}$$

and

$$\mathscr{G}_{22}(\varepsilon,\zeta) = \lambda^{-a+2b}\mathscr{G}_{22}(\lambda\varepsilon,\lambda^b\zeta). \tag{8.82}$$

Next we determine the behavior of these quantities along the coexistence line and its extension. For $\varepsilon < 0$ and $\zeta = 0$, and choosing $\lambda = |\varepsilon|^{-1}$, we get

$$\mathscr{G}_1(\varepsilon,0) = |\varepsilon|^{a-1}\mathscr{G}_1(-1,0), \tag{8.83}$$

$$\mathscr{G}_2(\varepsilon,0^\pm) = |\varepsilon|^{a-b}\mathscr{G}_2(-1,0^\pm), \tag{8.84}$$

$$\mathscr{G}_{11}(\varepsilon,0) = |\varepsilon|^{a-2}\mathscr{G}_{11}(-1,0), \tag{8.85}$$

$$\mathscr{G}_{22}(\varepsilon,0) = |\varepsilon|^{a-2b}\mathscr{G}_{22}(-1,0). \tag{8.86}$$

For $\varepsilon > 0$ and $\zeta = 0$, and choosing $\lambda = \varepsilon^{-1}$, we obtain

$$\mathscr{G}_{11}(\varepsilon,0) = \varepsilon^{a-2}\mathscr{G}_{11}(1,0), \tag{8.87}$$

$$\mathscr{G}_{22}(\varepsilon,0) = \varepsilon^{a-2b}\mathscr{G}_{22}(1,0). \tag{8.88}$$

Replacing these results into the expressions (8.71), (8.74) and (8.75), we reach the results

$$v_{\mathrm{G}} - v_{\mathrm{L}} \sim |\varepsilon|^{a-b}, \tag{8.89}$$

$$\kappa_T \sim |\varepsilon|^{a-2b}, \tag{8.90}$$

$$c_v \sim |\varepsilon|^{a-2}. \tag{8.91}$$

We see therefore that the Widom scaling theory predicts power laws for the critical behavior of these quantities. The result for the compressibility was obtained considering that the second term of (8.74) dominates over the first. Comparing these expressions with the (8.49), (8.50) and (8.52), we find the relations between the critical exponents and the indices a and b:

$$\alpha = -a + 2, \tag{8.92}$$

$$\beta = a - b, \tag{8.93}$$

$$\gamma = -a + 2b. \tag{8.94}$$

In doing the comparison, we have taken into account that along $\zeta = 0$ the variable ε is proportional to $T - T_c$.

For $\varepsilon = 0$ and $\zeta > 0$, and choosing $\lambda = \zeta^{-1/b}$, we get

$$\mathscr{G}_2(0, \zeta) = \zeta^{(a-b)/b}\mathscr{G}_2(0, 1). \tag{8.95}$$

Replacing into (8.73), we reach the result

$$v - v_c \sim \zeta^{(a-b)/b}, \tag{8.96}$$

where we have taken into account only the dominant term in (8.73) and we have approximated v_r by v_c. Comparing with expression (8.51), and taking into account that along $\varepsilon = 0$, ζ is proportional to $p - p_c$, we get the following relation between the exponent δ and the indices a and b

$$\frac{1}{\delta} = \frac{a-b}{b}. \tag{8.97}$$

Notice, however, that the exponent δ is defined by the behavior of $v - v_c$ (or $\rho - \rho_c$) versus $p - p_c$ along the critical isotherm and not along $\varepsilon = 0$. On the other hand, it is possible to argue that the critical behavior will be the same for both trajectories. In fact, it will be the same along any straight line that intercepts the critical point, that does not coincide with the coexistence line.

The four critical exponents α, β, γ, and δ are not all independent. We can choose only two as independent. Eliminating a and b we obtain several relations between them, that include the Rushbrooke relation

$$\alpha + 2\beta + \gamma = 2, \tag{8.98}$$

the Griffiths relation

$$\alpha + \beta(\delta + 1) = 2 \tag{8.99}$$

and the Widom relation

$$\gamma = \beta(\delta - 1). \tag{8.100}$$

The experimental values shown in Table 8.2 are in accordance with these relations, within the experimental errors. It is interesting to notice that the values of the exponents obtained from the van der Waals equation, $\alpha = 0$, $\beta = 1/2$, $\gamma = 1$ and $\delta = 3$ also satisfy these relations. The jump in the isochoric molar heat capacity c_v, predicted by the de van der Waals theory, must be understood as corresponding to $\alpha = 0$.

Replacing the result (8.83) into (8.72), we obtain

$$\frac{1}{2}(v_G + v_L) - v_r \sim |\varepsilon|^{1-\alpha} \tag{8.101}$$

where we used the relation $a = 2 - \alpha$. This result can be used to determine the critical exponent α from the measurements of v_G and v_L. Since the regular part v_r is linear with temperature, we see that the last term is a correction to the Cailletet-Mathias rule.

Problems

8.1 Determine the critical temperature T_c, the critical pressure p_c and the critical molar volume v_c corresponding to the Dieterici and Berthelot equations. Find also the critical compressibility factor $Z_c = p_c v_c / RT_c$.

8.2 Show that the critical behavior of the latent heat is related to the exponent β.

8.3 Show that the derivative of $g(T,p)$ along the coexistence line is well defined, that is, has the same value if we approach it from the liquid side or the vapor side. Determine its value at the critical point. Along a line perpendicular to the coexistence line the derivative has a jump. Determine this jump.

8.4 Determine the critical exponent related with the following functions

$$\ln|x|, \qquad x^{1/2}\ln|x|, \qquad ax + bx^{1/2}, \qquad ax^{-1} + bx^{-1/2}.$$

8.5 Show that the Dieterici and Berthelot equations of state provide the same critical exponent as the van der Waals equation.

8.6 Show that the coefficient of thermal expansion diverges at the critical point with the same exponent γ related to κ_T.

8.7 Demonstrate relation (8.55).

Chapter 9
Mixtures

9.1 Introduction

Mixtures of Pure Substances

The mixtures of pure substances are quite common in nature. Some mixtures are homogeneous others are heterogeneous. The atmospheric air is a homogeneous gaseous mixture of nitrogen, oxygen and other gases in smaller proportions. Oil is a natural mixture of hydrocarbons. The rocks, like granite, are solid heterogeneous mixtures. The homogeneous mixtures have uniform physical aspect and consists of a single thermodynamic phase. The heterogeneous mixtures, on the other hand, are formed by an aggregate of thermodynamic phase in coexistence. Due to gravity, the different phases of a heterogeneous liquid mixture are arranged one above the other, separated by menisci and sorted according to density. In a heterogeneous solid mixture the phases are displayed in a fragmented and intricate arrangement.

The separation of the components of a mixture to obtain pure compounds is a major problem. Once the separation is made, the pure substances may be used to obtain new compounds. Distillation is an example of a process of liquid separation by evaporation and condensation of vapor.

In this chapter and the next, we analyze mixtures of pure substances that do not lose their identities when they enter the composition of a mixture. That is, the chemical species in the mixture are the same as the pure substances used to make the mixture. They do not undergo any change, that is to say there is no chemical reactions within the mixture. Therefore, the number of moles of each pure substance remains unchanged before and after the mixture composition.

M.J. de Oliveira, *Equilibrium Thermodynamics*, Graduate Texts in Physics,
DOI 10.1007/978-3-662-53207-2_9

The chemical species comprising a mixture are generically called constituents of the mixture. In a molecular mixture, each constituent is represented by a type of molecule. If there is no chemical reaction, the number of chemical species corresponding to the number of components of the mixture. In mixtures in which chemical reactions occur, which will be treated in the last chapter, the number of components equals the number of chemical species minus the number of independent chemical reactions.

Solutions

If a mixture displays a single phase, that is, if it is homogeneous, it is also called a solution. In this case we say that the components are miscible. The component with the greatest concentration is conventionally called solvent and the others, of lower concentration, solutes. In a solution, we also say that the component of the greatest concentration dissolves the ones with lower concentration. If a mixture has more than one phase, that is, if it is a heterogeneous mixture, then each phase is a solution. In this case, however, each one of the coexisting solutions is called saturated solution. It is also possible that a component of the mixture is completely immiscible, forming a phase different from the others, which is not a proper solution.

A solution may be gaseous, liquid or solid. The gases are miscible in all proportions always forming a homogeneous mixture and therefore a single phase. A heterogeneous mixture can therefore have at most a single gas phase. The liquids dissolve gases, solids and liquids, sometimes in large quantities. Since the liquids are not fully miscible it is possible the occurrence of more than one liquid phase in a heterogeneous mixture. The solids also can dissolve gases, liquids and other solids, but the limits of miscibility are generally lower than that of liquids. Due to this reason, the number of solid phases in a heterogeneous mixture can be great.

9.2 Fundamental Relation

Thermodynamic State

An adequate description of the thermodynamic state of a mixture must contain quantities that determine the amount of each component. We may use for this purpose the mass, but the number of moles of each component is preferable. Thus, in the standard representation, the thermodynamic state of a mixture consisting of c components is defined by the extensive variables: entropy S, volume V, number of moles N_1 of the first component, number of moles N_2 of the second component, ..., number of moles of the N_c of the c-th component. The corresponding fundamental relation is given by the internal energy $U(S, V, N_1, N_2, \ldots, N_c)$ whose differential is

$$dU = TdS - pdV + \mu_1 dN_1 + \mu_2 dN_2 + \ldots + \mu_c dN_c, \tag{9.1}$$

where T is the temperature, p is the pressure, and μ_1, μ_2, ..., and μ_c are the chemical potentials of the components $1, 2, \ldots$, and c, respectively. Other representations can also be employed including the Helmholtz representation, the enthalpy representation and the Gibbs representation. The respective fundamental relations are given by the Helmholtz free energy $F(T, V, N_1, N_2, \ldots, N_c)$, by the enthalpy $H(S, p, N_1, N_2, \ldots, N_c)$ and by the Gibbs free energy $G(T, p, N_1, N_2, \ldots, N_c)$, which are obtained one from another by means of Legendre transformations.

The Gibbs representation is particularly useful in the study of mixtures. The differential of the Gibbs free energy is given by

$$dG = -SdT + Vdp + \mu_1 dN_1 + \mu_2 dN_2 + \ldots + \mu_c dN_c, \tag{9.2}$$

where

$$S = -\frac{\partial G}{\partial T}, \qquad V = \frac{\partial G}{\partial p}, \qquad \mu_i = \frac{\partial G}{\partial N_i}. \tag{9.3}$$

Being G an extensive function with respect to the variables N_i, we may write

$$G = \mu_1 N_1 + \mu_2 N_2 + \ldots + \mu_c N_c, \tag{9.4}$$

an expression that can be used to recover the fundamental relation $G(T, p, N_1, N_2, \ldots, N_c)$ in case the chemical potentials $\mu_i(T, p, N_1, N_2, \ldots, N_c)$ are known. From (9.2) and (9.4) one obtains the Gibbs-Duhem equation

$$-SdT + Vdp - N_1 d\mu_1 - N_2 d\mu_2 - \ldots - N_c d\mu_c = 0. \tag{9.5}$$

Molar Quantities

In many applications, we will have the opportunity to use molar quantities such as the molar Gibbs free energy g, defined by

$$g = \frac{G}{N}, \tag{9.6}$$

and the mole fraction x_i of the i-th component, defined by

$$x_i = \frac{N_i}{N}, \tag{9.7}$$

where

$$N = N_1 + N_2 + \ldots + N_c \tag{9.8}$$

is the total number of moles of the mixture. Notice that the c mole fractions are not independent because

$$x_1 + x_2 + \ldots + x_c = 1. \tag{9.9}$$

Equations (9.4) and (9.5) reduce to

$$g = \mu_1 x_1 + \mu_2 x_2 \ldots + \mu_c x_c \tag{9.10}$$

and

$$-sdT + vdp - x_1 d\mu_1 - x_2 d\mu_2 - \ldots - x_c d\mu_c = 0, \tag{9.11}$$

where $s = S/N$ is the molar entropy and $v = V/N$ is the molar volume. From (9.10) and (9.11), we conclude that

$$dg = -sdT + vdp + \mu_1 dx_1 + \mu_2 dx_2 + \ldots + \mu_c dx_c. \tag{9.12}$$

Notice, however, that the differentials dx_i are not independent but are tied by

$$dx_1 + dx_2 + \ldots + dx_c = 0. \tag{9.13}$$

Usually we eliminate dx_1 in (9.12), using (9.13), which allows us to write

$$dg = -sdT + vdp + (\mu_2 - \mu_1)dx_2 + \ldots + (\mu_c - \mu_1)dx_c \tag{9.14}$$

and to conclude that g can be considered as a function of temperature T, pressure p and mole fractions $x_2, x_3, \ldots, x_c$.

In some circumstances it is appropriate to use the activity z_i, introduced by Lewis and defined for each component by

$$z_i = e^{\mu_i/RT}. \tag{9.15}$$

9.3 Mixture of Ideal Gases

Helmholtz Free Energy

Consider a mixture of c distinct ideal gases in a vessel of volume V at temperature T. We postulate that the Helmholtz free energy of the mixture $F(T, V, N_1, N_2, \ldots, N_c)$ is equal to the sum of the Helmholtz free energy of each gas as if each one of them were alone in the vessel, that is, as if each one were confined in a vessel of volume V at temperature T. That is,

$$F = \sum_{i=1}^{c} N_i f_i(T, \frac{V}{N_i}), \tag{9.16}$$

where $f_i(T, v_i)$ is the molar Helmholtz free energy of the i-th component alone, given by

$$f_i(T, v_i) = -a_i T - c_i T \ln \frac{T}{T_{0i}} - RT \ln \frac{v_i}{v_{0i}}, \tag{9.17}$$

where a_i, c_i, T_{0i} and v_{0i} are characteristic constants of the ideal gas i.

The pressure $p = -\partial F/\partial V$ is thus

$$p = \sum_{i=1}^{c} \frac{N_i RT}{V} = \frac{NRT}{V}. \tag{9.18}$$

The partial pressure p_i, which is the pressure exerted by the i-th ideal gas if it were alone in the vessel of volume V, is given by $p_i = -\partial f_i/\partial v_i$ or

$$p_i = \frac{RT}{v_i} = \frac{N_i RT}{V}. \tag{9.19}$$

Therefore,

$$p = \sum_{i=1}^{c} p_i, \tag{9.20}$$

which is Dalton law: the pressure of an ideal gas mixture is the sum of the partial pressures.

Gibbs Free Energy

The Gibbs free energy $G(T, p, N_1, N_2, \ldots, N_c)$ of an ideal gas mixture is obtained from the Legendre transformation of the free energy F, given by (9.16). Making the transformation, we get

$$G = \sum_{i=1}^{c} N_i \{g_i + RT \ln \frac{N_i}{N}\}, \tag{9.21}$$

where g_i is the molar Gibbs free energy of the i-th component alone, given by

$$g_i = -b_i T - (c_i + R) T \ln \frac{T}{T_{0i}} + RT \ln \frac{p}{p_{0i}}, \tag{9.22}$$

where $b_i = a_i - R$ and $p_{0i} = RT_{0i}/v_{0i}$, and is related with $f_i(T, v_i)$ by means of a Legendre transformation.

The entropy $S = -\partial G/\partial T$, as a function of T and p, is given by

$$S = \sum_{i=1}^{r} N_i \{s_i - R \ln \frac{N_i}{N}\}, \tag{9.23}$$

where $s_i = -\partial g_i/\partial T$ is the molar entropy of the i-th component alone, given by

$$s_i = a_i + c_i + (c_i + R) \ln \frac{T}{T_{0i}} - R \ln \frac{p}{p_{0i}}. \tag{9.24}$$

The chemical potential $\mu_i = \partial G/\partial N_i$ of the i-th component is

$$\mu_i = g_i + RT \ln \frac{N_i}{N}. \tag{9.25}$$

It is worth noting that the molar Gibbs free energy $g = G/N$, the molar entropy $s = S/N$ and the chemical potential can be written in the forms

$$g = \sum_{i=1}^{c} x_i \{g_i + RT \ln x_i\}, \tag{9.26}$$

$$s = \sum_{i=1}^{c} x_i \{s_i - R \ln x_i\} \tag{9.27}$$

and

$$\mu_i = g_i + RT \ln x_i, \tag{9.28}$$

where $x_i = N_i/N$. Recall that the mole fractions x_i are not independent but are related by (9.9).

The activity z_i of the i-th ideal gas is connected to the partial pressure p_i by the linear relation

$$p_i = p_i^0 z_i \tag{9.29}$$

where p_i^0 depends only on temperature. This relation is obtained by noting that for an ideal gas mixture $p_i/p = N_i/N = x_i$ and by using (9.22) and (9.28).

Entropy of Mixing

Let us determine what is called the entropy of mixing. Consider a vessel divided into c compartments separated by diathermal and mobile walls. Each compartment contains a different type of ideal gas and the i-th compartment contains N_i moles of ideal gas of type i. As the walls that separate the compartments are diathermal and mobile, all gases have the same temperature T and the same pressure p. Next the inner walls are removed and the gases mix. The temperature and pressure of the ideal gas mixture does not change, but the entropy increases. The entropy of mixing is the difference between the entropy of the mixed gases and the entropy of the gases isolated, within their respective compartments, at the same temperature and pressure.

The total entropy of the isolated gases, each in its compartment, is simply

$$S_I = \sum_{i=1}^{c} N_i s_i, \tag{9.30}$$

where s_i is the molar entropy of the i-th component. For an ideal gas mixture, the entropy is

$$S_F = \sum_{i=1}^{c} N_i \{s_i - R \ln \frac{N_i}{N}\}, \tag{9.31}$$

so that

$$\Delta S = S_F - S_I = -R \sum_{i=1}^{c} N_i \ln \frac{N_i}{N}. \tag{9.32}$$

As $N_i/N \leq 1$, then $\Delta S \geq 0$.

The variation in the Gibbs free energy is

$$\Delta G = RT \sum_{i=1}^{c} N_i \ln \frac{N_i}{N} = -T\Delta S, \tag{9.33}$$

from which one concludes that the change in the enthalpy is zero, since $\Delta H = \Delta G + T\Delta S = 0$. We conclude also that the volume change is zero. Indeed, since $\Delta V = \partial \Delta G/\partial p = 0$ and being ΔG independent of p, then $\Delta V = 0$. The variation in the internal energy is also zero, since $\Delta U = \Delta H - p\Delta V = 0$.

Suppose that the gases are all equal, for example, all of type 1. In this case there should be no increase in entropy. The initial entropy is

$$S_I = \sum_{i=1}^{c} N_i s_1. \tag{9.34}$$

The final entropy, however, is

$$S_F = N s_1, \tag{9.35}$$

since the vessel is filled by a single component, that has N moles. As $N = \sum_i N_i$, then $S_F = S_I$ and there is no change in entropy as expected.

9.4 Dilute Solutions

Raoult and Henry Laws

Let us examine here the properties of dilute solutions. We consider a mixture of c components such that $c - 1$ components have very small concentrations. We adopt the component 1 as the solvent and the other components as solutes. The sum of the mole fractions of solutes is denoted by x, that is,

$$x = x_2 + x_3 + \ldots + x_c. \tag{9.36}$$

For a dilute solution $x << 1$ so that the mole fraction of the solvent $x_1 = 1 - x$ is close to unit.

Denoting by $\bar{\rho} = N/V$ the total number of moles of the solution per unit volume and by $\bar{\rho}_i = N_i/V$ the number of moles of the i-th component per unit volume, then the condition $x << 1$ implies $\bar{\rho}_2 + \bar{\rho}_3 + \ldots + \bar{\rho}_c << \bar{\rho}$. That is to say, in a dilute solution the number of moles per unit volume of the solutes must be small. Note, however, that the number of moles per unit volume of the solvent need not necessarily be small, unlike what happens with a mixture of ideal gases, in which all the components must have low density. On the other hand, a mixture of ideal gases need not be diluted.

We now examine the behavior of the molar Gibbs free energy $g(T, p, x_2, x_3, \ldots, x_c)$ for small values of x. We are considering g to be a function of the mole fractions of the solutes and not of the solvent, because the mole fractions can not all be independent. If we use (9.10), we see that the determination of g is reduced to the determination of the chemical potentials μ_i. Proceeding according to Lewis, we start with two assumptions concerning the behavior of chemical potentials for dilute solutions. The first hypothesis states that the chemical potential of the solvent reduces the chemical potential of pure solvent in the limit $x \to 0$. The second hypothesis says that the activities $z_i = \exp\{\mu_i/RT\}$ of the solutes are proportional to their mole fractions, that is,

$$z_i = \kappa_i x_i, \tag{9.37}$$

where the constant κ_i depends on the pressure, temperature and nature of the solvent and solute, but not of the mole fractions. We call this expression Henry law for dilute solutions. The chemical potential of the solute behaves, therefore, according to

$$\mu_i = \mu_i^\circ + RT \ln x_i, \tag{9.38}$$

where $\mu_i^\circ = RT \ln \kappa_i$.

To determine the chemical potential of the solvent, we use the Gibbs-Duhem equation (9.11). But before, we see that for T and p constant, we have $d\mu_i = RTdx_i/x_i$ for $i = 2, 3, \ldots, c$ which, replaced in the Gibbs-Duhem equation, gives

$$(1 - x)d\mu_1 + RTdx = 0, \tag{9.39}$$

because $x_1 = 1 - x$ and $x = x_2 + x_3 + \ldots + x_c$. Integrating this equation, we get

$$\mu_1 = g_1 + RT \ln(1 - x), \tag{9.40}$$

where we take into account that in the limit $x \to 0$, the chemical potential of the solvent reduces to the chemical potential of pure solvent, which is identified with the molar Gibbs free energy of the pure solvent, denoted by g_1. The activity $z_1 = \exp\{\mu_1/RT\}$ of the solvent is given therefore by

$$z_1 = z_1^*(1 - x), \tag{9.41}$$

that is, the activity of the solvent of diluted solutions is proportional to the mole fraction of the solvent $x_1 = 1 - x$. Moreover, the constant of proportionality $z_1^* = \exp\{g_1/RT\}$ is identified with the activity of the pure solvent. We call this expression Raoult law for dilute solutions.

We remark that the Raoult law concerns the solvent (whose mole fraction is close to unity), while Henry law concerns the solutes (whose mole fractions approach zero). If the dilute solution is in coexistence with its vapor, then the equilibrium condition implies that the activities of the components in the solution should be equal to the activities of the components in the vapor phase. If, in addition, the vapor can be considered an ideal gas mixture, then, according to (9.29), the activities z_i are proportional to the partial pressures p_i of the gases. Therefore, from Raoult and Henry laws, we can write

$$p_1 = p_1^*(1 - x), \tag{9.42}$$

for the solvent, and

$$p_i = k_i x, \tag{9.43}$$

for the solutes, where p_1^*, the vapor pressure of the pure solvent, and k_i depend only on temperature. These relations constitute the original forms of Raoult and Henry laws, respectively.

Replacing, finally, the expressions for the chemical potentials in (9.10), we obtain the explicit form for the molar Gibbs free energy of a dilute solution,

$$g = g_1(1 - x) + \sum_{i=2}^{c} \mu_i^\circ x_i + RT(1 - x)\ln(1 - x) + RT \sum_{i=2}^{c} x_i \ln x_i. \tag{9.44}$$

The chemical potential of the solvent can be written as

$$\mu_1 = g_1 - RTx, \tag{9.45}$$

since for small values of x we can approximate $\ln(1 - x)$ by $-x$. Importantly, the chemical potential μ_1, both in the form (9.40) or as (9.45), does not depend on the characteristics of the solutes, but only on the sum of their mole fractions. This particularity provides the emergency of so-called colligative properties, which are those properties that depend on the number of moles of the solutes but do not depend on the type of solutes present in the solution. Typical examples of colligative properties are:

(a) osmotic pressure;
(b) elevation of the boiling temperature of liquid solutions with non-volatile solutes (the vapor has a single component: the solvent), and
(c) lowering of the fusion temperature of the solids immiscible in solute (the solid phase contains a single component: the solvent).

Osmotic Pressure

Consider a system consisting of two compartments separated by a rigid and diathermal wall so that, at equilibrium, the temperature will be equal, but not the pressure. On one side there is a dilute solution and on the other there is only the pure substance number 1, the solvent. The wall is permeable but only to the component 1. The wall being permeable to the solvent, in equilibrium, the chemical potential of the solvent in solution should be equal to the chemical potential of the pure solvent, that is,

$$\mu_1(T,p,x_2,x_3,\ldots,x_c) = g_1(T,p_0), \tag{9.46}$$

where p_0 is the pressure of the pure solvent. The osmotic pressure Π is defined as the difference in pressure between the two sides of the wall, that is, $\Pi = p - p_0$.

If the solution is diluted, then the chemical potential of the solvent is given by (9.45) which, replaced in (9.46), yields

$$g_1(T,p) - RTx = g_1(T,p_0). \tag{9.47}$$

But for small values of $\Pi = p - p_0$, we may approximate $g_1(T,p)$ by

$$g_1(T,p) = g_1(T,p_0) + v_1(T,p_0)(p - p_0), \tag{9.48}$$

where $v_1 = \partial g_1/\partial p$ is the molar volume of the pure solvent. Since the solution is diluted, v_1 can be replaced by the molar volume v of the solution. Therefore,

$$RTx = \Pi v, \tag{9.49}$$

or

$$\Pi = \bar{\rho}_{\text{solute}} RT, \tag{9.50}$$

which is the van 't Hoff equation, where $\bar{\rho}_{\text{solute}} = x/v = (N_2 + N_3 + \ldots + N_c)/V$ is the total number of moles of solute per unit volume (molarity). For dilute solutions, the osmotic pressure is thus independent of the nature of the solvent, depending only on the molarity of the solute.

The experimental measurement of the osmotic pressure can be used for the determination of molecular weight M of the molecules, especially macromolecules. For a solution with only one type of solute $\bar{\rho}_{\text{solute}} = \rho_{\text{solute}}/M$, where ρ_{solute} is the density of the solute (mass per unit volume). Therefore,

$$M = \rho_{\text{solute}} \frac{RT}{\Pi}, \tag{9.51}$$

so that the experimental measurements of pressure and density provide the molecular weight of the solute. As the above equation should be valid for dilute solutions then it is convenient to plot the right side of (9.51) as a function of density for various values of ρ_{solute} and make the extrapolation $\rho_{\text{solute}} \to 0$.

Lowering the Melting Point

We will examine here the lowering of the melting temperature of a pure substance when certain impurities are dissolved in it. Suppose that the impurities and the pure substance are miscible in the liquid phase but not on the solid phase. Thus we should examine the coexistence of a liquid phase, which is a dilute solution, and a solid phase consisting only of the pure solvent. The liquid phase is a dilute solution whose chemical potential of the solvent is given by (9.45), that is,

$$\mu_1 = g_1^L(T, p) - RTx, \tag{9.52}$$

where $g_1^L(T, p)$ is the molar Gibbs free energy of the pure solvent and x is the sum of mole fractions of the solutes in the liquid phase. Since we are assuming that the solid phase is composed only by the solvent, then the chemical potential of the solvent in the solid phase is independent of x and is equal to the molar Gibbs free energy of the solid which we denote by $g_1^S(T, p)$. The equilibrium condition is then

$$g_1^L(T, p) - RTx = g_1^S(T, p), \tag{9.53}$$

so that

$$RTx = g_1^L(T, p) - g_1^S(T, p). \tag{9.54}$$

Let T_0 be the temperature for which $g_1^L(T_0, p) = g_1^S(T_0, p)$, which is the melting temperature of the pure solid. Then for small differences $T - T_0$ we have

$$g_1^L(T, p) - g_1^S(T, p) = -[s_1^L(T_0, p) - s_1^S(T_0, p)](T - T_0), \tag{9.55}$$

or

$$g_1^L(T, p) - g_1^S(T, p) = -\frac{\ell_f}{T_0}(T - T_0), \tag{9.56}$$

where $\ell_f = T_0(s_1^L - s_1^S)$ is the molar fusion latent heat of the pure solid. Therefore,

$$T_0 - T = \frac{RT_0^2}{\ell_f}x, \tag{9.57}$$

which is the van 't Hoff formula for the lowering of the fusion temperature of the solvent. The coefficient RT_0^2/ℓ_f depends only on the properties of the solvent and does not depend on the nature of the solute, but only on its number of moles, and is known as the cryoscopic constant. Notice that this expression is valid as long as the liquid solution coexist with the pure solid.

If the solute is composed by a single substance, the measurement of the decrease in temperature $\Delta T = T_0 - T$ can be used to determine the molar mass of the solute M_2 from the molar mass of the solvent M_1. If the mole fraction x of the solute is small then $x = N_2/N_1$. But

$$\frac{N_2}{N_1} = \frac{m_2}{m_1}\frac{M_1}{M_2}, \tag{9.58}$$

where m_1 and m_2 are the masses of the solvent and solute, respectively. This equation, combined with the van 't Hoff formula (9.57), allows the determination of M_2 from M_1.

9.5 General Solutions

Ideal Solutions

So far we have obtained two expressions for the Gibbs free energy. One related to the mixture of ideal gases, valid in the regime of low densities of all components, and another related to the dilute solutions, valid in the regime of low mole fractions of solutes. We now develop expressions of the Gibbs free energy for solutions that are not necessarily dilute or a mixture of ideal gases.

Suppose that N_1 moles of a pure substance 1, N_2 moles of a pure substance 2, ... and N_c moles of a pure substance c, are used to form a mixture. Initially, all pure substances which are isolated have the same temperature T and the same pressure p. The process of formation of the mixture is such that the final mixture has the same temperature and the same pressure of the isolated substances. In this process, the change in Gibbs free energy ΔG is given by

$$\Delta G = G(T,p,N_1,N_2,\ldots,N_c) - \sum_{i=1}^{c} N_i g_i(T,p), \tag{9.59}$$

where $G(T,p,N_1,N_2,\ldots,N_c)$ is the Gibbs free energy of the mixture and $g_i(T,p)$ is the molar Gibbs free energy of the i-th pure substance. Similarly, we define the variation in entropy ΔS, the volume change ΔV, and the enthalpy change ΔH.

In the process of formation of the mixture, there will be certainly an increase in the entropy and in general one expects a nonzero variation of the enthalpy and of other thermodynamic properties. In an ideal mixture, or ideal solution, we assume that the variation in entropy is the same as that occurring in a mixture of ideal gases, at constant temperature and pressure, that is

$$\Delta S = -R\sum_{i=1}^{c} N_i \ln\frac{N_i}{N}, \tag{9.60}$$

and that the variation in enthalpy vanishes, that is, $\Delta H = 0$. As the mixture of the substances is an isothermal process, then $\Delta H = \Delta G + T\Delta S$, so that, for ideal solutions,

$$\Delta G = -T\Delta S. \tag{9.61}$$

A significant consequence is that there is no variation of volume nor variation of internal energy when pure substances are mixed to form an ideal solution. In fact, deriving both sides of (9.61) with respect to pressure and taking into account that ΔS is independent on p, we see that $\Delta V = \partial\Delta G/\partial p = 0$. Since p is constant then $\Delta U = \Delta H - p\Delta V = 0$.

Replacing (9.61) in (9.59) and using (9.60), one obtains the following expression for the Gibbs free energy of an ideal solution

$$G = \sum_{i=1}^{c} N_i\{g_i + RT\ln\frac{N_i}{N}\}, \tag{9.62}$$

which is the fundamental relation in the Gibbs representation. From it we found the entropy of the ideal solution $S = -\partial G/\partial T$, given by

$$S = \sum_{i=1}^{c} N_i\{s_i - R\ln\frac{N_i}{N}\}, \tag{9.63}$$

where $s_i = -\partial g_i/\partial T$ is the molar entropy of the i-th pure substance.

Since the enthalpy, the volume and the internal energy do not vary in the process of formation of an ideal solution then

$$H = \sum_{i=1}^{c} N_i h_i, \qquad V = \sum_{i=1}^{c} N_i v_i, \qquad U = \sum_{i=1}^{c} N_i u_i, \tag{9.64}$$

where $h_i = g_i + Ts_i$, $v_i = \partial g_i/\partial p$ and $u_i = h_i - pv_i$ are the molar enthalpy, the molar volume and the molar energy of the i-th component in the pure state. The chemical potential $\mu_i = \partial G/\partial N_i$ of the i-th component is given by

$$\mu_i = g_i + RT\ln\frac{N_i}{N}. \tag{9.65}$$

The expressions for the molar Gibbs free energy $g = G/N$ and for the molar entropy $s = S/N$ are given by

$$g = \sum_{i=1}^{c} x_i\{g_i + RT\ln x_i\}, \tag{9.66}$$

$$s = \sum_{i=1}^{c} x_i\{s_i - R\ln x_i\}, \tag{9.67}$$

and for the enthalpy, volume and molar energy by

$$h = \sum_{i=1}^{c} x_i h_i, \qquad v = \sum_{i=1}^{c} x_i v_i, \qquad u = \sum_{i=1}^{c} x_i u_i. \tag{9.68}$$

The i-th chemical potential is given by

$$\mu_i = g_i + RT \ln x_i. \tag{9.69}$$

Remember that the mole fractions x_i are not independent but are linked by (9.9). From (9.69), we see that the activity $z_i = e^{\mu_i/RT}$ for the i-th component is related with the mole fraction by means of

$$z_i = z_i^* x_i, \tag{9.70}$$

where $z_i^* = e^{g_i/RT}$ is the activity of the pure component. Therefore, for ideal solutions, if any component is considered to be a solvent (x_i close to unit), Raoult law is fulfilled. If on the other hand one component is considered to be a solute (x_i close to zero), Henry law is satisfied with $\kappa_i = z_i^*$.

In general, we do not expect real solutions to behave as ideal solutions. However, certain molecular solutions behaves approximately as ideal solutions. In particular a mixture of ideal gases is an ideal solution. It suffices to compare the equations for the Gibbs free energies of an ideal gas mixture and of an ideal mixture, given by (9.21) and (9.62), respectively.

Hildebrand Regular Solutions

A Hildebrand regular solution is defined as one whose change in entropy is the same as that of an ideal solution. This means that when mixing several pure substances at the same temperature and pressure, the entropy change ΔS is given by (9.60). However, unlike what happens in an ideal solution, the change in enthalpy ΔH in a Hildebrand regular solution does not vanish. Since ΔS is independent of temperature, it follows that ΔH should also be independent of temperature. Indeed, deriving $\Delta H = \Delta G + T\Delta S$ with respect to temperature, we get

$$\frac{\partial}{\partial T}\Delta H = \frac{\partial}{\partial T}\Delta G + \Delta S + T\frac{\partial}{\partial T}\Delta S = 0, \tag{9.71}$$

because ΔS is independent on T and $\Delta S = -\partial \Delta G/\partial T$.

If we demand that the mixture is formed without change of volume, then ΔH should be independent of pressure. In fact, deriving $\Delta H = \Delta G + T\Delta S$ with respect to pressure,

$$\frac{\partial}{\partial p}\Delta H = \frac{\partial}{\partial p}\Delta G + T\frac{\partial}{\partial p}\Delta S = \Delta V, \tag{9.72}$$

since ΔS does not depend on p and $\Delta V = \partial \Delta G/\partial p$. Therefore, if $\Delta V = 0$, then ΔH will be independent of pressure. In this case, $\Delta H = \Delta U$, since $\Delta H = \Delta U + p\Delta V$.

Replacing $\Delta G = \Delta H - T\Delta S$ in expression (9.59) and using the notation $\widetilde{G} = \Delta H$, then the fundamental relation $G(T, p, N_1, N_2, \ldots, N_c)$ of a Hildebrand regular mixture has the form

$$G = \widetilde{G} + \sum_{i=1}^{c} N_i g_i + RT\sum_{i=1}^{c} N_i \ln\frac{N_i}{N}, \tag{9.73}$$

where $\widetilde{G}$ depends on the numbers of moles of each component, can depend on pressure, but not on the temperature, as shown in (9.71).

As the function $\widetilde{G}(p, N_1, N_2, \ldots, N_c)$ is extensive, we may write

$$\widetilde{G}(p, N_1, N_2, \ldots, N_c) = N\tilde{g}(p, x_1, x_2, \ldots, x_c), \tag{9.74}$$

where $x_i = N_i/N$ are the mole fractions and $N = \sum_i N_i$. Therefore the molar Gibbs free energy $g = G/N$ of a Hildebrand regular solution is given by

$$g = \tilde{g} + \sum_{i=1}^{c} x_i\{g_i + RT\ln x_i\}, \tag{9.75}$$

where $\tilde{g}$ does not depend on T.

Excess Quantities

In the study of real solutions it is convenient to use as a reference system an ideal solution with the same composition. Thus, we introduce the thermodynamic excess quantities, defined as the differences of the thermodynamic quantities from the ideal forms. The excess Gibbs free energy G^{E} is defined by

$$G = G^{\mathrm{E}} + \sum_{i=1}^{c} N_i\{g_i + RT\ln\frac{N_i}{N}\}. \tag{9.76}$$

Clearly, for an ideal solution, $G^{\mathrm{E}} = 0$. Comparing this expression with (9.73), we see that the Hildebrand regular solution is the one such that G^{E} does not depend on temperature but only on pressure and number of moles.

Other excess quantities are defined in an analogous way and can be obtained directly form G^E. Thus, the excess entropy is given by $S^E = -\partial G^E/\partial T$ and the excess enthalpy by $H^E = G^E + TS^E$. The excess chemical potentials $\mu_i^E = \partial G^E/\partial N_i$ are related to the chemical potentials μ_i by

$$\mu_i = \mu_i^E + g_i + RT \ln \frac{N_i}{N}. \tag{9.77}$$

The excess Gibbs free energy G^E is related to the variation of the Gibbs free energy in the formation of a mixture ΔG by

$$\Delta G = G^E + \sum_{i=1}^{c} N_i RT \ln \frac{N_i}{N}. \tag{9.78}$$

From this relation we conclude that $H^E = \Delta H$, that is, the excess enthalpy coincides with the variation in enthalpy in the formation of a mixture.

Taking into account that $G^E(T, p, N_1, N_2, \ldots, N_c)$ is an extensive function we may write

$$G^E(T, p, N_1, N_2, \ldots, N_c) = Ng^E(T, p, x_1, x_2, \ldots, x_c). \tag{9.79}$$

Therefore, the molar Gibbs free energy $g = G/N$ of any solution can be written in the form

$$g = g^E + \sum_{i=1}^{c} x_i \{g_i + RT \ln x_i\}. \tag{9.80}$$

The chemical potentials are given by

$$\mu_i = \mu_i^E + g_i + RT \ln x_i. \tag{9.81}$$

Substituting (9.80) and (9.81) into (9.10), we conclude that

$$g^E = \sum_{i}^{c} \mu_i^E x_i. \tag{9.82}$$

Replacing (9.81) in the Gibbs-Duhem equation in the form given by (9.11), we obtain the Gibbs-Duhem-Margules equation

$$\sum_{i=1}^{c} x_i d\mu_i^E = 0, \tag{9.83}$$

valid at T and p constant.

In the limit of dilute mixtures, the excess quantities must have certain properties that we will examine. Comparing the expression (9.81) with the expression of the chemical potential of the solvent, given by (9.40), and with the chemical potentials of the solutes, given by (9.38), we see that in the limit $x_i \to 1$, the excess chemical potential μ_i^{E} vanishes and the others, μ_j^{E}, approaches $\mu_j^\circ - g_j$. Using these results in (9.82), we conclude that g^{E} also vanish in this limit.

The knowledge of specific forms of the excess Gibbs free energy g^{E} allows the determination of the intrinsic properties of the mixtures, as long as we know the properties of the pure substances. In fact, if g^{E} and g_i were known it is possible to recover g by (9.80). To set up g^{E} we should take into account that $g^{\mathrm{E}} \to 0$ when $x_i \to 1$ for any component i. The simplest form of g^{E} that meets this property is

$$g^{\mathrm{E}} = \frac{1}{2}\sum_{i,j} w_{ij}x_ix_j, \tag{9.84}$$

where w_{ij} depends only on T and p, and $w_{ij} = w_{ji}$ and $w_{ii} = 0$.

We remark that g^{E} can be a singular function of the mole fractions. For example, in the vicinity of a critical point, we expect that the thermodynamic quantities have singular behavior. In particular, the excess Gibbs free energy must by a nonanalytic at the critical point. However, in most cases the systems are not found inside the critical region and can therefore be described by functions g^{E} that are analytic in x_i, as is the case of that given by (9.84), which we call regular forms. The mixtures that are described by such functions are called regular mixtures. Although the regular forms for g^{E} do not provide a precise description at the critical region, they allows us, on the other hand, a qualitative description of this region.

Problems

9.1 Show that if $G(T, p, N_1, N_2, \ldots, N_c)$ is an extensive function, that is, homogeneous function of first order in $N_1, N_2, \ldots, N_c$, then

$$G = \mu_1 N_1 + \mu_2 N_2 + \ldots + \mu_c N_c.$$

9.2 From $H = G + TS$, $V = \partial G/\partial p$ and $U = H - pV$, and using the expressions (9.21) and (9.23) for G and S, show that for an ideal gas mixture the enthalpy, the volume and the internal energy are given by

$$H = \sum_{i=1}^{c} N_i h_i, \qquad V = \sum_{i=1}^{c} N_i v_i, \qquad U = \sum_{i=1}^{c} N_i u_i,$$

where h_i is the molar enthalpy, v_i is the molar volume and u_i is the molar energy of the i-th component, alone. The enthalpy, the volume and the energy of a mixture of ideal gases are therefore equal to the sum of the enthalpies, the volumes and the energy of the gases isolated, at the same temperature and pressure of the mixture. Or, equivalently, the enthalpy, the volume and the internal energy do not change when ideal gases are mixed at constant temperature and pressure.

9.3 Suppose that a vessel of volume V has r partitions separated by rigid and diathermal walls containing distinct gases. The i-th partition has volume V_i and contains N_i moles of a gas of type i. Determine the change in the Helmholtz free energy and of the entropy after the removal of the walls. Notice that initially the gases are at the same temperature, since the walls are diathermal, but not necessarily at the same pressure. Repeat the problem for the case in which all partitions have the same type of gas.

9.4 Derive a formula similar to the van 't Hoff one (9.57) for the elevation of the boiling point of a liquid by the addition of non-volatile solutes. Taking into account that the vapor contains only the solvent as its sole component, the chemical potential of the solvent in the gaseous phase is independent of the mole fractions of the solutes. Show that how it can be used to determine the ratio between the molecular mass of the solvent and of the solute.

9.5 Show that a regular solution fulfills the Raoult and Henry laws. Determine the quantity κ_i which appears in the Henry law.

Chapter 10
Binary Mixtures

10.1 Fundamental Relation

Substances that are Binary Mixtures

In this chapter we study mixtures of two components, called binary mixtures which are quite common in nature, both the homogeneous and heterogeneous. Many minerals, when not pure substances are homogeneous solid binary mixtures such as olivine (magnesium silicate and iron silicate) and amalgam (solid solution of mercury and silver). The manufactured alloys, such as bronze (copper and tin) and brass (copper and zinc) are also solid binary mixtures. Alcohol is usually found in association with water, forming a homogeneous binary mixture.

We focus our study on the phase equilibria of binary mixtures. We initially examine the liquid-vapor coexistence of mixtures of liquids which are miscible in all proportions. Such mixtures may or may not have an azeotropic point. Next, we examine the solid-liquid transition of mixtures of substances that are completely miscible in both the solid phase and liquid phase.

Later, we pass to the study of mixtures of partially miscible liquid. In this case it is common the occurrence of the coexistence of two liquids of different compositions. The variation in temperature or pressure may lead the system to a critical point where both coexisting liquids become identical. We consider, next, the mixtures of substances partially miscible in the solid phase but fully miscible liquid phase. The partial solubility provides the appearance of two solid phases in coexistence. If the pure solid substances have the same crystalline structure, then the two solid solutions also have the same crystalline structure but different compositions. In this case, their coexistence may disappear at a critical point at which the two solid phases become equal.

It is still possible, the occurrence of the melting of the two solid solutions while in coexistence. In this case we will be facing the equilibrium of three phases, two solid phases and the liquid phase, and we will have the opportunity to examine systems

M.J. de Oliveira, *Equilibrium Thermodynamics*, Graduate Texts in Physics,
DOI 10.1007/978-3-662-53207-2_10

with eutectic point. Other types of phase equilibria may occur, but in this chapter, we limit ourselves to the study of equilibrium types mentioned above.

Thermodynamic State

In the representation of internal energy, the thermodynamic state of a binary mixture is defined by the extensive variables: entropy S, volume V, number of moles N_1 of the first component and number of moles N_2 of the second component. The fundamental relation is given by the corresponding internal energy $U(S, V, N_1, N_2)$, whose differential is

$$dU = TdS - pdV + \mu_1 dN_1 + \mu_2 dN_2, \tag{10.1}$$

where T is the temperature, p is the pressure, and μ_1 and μ_2 are the chemical potentials of components 1 and 2, respectively. Other representations can be used in the study of the binary mixtures, but the most convenient is the Gibbs representation. The fundamental relation is given by the Gibbs free energy $G(T, p, N_1, N_2)$, whose differential is

$$dG = -SdT + Vdp + \mu_1 dN_1 + \mu_2 dN_2, \tag{10.2}$$

where

$$S = -\frac{\partial G}{\partial T}, \qquad V = \frac{\partial G}{\partial p}, \qquad \mu_1 = \frac{\partial G}{\partial N_1}, \qquad \mu_2 = \frac{\partial G}{\partial N_2}. \tag{10.3}$$

The activities z_1 and z_2 of the components 1 and 2 are defined by

$$z_1 = e^{\mu_1/RT} \qquad \text{and} \qquad z_2 = e^{\mu_2/RT}. \tag{10.4}$$

Being G extensive we may write

$$G = \mu_1 N_1 + \mu_2 N_2, \tag{10.5}$$

which can be uses to recover the fundamental relation $G(T, p, N_1, N_2)$ in the case the chemical potentials $\mu_1(T, p, N_1, N_2)$ and $\mu_2(T, p, N_1, N_2)$ are known. From (10.2) and (10.5) we get the Gibbs-Duhem equation

$$-SdT + Vdp - N_1 d\mu_1 - N_2 d\mu_2 = 0. \tag{10.6}$$

Molar Quantities

We will make frequent use of molar quantities such as the molar Gibbs free energy g defined by $g = G/N$ and the mole fractions x_1 and x_2 of the components, defined by $x_1 = N_1/N$ and $x_2 = N_2/N$, where $N = N_1 + N_2$ is the total number of moles of the mixture. Note that the mole fractions are not independent since $x_1 + x_2 = 1$.

Dividing both members of (10.5) and (10.6) by N, they reduce to

$$g = \mu_1 x_1 + \mu_2 x_2 \tag{10.7}$$

and

$$-sdT + vdp - x_1 d\mu_1 - x_2 d\mu_2 = 0, \tag{10.8}$$

where $s = S/N$ is the molar entropy and $v = V/N$ molar volume. From (10.7) and (10.8) we conclude that

$$dg = -sdT + vdp + \mu_1 dx_1 + \mu_2 dx_2. \tag{10.9}$$

Notice however that the differentials dx_i are not independent but are linked by

$$dx_1 + dx_2 = 0. \tag{10.10}$$

Usually we eliminate dx_1 in (10.9), using (10.10), which permits us to write

$$dg = -sdT + vdp + (\mu_2 - \mu_1)dx_2. \tag{10.11}$$

Defining μ and x by

$$\mu = \mu_2 - \mu_1 \qquad \text{and} \qquad x = x_2, \tag{10.12}$$

respectively, then, according to (10.11), we have

$$dg = -sdT + vdp + \mu dx, \tag{10.13}$$

so that the molar Gibbs free energy $g(T, p, x)$ can be considered a function of temperature T, pressure p and mole fraction of the second component x. The equations of state are

$$\mu = \frac{\partial g}{\partial x}, \qquad s = -\frac{\partial g}{\partial T}, \qquad v = \frac{\partial g}{\partial p}. \tag{10.14}$$

Given $g(T,p,x)$, the fundamental relation in the Gibbs representation can be recovered by

$$G(T,p,N_1,N_2) = Ng(T,p,x), \tag{10.15}$$

where $x = N_2/(N_1 + N_2)$. From this equation we get the following relations,

$$\mu_1 = g - x\mu \tag{10.16}$$

and

$$\mu_2 = g + (1-x)\mu, \tag{10.17}$$

between the chemical potentials $\mu_1 = \partial G/\partial N_1$ and $\mu_2 = \partial G/\partial N_2$ of the components of the mixture and the chemical potential μ. It suffices to take into account that $\partial x/\partial N_1 = -x/N$ and that $\mu_2 = \partial G/\partial N_2$.

It is convenient to define the relative activity ζ of the component 2 as $\zeta = z_2/(z_1 + z_2)$, which can be written in the form

$$\zeta = \frac{1}{1 + e^{-\mu/RT}}. \tag{10.18}$$

10.2 Gibbs Free Energy

Dilute Mixtures

In the applications that we will consider below we will use specific forms for the Gibbs free energy. In the preceding chapter we have seen some forms that are applied to a mixture of ideal gases and to other systems called ideal and regular solutions. Next we will write explicitly the various expressions for the case of a binary mixture. We consider first the Gibbs free energy of a diluted mixture and after we consider the ideal mixture and regular mixture forms. These expressions should be understood as approximations to the free energy. Although approximate they adequately describe it in many cases, the properties of the mixtures, in particular the phase coexistence.

The molar Gibbs free energy of a dilute mixture was presented in the previous section and reduces, for the present case of two components, to the following expression

$$g = g_1(1-x) + \mu_2^\circ x + RT(1-x)\ln(1-x) + RTx\ln x, \tag{10.19}$$

where g_1 is the molar Gibbs free energy of the pure solvent, the component 1 of the mixture, and μ_2° depends only on the temperature and pressure. The mole fraction x is related to the component 2, the solute. The expression above is valid for $x << 1$.

The chemical potential $\mu = \partial g/\partial x$ is given by

$$\mu = -g_1 + \mu_2^\circ - RT\ln(1-x) + RT\ln x. \tag{10.20}$$

From μ and g, we obtain the chemical potential of the solvent $\mu_1 = g - x\mu$, given by

$$\mu_1 = g_1 + RT\ln(1-x), \tag{10.21}$$

which can also be written in the form

$$\mu_1 = g_1 - RTx \tag{10.22}$$

taking into account that $x << 1$. The chemical potential of the solute, $\mu_2 = \mu + \mu_1$, is given by

$$\mu_2 = \mu_2^\circ + RT\ln x. \tag{10.23}$$

The activities of the solvent z_1 and solute z_2 are obtained by exponentiation of both sides of (10.21) and (10.23) and are therefore given by

$$z_1 = z_1^o(1-x) \qquad \text{and} \qquad z_2 = \kappa_2 x, \tag{10.24}$$

where $z_1^o = \exp\{g_1/RT\}$ is the activity of the pure solvent and $\kappa_2 = \exp\{\mu_2^\circ/RT\}$. These two expressions are the Raoult and Henry laws, respectively.

Ideal Mixtures

The expression for the molar Gibbs free energy g of an ideal mixture of several components, seen in the previous chapter, is reduced to the following expression for the case of two components:

$$g = xg_2 + (1-x)g_1 + xRT\ln x + (1-x)RT\ln(1-x). \tag{10.25}$$

where g_1 and g_2 are the molar Gibbs free energies of the pure substances 1 and 2, respectively, and x is the mole fraction of component 2.

The molar entropy $\partial g/\partial T$ is given by

$$s = xs_2 + (1-x)s_1 - xR\ln x - (1-x)R\ln(1-x), \tag{10.26}$$

where $s_1 = \partial g_1/\partial T$ and $s_2 = \partial g_2/\partial T$ are the molar entropies of pure substances 1 and 2, respectively. The chemical potential μ is given by

$$\mu = g_2 - g_1 + RT \ln \frac{x}{1-x}. \tag{10.27}$$

while the chemical potentials μ_1 and μ_2 of each of the components are given by

$$\mu_1 = g_1 + RT \ln(1-x) \tag{10.28}$$

and

$$\mu_2 = g_2 + RT \ln x. \tag{10.29}$$

If the i-th pure substance is an ideal gas then the corresponding free energy $g_i(T, p)$ is known and is given by

$$g_i(T, p) = -b_i T - (c_i + R) T \ln \frac{T}{T_{0i}} + RT \ln \frac{p}{p_{0i}}, \tag{10.30}$$

where b_i, c_i, T_{0i} and p_{0i} are constants that characterize the i-th ideal gas. The corresponding molar entropy $s_i = \partial g_i/\partial T$ is

$$s_i(T, p) = a_i + c_i + (c_i + R) \ln \frac{T}{T_{0i}} - R \ln \frac{p}{p_{0i}}. \tag{10.31}$$

Regular Mixtures

The molar Gibbs free energy $g(T, p, x)$ of a mixture of two components is given by

$$g = g^{\mathrm{E}} + (1-x) g_1 + x g_2 + RT(1-x) \ln(1-x) + RTx \ln x, \tag{10.32}$$

where $g^{\mathrm{E}}(T, p, x)$ is the molar excess Gibbs free energy, that depends on temperature, pressure and mole fraction x of component 2. From it we get the chemical potential $\mu = \partial g/\partial x$, given by

$$\mu = \mu^{\mathrm{E}} - g_1 + g_2 + RT \ln \frac{x}{1-x}, \tag{10.33}$$

where $\mu^{\mathrm{E}} = \partial g^{\mathrm{E}}/\partial x$. The chemical potentials μ_1 and μ_2 of the components are obtained by the relations (10.16) and (10.17) and are

$$\mu_1 = \mu_1^{\mathrm{E}} + g_1 + RT \ln(1-x) \tag{10.34}$$

and

$$\mu_2 = \mu_2^{\mathrm{E}} + g_2 + RT \ln x, \tag{10.35}$$

where the excess chemical potential are given by

$$\mu_1^{\mathrm{E}} = g^{\mathrm{E}} - x\mu^{\mathrm{E}} = g^{\mathrm{E}} - x\frac{\partial g^{\mathrm{E}}}{\partial x} \tag{10.36}$$

and

$$\mu_2^{\mathrm{E}} = g^{\mathrm{E}} + (1-x)\mu^{\mathrm{E}} = g^{\mathrm{E}} + (1-x)\frac{\partial g^{\mathrm{E}}}{\partial x}. \tag{10.37}$$

In order to describe the intrinsic properties of binary solutions, we must seek specific forms of g^{E}. A large class of binary mixtures is well described by the excess Gibbs free energies g^{E} that are analytic functions of x, which we call regular forms. The mixtures are described by such regular functions are called regular mixtures. These forms must be constructed so that $g^{\mathrm{E}} \to 0$ when $x \to 0$ and when $x \to 1$, since, taking into account that the function g^{E} describes the deviations of the real solution from the ideal behavior, it must vanish when the solution is reduced to a pure substance. We assume, therefore, that g^{E} is the product of $x(1-x)$ with a function of x which is finite at $x = 0$ and $x = 1$. Under these conditions the simplest regular form is

$$g^{\mathrm{E}} = wx(1-x), \tag{10.38}$$

where w depends only on T and p. Therefore,

$$g = wx(1-x) + (1-x)g_1 + xg_2 + RT(1-x)\ln(1-x) + RTx\ln x. \tag{10.39}$$

From g^{E} given by (10.38) we obtain the simple regular equations

$$\mu_1^{\mathrm{E}} = wx^2 \qquad \text{and} \qquad \mu_2^{\mathrm{E}} = w(1-x)^2, \tag{10.40}$$

so that the chemical potentials μ_1 and μ_2 of each component are

$$\mu_1 = wx^2 + g_1 + RT\ln(1-x) \tag{10.41}$$

and

$$\mu_2 = w(1-x)^2 + g_2 + RT\ln x. \tag{10.42}$$

A regular form particularly useful is

$$g^{\mathrm{E}} = x(1-x)\{a + b(1-2x) + c(1-2x)^2\}, \tag{10.43}$$

which leads as to the Margules equations:

$$\mu_1^{\mathrm{E}} = x^2\{(a+b) + 2(b+c)(1-2x) + 3c(1-2x)^2\} \tag{10.44}$$

and

$$\mu_2^{\mathrm{E}} = (1-x)^2\{(a-b) + 2(b-c)(1-2x) + 3c(1-2x)^2\}. \tag{10.45}$$

They are reduced to the simple regular equations when $b = c = 0$.

Another regular form is

$$g^{\mathrm{E}} = \frac{x(1-x)}{b(1-x) + ax}, \tag{10.46}$$

which yields the van Laar equations

$$\mu_1^{\mathrm{E}} = \frac{ax^2}{[b(1-x) + ax]^2} \tag{10.47}$$

and

$$\mu_2^{\mathrm{E}} = \frac{b(1-x)^2}{[b(1-x) + ax]^2}. \tag{10.48}$$

These equations are reduced to the simple regular equations when $b = a$.

10.3 Phase Transition

Two Phase Coexistence

The coexistence of two phases in a binary mixture is described by a linear segment of $g(T, p, x)$ as a function of x, as shown in Fig. 10.1. To show this result, let us consider a process in which the concentration of a mixture, which displays two coexisting phases, are changed, keeping the temperature and pressure constant. Under these conditions, one phase grows over the other which decreases, changing the proportion of each coexisting phase. However, the mole fractions of the components in each phase remain unchanged.

Denoting by N_A and N_B the number of moles of each phase, then the fraction in number of moles of phase A is $r_A = N_A/N$ and of phase B is $r_B = N_B/N$, where

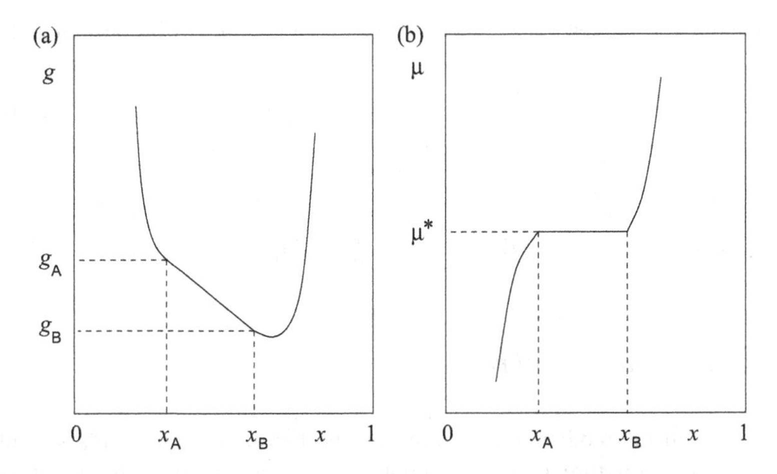

Fig. 10.1 (**a**) Molar Gibbs free energy g as a function of the mole fraction x for T and p constant. (**b**) Chemical potential $\mu = \partial g/\partial x$ as a function of the mole fraction x for T and p constant

N is the total number of moles. Denoting by x_A and by x_B the mole fractions of component 2 in phases A and B, respectively, then the number of mole of component 2 in phase A is $x_A N_A$ and in phase B is $x_B N_B$, so that the total number of moles of component 2 in the mixture is $N_2 = x_A N_A + x_B N_B$. But the mole fraction of component 2 in the mixture is $x = N_2/N$ and therefore

$$x = x_A r_A + x_B r_B, \tag{10.49}$$

which is the lever rule. Since $r_A + r_B = 1$, we obtain from this rule the relations

$$r_A = \frac{x - x_B}{x_A - x_B} \qquad \text{and} \qquad r_B = \frac{x_A - x}{x_A - x_B}. \tag{10.50}$$

Since x_A and x_B remain unchanged, the same happens with the molar Gibbs free energy $g_A^* = g(T, p, x_A)$ and $g_B^* = g(T, p, x_B)$ of each phase because the temperature and pressure are constant. The molar Gibbs free energy g of the mixture is then

$$g = r_A g_A^* + r_B g_B^*. \tag{10.51}$$

Substituting (10.50) in (10.51), we get

$$g = \frac{x_B - x}{x_B - x_A} g_A^* + \frac{x - x_A}{x_B - x_A} g_B^*, \tag{10.52}$$

and therefore g is linear in x in the range $x_A \le x \le x_B$ as g_A^* and g_B^* do not depend on x but only of T and p.

In the range $x_A \leq x \leq x_B$, the chemical potential $\mu = \partial g/\partial x$ is constant and is

$$\mu^* = \frac{g_B^* - g_A^*}{x_B - x_A}, \tag{10.53}$$

from which we get

$$g_A^* - \mu^* x_A = g_B^* - \mu^* x_B. \tag{10.54}$$

Thermodynamic Equilibrium

In a system in thermodynamic equilibrium the thermodynamic fields should be uniform. For a mixture of two components, the condition of thermodynamic equilibrium means that the temperature T, the pressure p, the chemical potential μ_1 of component 1 and the chemical potential μ_2 of component 2 must have the same value anywhere in the system. In particular, for two phases A and B in coexistence, the condition of equilibrium is given by

$$T_A = T_B, \qquad p_A = p_B, \qquad \mu_{1A} = \mu_{1B}, \qquad \mu_{2A} = \mu_{2B}. \tag{10.55}$$

where T_A, p_A, μ_{1A} and μ_{2A} denote the fields in phase A and T_B, p_B, μ_{1B} and μ_{2B} denote the fields in phase B.

Let us imagine that the Gibbs free energy of the thermodynamic phases A and B are described by the functions $g_A(T, p, x)$ an $g_B(T, p, x)$, respectively. If these functions are two distinct branches, then, the Gibbs free energy of the mixture $g(T, p, x)$ is obtained by the construction of a double tangent between the two branches, that is, by the construction of a linear segment that connect the two points of tangency (x_A, g_A^*) and (x_B, g_B^*) and having slope μ^* given by (10.53). Taking into account that $g_A^* = g_A(T, p, x_A)$, that $g_B^* = g_A(T, p, x_B)$, and that

$$\mu^* = \mu_A(T, p, x_A) = \mu_B(T, p, x_B), \tag{10.56}$$

where $\mu_A = \partial g_A/\partial x$ and $\mu_B = \partial g_B/\partial x$, then, (10.54) reads

$$g_A(T, p, x_A) - \mu^* x_A = g_B(T, p, x_B) - \mu^* x_B. \tag{10.57}$$

Equations (10.56) and (10.57) are, therefore, the conditions of thermodynamic equilibrium and can be use to determine x_A and x_B as functions of T and p.

Using (10.16), it is easy to see that (10.57) is equivalent to equation

$$\mu_{1A}(T, p, x_A) = \mu_{1B}(T, p, x_B). \tag{10.58}$$

Adding (10.56) and (10.57) and using (10.17), we see that the resulting equation is equivalent to equation

$$\mu_{2A}(T, p, x_A) = \mu_{2B}(T, p, x_B). \tag{10.59}$$

These two equations are the conditions of thermodynamic equilibrium stated above and may equivalently be used in the place of (10.56) and (10.57) to determine x_A and x_B.

Condensation and Boiling Lines

As an example of phase transition, we examine the liquid-vapor transition of the water-methanol mixture whose diagram is shown in Fig. 10.2. The water and methanol are miscible in all proportions forming a liquid solution. Introducing heat in a mixture of completely miscible liquid, at constant pressure, the temperature increases and the mixture passes to the gaseous state. However, unlike what happens with a pure substance, the transition does not occur in a single temperature, but a range of temperatures. The transition begins at the boiling temperature and ends at the condensation temperature. These two temperatures depend on the mole fraction x of one component, as can be seen in Fig. 10.2a. In this graph, the line that marks the threshold of boiling is called boiling or bubble line and the line that marks the threshold of condensation is called condensation or dew line. Similarly if the temperature is maintained constant, the phase transition occurs between two pressures, as can be seen in Fig. 10.2b.

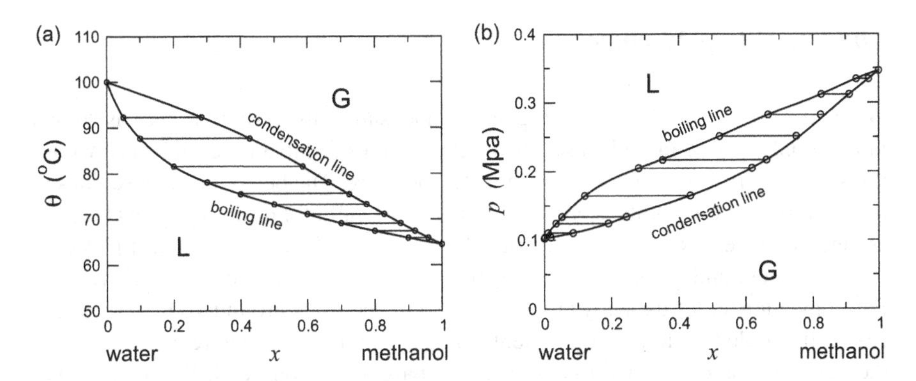

Fig. 10.2 Boiling and condensation lines for water-methanol mixture, where x is the mole fraction of methanol. (**a**) Diagram temperature versus mole fraction, at the pressure of 1 atm. (**b**) Diagram pressure versus mole fraction, at the temperature of 100 °C. The horizontal segments are the *tie lines* and represent the coexistence between the liquid (L) and gas (G) phases. *Source*: LB

To understand the graphs in Fig. 10.2, we recall that the coexistence of two phases, for T and p fixed, occurs at a well defined value of the chemical potential $\mu^*(T,p)$. The two phases in coexistence have also well defined mole fractions $x_L(T,p)$ and $x_G(T,p)$. In a graph of T versus x at constant pressure, $x_L(T,p)$ and $x_G(T,p)$ constitute the boiling and condensation lines, respectively. The horizontal line segments that connect two points on these lines are tie lines shown in Fig. 10.2a. Along the tie line, not just the pressure p and temperature T are constant, but the chemical potential μ as well.

If the mixture is in a state corresponding to a point on the tie line, the fraction in number of moles r_L of the liquid phase and the fraction in number of moles r_G of the vapor are given by

$$r_L = \frac{x - x_G}{x_L - x_G} \qquad \text{and} \qquad r_G = \frac{x_L - x}{x_L - x_G}. \tag{10.60}$$

Therefore, the fraction in number of moles of one phase is proportional to the distance between the point under consideration and the end point of the tie line corresponding to the other phase.

Heating the solution at constant pressure until a temperature between the boiling and condensation temperatures, the mixture will be found in a state of coexistence of liquid and vapor whose compositions are distinct from one another and distinct from the mixture. One phase is richer in one component than the other. In Fig. 10.2 we can see that the vapor is rich in methanol, component 2, as the mole fraction x_G of the component 2 in the gas phase is greater than the mole fraction x_L of component 2 in the vapor. The method of enrichment of one of the components of a solution by boiling is called distillation.

Limit of Dilute Mixture

When $x = 0$ or $x = 1$, the boiling and condensation lines should meet as shown in Fig. 10.2 because in both cases the mixture is reduced to a pure substance whose transition liquid-vapor occurs at a single temperature. The behavior of these lines in the vicinity of these points is obtained from results valid for dilute solutions because around them the mixture becomes diluted. Around $x = 0$, the component 1 plays the role of solvent and component 2, the solute; around $x = 1$, the roles are inverted. For convenience, let us analyze the behavior of the lines around $x = 0$, in which case x, the mole fraction of component 2, is very small. Furthermore, let us consider the general case of a transition between a phase A and phase B. We assume that phase A occurs at temperatures higher than phase B.

According to Raoult law, in the form given by (10.22), the chemical potential of the solvent in dilute solutions is given by

$$\mu_1^A = g_1^A - RTx_A, \qquad \text{and} \qquad \mu_1^B = g_1^B - RTx_B, \tag{10.61}$$

where $g_1^{\rm A}$ and $g_1^{\rm B}$ are the molar Gibbs free energies of the pure solvent in phases A and B, respectively, and depend only on temperature and pressure. According to Henry law the activity of the solute of dilute solutions is given by

$$z_2^{\rm A} = \kappa_2^{\rm A} x_{\rm A}, \qquad \text{and} \qquad z_2^{\rm B} = \kappa_2^{\rm B} x_{\rm B}, \tag{10.62}$$

where $\kappa_2^{\rm A}$ and $\kappa_2^{\rm B}$ depend only on temperature and pressure.

The condition of thermodynamic equilibrium means that the chemical potentials of the solvent and the solute in each phase are identical. Therefore,

$$g_1^{\rm A} - RTx_{\rm A} = g_1^{\rm B} - RTx_{\rm B} \tag{10.63}$$

or

$$x_{\rm A} - x_{\rm B} = \frac{\Delta g_1}{RT}, \tag{10.64}$$

where $\Delta g_1 = g_1^{\rm A} - g_1^{\rm B}$. And

$$\kappa_2^{\rm B} x_{\rm B} = \kappa_2^{\rm A} x_{\rm A}. \tag{10.65}$$

From these equations we get

$$x_{\rm A} = \frac{\Delta g_1}{RT}\frac{\kappa_2^{\rm B}}{\kappa_2^{\rm B} - \kappa_2^{\rm A}} \qquad \text{and} \qquad x_{\rm B} = \frac{\Delta g_1}{RT}\frac{\kappa_2^{\rm A}}{\kappa_2^{\rm B} - \kappa_2^{\rm A}}, \tag{10.66}$$

which are the equations describing the lines that determine the thresholds of the phase transition. Note that if the pure substances are immiscible in one phase, the results above yield the van't Hoff formula. Suppose that the miscibility occurs only at the phase A. Then, $\kappa_2^{\rm A} = 0$, which gives $x_{\rm A} = \Delta g_1/RT$ from which we obtain van't Hoff formula.

Expanding Δg_1 around the transition pressure $p_1^*(T)$ of the pure solvent and taking into account that $\Delta g_1 = 0$ when $p = p_1^*$ then

$$x_{\rm L} = a(p - p_1^*) \qquad \text{and} \qquad x_{\rm G} = b(p - p_1^*), \tag{10.67}$$

where a and b depend only on temperature. Therefore the lines that determine the thresholds of the transition in the diagram pressure versus mole fraction are linear near $x = 0$ and have slopes of the same sign because $a/b = \kappa_2^{\rm A}/\kappa_2^{\rm B} > 0$, as seen in Fig. 10.2b. The same behavior holds in the diagram temperature versus mole fraction, as seen in Fig. 10.2a, what can be verified by the expansion of Δg_1 around the transition temperature of the pure solvent.

10.4 Completely Miscible Substances

Phase Coexistence

Here we examine liquid-vapor and solid-liquid transitions in binary mixtures of pure substances which are miscible in all proportions. When mixed, such substances form a single liquid solution or a single solid solution. Examples of pairs of completely miscible liquids are water and methanol, water and ethanol, benzene and ethanol, acetone and chloroform. Heating a liquid solution, it turns into vapor, which is also a solution. Many pairs of pure solids also are miscible in all proportions. Examples of solid solutions are: olivine (magnesium silicate and iron silicate), silicon and germanium, silver and gold, bismuth and antimony. Such solid solutions become liquid solutions at temperatures sufficiently high.

Let us suppose that, when isolated, the pure substances 1 and 2 undergo a transition from phase A to phase B, along the line of coexistence $p_1^*(T)$ and $p_2^*(T)$, respectively. The molar Gibbs free energy of the pure substances $g_1(T,p)$ and $g_2(T,p)$ are considered to have two branches each corresponding to the two phases, that is,

$$g_i(T,p) = \begin{cases} g_i^{\mathrm{A}}(T,p), & \text{phase A}, \\ g_i^{\mathrm{B}}(T,p), & \text{phase B}, \end{cases} \tag{10.68}$$

for $i = 1,2$. The coexistence line $p_i^*(T)$ for each isolated pure substance is determined by imposing the equality $g_i^{\mathrm{A}}(T,p) = g_i^{\mathrm{B}}(T,p)$.

We imagine that the molar Gibbs free energy of the mixture $g(T,p,x)$, where x represents the mole fraction of component 2, is constituted also by two branches corresponding to the two phases, that is,

$$g(T,p,x) = \begin{cases} g^{\mathrm{A}}(T,p,x), & \text{phase A}, \\ g^{\mathrm{B}}(T,p,x), & \text{phase B}. \end{cases} \tag{10.69}$$

Let us assume that each phase can be approximated by a regular solution so that the molar Gibbs free energies g^{A} and g^{B} of each phase are of the form (10.32). The chemical potentials of the components 1 and 2 of the phase A are thus of the form (10.34) and (10.35), that is,

$$\mu_1^{\mathrm{A}}(x) = \mu_1^{\mathrm{EA}}(x) + g_1^{\mathrm{A}} + RT\ln(1-x) \tag{10.70}$$

and

$$\mu_2^{\mathrm{A}}(x) = \mu_2^{\mathrm{EA}}(x) + g_2^{\mathrm{A}} + RT\ln x. \tag{10.71}$$

Similarly, the chemical potentials of the components 1 and 2 of phase B are also of the form (10.34) and (10.35), that is,

$$\mu_1^{\rm B}(x) = \mu_1^{\rm EB}(x) + g_1^{\rm B} + RT\ln(1-x) \tag{10.72}$$

and

$$\mu_2^{\rm B}(x) = \mu_2^{\rm EB}(x) + g_2^{\rm B} + RT\ln x, \tag{10.73}$$

where $\mu_i^{\rm EA}$ and $\mu_i^{\rm EB}$ are the excess chemical potentials of the components in phases A and B, respectively.

If there is phase coexistence, the mole fraction of component 2 in phase A will be denoted by $x_{\rm A}$ and in phase B by $x_{\rm B}$. According to (10.58) and (10.59), the coexistence of phases is obtained by imposing the equality of the chemical potentials $\mu_i^{\rm A}(x_{\rm A}) = \mu_i^{\rm B}(x_{\rm B})$ or

$$\mu_1^{\rm EA}(x_{\rm A}) + g_1^{\rm A} + RT\ln(1-x_{\rm A}) = \mu_1^{\rm EB}(x_{\rm B}) + g_1^{\rm B} + RT\ln(1-x_{\rm B}) \tag{10.74}$$

and

$$\mu_2^{\rm EA}(x_{\rm A}) + g_2^{\rm A} + RT\ln x_{\rm A} = \mu_2^{\rm EB}(x_{\rm B}) + g_2^{\rm B} + RT\ln x_{\rm B}, \tag{10.75}$$

which we write in the form

$$\ln\frac{1-x_{\rm B}}{1-x_{\rm A}} = \frac{1}{RT}\{\Delta g_1 + \mu_1^{\rm EA}(x_{\rm A}) - \mu_1^{\rm EB}(x_{\rm B})\} \tag{10.76}$$

and

$$\ln\frac{x_{\rm B}}{x_{\rm A}} = \frac{1}{RT}\{\Delta g_2 + \mu_2^{\rm EA}(x_{\rm A}) - \mu_2^{\rm EB}(x_{\rm B})\}, \tag{10.77}$$

where $\Delta g_1 = g_1^{\rm A} - g_1^{\rm B}$ and $\Delta g_2 = g_2^{\rm A} - g_2^{\rm B}$. These two expressions permit to determine $x_{\rm A}$ and $x_{\rm B}$ as functions of pressure and temperature and constitute the lines that determine the thresholds of the transition between A and B.

Liquid-Vapor Equilibrium

To determine the lines that give the threshold of the transition in the diagram p versus x or in the diagram T versus x, we must know how the free energies $g_i(T,p)$ of the pure substances depend on T and p. We consider initially the liquid-vapor transition. We assume that the gas corresponding to the pure substances are ideal gases so that

$$g_i^G = RT\ln p + \gamma_i, \tag{10.78}$$

where γ_i depends only on temperature and that the liquids corresponding to the pure substances are described by

$$g_i^L = RT \ln p_i^* + \gamma_i, \tag{10.79}$$

where p_i^* is the vapor pressure of the i-th pure substance and depends only on temperature.

From these expressions we have $\Delta g_i = g_i^{\mathrm{G}} - g_i^{\mathrm{L}} = \ln(p/p_i^*)$ which replaced in (10.76) and (10.77) yields

$$\ln \frac{1 - x_{\mathrm{L}}}{1 - x_{\mathrm{G}}} = \ln \frac{p}{p_1^*} - \frac{1}{RT}\mu_1^{\mathrm{EL}}(x_{\mathrm{L}}) \tag{10.80}$$

and

$$\ln \frac{x_{\mathrm{L}}}{x_{\mathrm{G}}} = \ln \frac{p}{p_2^*} - \frac{1}{RT}\mu_2^{\mathrm{EL}}(x_{\mathrm{L}}), \tag{10.81}$$

where we have taken into account that the excess chemical potentials of the pure substances in vapor phase vanish, $\mu_i^{\mathrm{EG}} = 0$, because the vapor is a mixture of ideal gases. These equations describe the condensation and boiling lines in the diagram pressure versus mole fraction.

They can also be written in the form

$$p(1 - x_{\mathrm{G}}) = p_1^*(1 - x_{\mathrm{L}}) \exp\{\frac{1}{RT}\mu_1^{\mathrm{EL}}(x_{\mathrm{L}})\} \tag{10.82}$$

and

$$p x_{\mathrm{G}} = p_2^* x_{\mathrm{L}} \exp\{\frac{1}{RT}\mu_2^{\mathrm{EL}}(x_{\mathrm{L}})\}. \tag{10.83}$$

Adding these two equations, we get p as a function of x_{L}, which constitutes the boiling line

$$p = p_1^*(1 - x_{\mathrm{L}}) \exp\{\frac{1}{RT}\mu_1^{\mathrm{EL}}(x_{\mathrm{L}})\} + p_2^* x_{\mathrm{L}} \exp\{\frac{1}{RT}\mu_2^{\mathrm{EL}}(x_{\mathrm{L}})\}. \tag{10.84}$$

The condensation line, p as a function of x_{G}, is obtained in implicit form from the boiling line and using one of the two equations (10.82) or (10.83). Alternatively, dividing both (10.82) and (10.83), we get

$$\frac{1}{x_{\mathrm{G}}} = 1 + \frac{p_1^*}{p_2^*}(\frac{1}{x_{\mathrm{L}}} - 1) \exp\{\frac{1}{RT}[\mu_1^{\mathrm{EL}}(x_{\mathrm{L}}) - \mu_2^{\mathrm{EL}}(x_{\mathrm{L}})]\}, \tag{10.85}$$

which determined x_{G} from x_{L}.

To determine the condensation and boiling lines in the diagram temperature versus mole fraction we should explicit the dependence of the vapor pressure p_i^* of the pure substance with temperature. To this end we use the following equation

$$\ln \frac{p}{p_i^*} = \frac{\ell_i}{R}\left(\frac{1}{T} - \frac{1}{T_i^*}\right) \tag{10.86}$$

where T_i^* is the transition temperature of the i-th pure substance, at pressure p. This equation can be derived with the help of the Clausius-Clapeyron equation, assuming that the boiling latent heat ℓ_i is a constant and that the volume of the vapor is much greater than the volume of the liquid. Substituting this result in (10.80) and (10.81), we get

$$\ln \frac{1 - x_\text{L}}{1 - x_\text{G}} = \frac{\ell_1}{R}\left(\frac{1}{T} - \frac{1}{T_1^*}\right) - \frac{1}{RT}\mu_1^\text{EL}(x_\text{L}) \tag{10.87}$$

and

$$\ln \frac{x_\text{L}}{x_\text{G}} = \frac{\ell_2}{R}\left(\frac{1}{T} - \frac{1}{T_2^*}\right) - \frac{1}{RT}\mu_2^\text{EL}(x_\text{L}), \tag{10.88}$$

If the liquid phase is an ideal solution, the excess chemical potentials of the substances in this phase are also zero, $\mu_i^\text{EL} = 0$, so that (10.80) and (10.81) become

$$\frac{1 - x_\text{L}}{1 - x_\text{G}} = \frac{p}{p_1^*} \quad \text{and} \quad \frac{x_\text{L}}{x_\text{G}} = \frac{p}{p_2^*}, \tag{10.89}$$

and can be solved for x_L and x_G, with the results

$$x_L = \frac{p - p_1^*}{p_2^* - p_1^*} \quad \text{and} \quad x_G = \frac{p_2^*}{p}\frac{p - p_1^*}{p_2^* - p_1^*}. \tag{10.90}$$

The boiling and condensation lines are thus

$$p = p_2^* x_L + p_1^*(1 - x_L) \tag{10.91}$$

and

$$\frac{1}{p} = \frac{x_G}{p_2^*} + \frac{1 - x_G}{p_1^*}. \tag{10.92}$$

We see that p is linear in x along the boiling line and that $1/p$ is linear in x along the condensation line. The two properties are observed experimentally, for example, in the system heptane-hexane, as can be seen in Fig. 10.3a, which can then be considered as an ideal mixture.

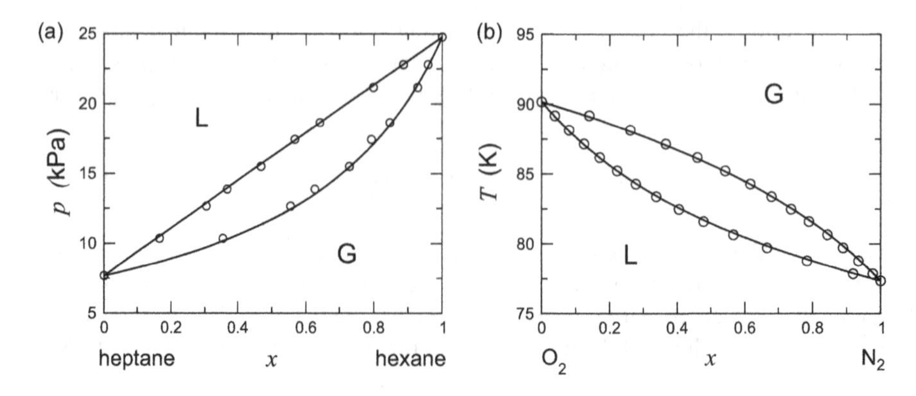

Fig. 10.3 (**a**) Liquid-vapor transition of the system heptane-hexane, at the temperature of 30 °C. Diagram p versus mole fraction x of hexane. *Source*: LB. (**b**) Liquid-vapor transition of the system oxygen-nitrogen, at the pressure of 1 atm. Diagram temperature T versus mole fraction x of nitrogen. The *circles* represent the experimental data obtained by Dodge and Dunbar [130]

The ideal behavior displayed by the system heptane-hexane is also presented by other pairs of similar pure substances. However, in general, we expect deviations from the ideal behavior, that can be relevant such as in azeotropes or small such as in the oxygen-nitrogen system shown if Fig. 10.3b. These deviations can be described by means of the regular forms for the excess quantities. Using (10.87) and (10.88), we can describe the condensation and boiling lines assuming that the liquid solution is a simple regular solution, that is, such that

$$\mu_1^{\mathrm{EL}} = w x_{\mathrm{L}}^2 \qquad \text{and} \qquad \mu_2^{\mathrm{EL}} = w(1 - x_{\mathrm{L}})^2. \tag{10.93}$$

Using the oxygen and nitrogen latent heats, $\ell_1/R = 820$ K and $\ell_2/RT = 670$ K, respectively, and the value $w/R = 12$ K, we get the continuous line shown in Fig. 10.3b.

Azeotropy

The results obtained so far show boiling and condensation lines that are strictly monotonic, as seen in Figs. 10.2 and 10.3. However, these lines may have a maximum or a minimum as seen in Fig. 10.4. At this point, called azeotropic, the boiling and condensation lines meet so that in the azeotropic point $x_{\mathrm{L}} = x_{\mathrm{G}} = x_{\mathrm{az}}$. A binary mixture, with a composition corresponding to the azeotropic point, passes from the liquid to the vapor phase, at constant pressure, at a unique temperature and in this sense it resembles a pure substance. However, an azeotropic mixture distinguishes from a pure substance because the composition of the azeotropic point

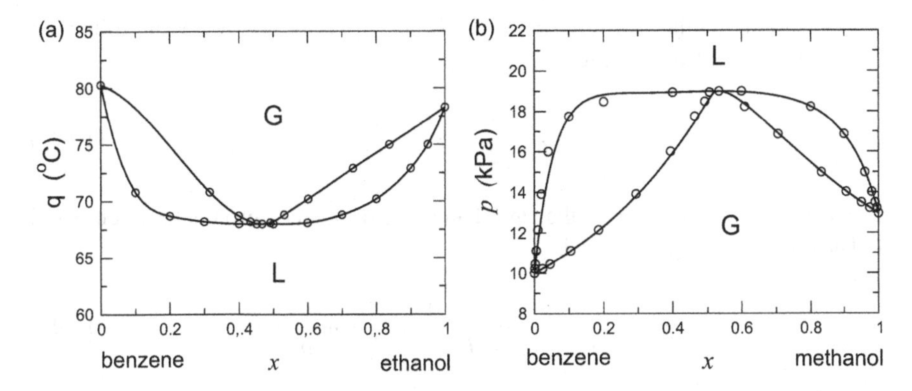

Fig. 10.4 (**a**) Liquid-vapor transition for the system benzene-ethanol in the diagram temperature versus mole fraction x of ethanol, at the pressure of 1 atm. The azeotropic point occurs at $x_{\text{az}} = 0.45$. (**b**) Liquid-vapor transition for the system benzene-methanol in the diagram pressure versus mole fraction x of methanol, at the temperature of 20 °C. The azeotropic point occurs at $x_{\text{az}} = 0.52$. *Source*: LB

depends on pressure. The properties of mixtures related to the azeotropic point are known as Gibbs-Konovalov laws.

There are many pairs of liquids that displays an azeotropic point including benzene-ethanol and benzene-methanol, whose diagrams are shown in Fig. 10.4, chloroform-acetone, water-ethanol, etc. At the pressure of 1 atm, the azeotropic point of the mixture water-ethanol occurs when the mole fraction of ethanol $x_{\text{az}} = 0.894$. Decreasing the pressure, the azeotropic point of the system water-ethanol increases, reaching the value $x_{\text{az}} = 0.98$ at the pressure of 0.13 atm. An important consequence of the existence of the azeotropic point is that the distillation process stops working when one reaches this point. A mixture water-ethanol that is poor in ethanol may become richer in ethanol by distillation until the concentration reaches the azeotropic point.

At the azeotropic point the coexistence line p versus μ, at constant T, reaches a maximum or a minimum and in both cases $dp/d\mu = 0$. From the Clausius-Clapeyron equation

$$\frac{dp}{d\mu} = \frac{x_{\text{G}} - x_{\text{L}}}{v_{\text{G}} - v_{\text{L}}}, \tag{10.94}$$

valid at constant temperature, we see that $(x_{\text{G}} - x_{\text{L}}) \to 0$ at the azeotropic point, that is, the length of the tie lines vanishes.

To determine the azeotropic point it suffices to impose $x_{\text{G}} = x_{\text{L}} = x_{\text{az}}$ in (10.85) since at this point x_{G} and x_{L} are equal. We get then

$$\mu_1^{\text{EL}}(x_{\text{az}}) - \mu_2^{\text{EL}}(x_{\text{az}}) = RT \ln \frac{p_2^*}{p_1^*}. \tag{10.95}$$

For the simple regular form $\mu_1^{\mathrm{E}} = wx^2$ and $mu_2^{\mathrm{E}} = w(1-x)^2$ we get the following result for the azeotropic point

$$x_{\mathrm{az}} = \frac{1}{2} - \frac{RT}{2w} \ln \frac{p_1^*}{p_2^*}. \tag{10.96}$$

The azeotropic point exists if the right hand side of this equation is between 0 and 1. The corresponding pressure is given by

$$p_{\mathrm{az}} = \sqrt{p_1^* p_2^*} \exp\{\frac{w}{RT}\frac{1}{4} + \frac{RT}{4w}(\ln \frac{p_1^*}{p_2^*})^2\}. \tag{10.97}$$

Other regular forms can also predict azeotropic points. The continuous lines shown in Fig. 10.4b were determined from (10.84) and (10.85) and by using the Margules regular form given by (10.44) and (10.45), with the following parameters: $a/RT = 2.04$, $b/RT = 0.28$ and $c/RT = 0.30$, and $T = 293.15$ K.

Solid-Liquid Equilibrium

We examine now the solid-liquid equilibrium of substances that are completely miscible both in the solid and liquid phases or, equivalently, the coexistence of a solid solution and a liquid solution. Complete miscibility occurs, for example, in mixtures of substances which in the solid phase has the same crystalline structure. The solid solutions can be of type substitutional or interstitial. In the first type, the atoms of one species are replaced by atoms of the other species in the vertices of the crystalline structure, which remains unchanged. In the second type, the atoms of the solute are inserted in the interstices between the atoms of the solvent. The complete solubility can occur only on the substitutional type. Examples of binary mixtures whose components have the same crystalline structure and are miscible in all proportions include: silver and gold, gold and copper (above 410 °C), gold and nickel (above 810 °C), gold an platinum (above 1260 °C), bismuth and antimony, cadmium and magnesium (above 253 °C), chromium and molybdenum (above 880 °C), copper and nickel, copper and platinum (above 816 °C), germanium and silicon, selenium and tellurium, magnesium silicate and iron silicate (olivine), and Al_2O_3 and Cr_2O_3.

Raising the temperature of these systems, they melt giving rise to a liquid solution, which coexist with the solid solution. We denote by x_{S} and x_{L} the mole fractions of component 2 in the solid and liquid phases, respectively. The mole fractions x_{S} and x_{L} as functions of temperature or pressure constitute the solid line (solidus) and the liquid line (liquidus). To determine them, we start from (10.76) and (10.77), that is,

$$\ln \frac{1-x_{\mathrm{S}}}{1-x_{\mathrm{L}}} = \frac{1}{RT}\{\Delta g_1 + \mu_1^{\mathrm{EL}}(x_{\mathrm{L}}) - \mu_1^{\mathrm{ES}}(x_{\mathrm{S}})\} \tag{10.98}$$

and

$$\ln \frac{x_S}{x_L} = \frac{1}{RT}\{\Delta g_2 + \mu_2^{EL}(x_L) - \mu_2^{ES}(x_S)\}, \tag{10.99}$$

where $\Delta g_1 = g_1^L - g_1^S$ and $\Delta g_2 = g_2^L - g_2^S$. These two expressions allows the determination of the solidus and liquidus. The lines can be obtained as long as Δg_1 and Δg_2 are known as functions of temperature and pressure. Let us consider p constant and use the approximation

$$\Delta g_1 = \frac{\ell_1}{T_1^*}(T_1^* - T) \qquad \text{and} \qquad \Delta g_2 = \frac{\ell_1}{T_2^*}(T_2^* - T), \tag{10.100}$$

where T_1^* and T_2^* are the melting temperatures and ℓ_1 and ℓ_2 are the melting latent heat of the pure substances. Substituting these expression in the previous equations, we get

$$\ln \frac{1 - x_S}{1 - x_L} = \frac{\ell_1}{R}(\frac{1}{T} - \frac{1}{T_1^*}) + \frac{1}{RT}\{\mu_1^{EL}(x_L) - \mu_1^{ES}(x_S)\} \tag{10.101}$$

and

$$\ln \frac{x_S}{x_L} = \frac{\ell_2}{R}(\frac{1}{T} - \frac{1}{T_2^*}) + \frac{1}{RT}\{\mu_2^{EL}(x_L) - \mu_2^{ES}(x_S)\}, \tag{10.102}$$

which permit determine the liquidus and solidus in the diagram temperature versus mole fraction.

As an example of the solid-liquid transition we show in Fig. 10.5 the diagram T versus x of the system germanium-silicon. Germanium and silicon are completely miscible both in the liquid and solid phases. Both have the same type of crystalline structure and their atoms similar sizes and the solid solution is of the type

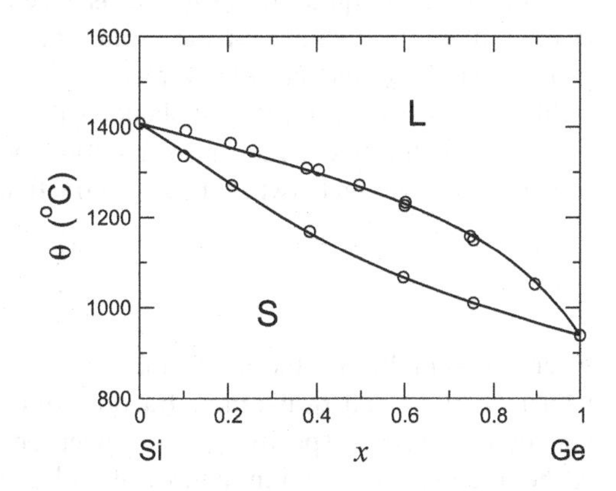

Fig. 10.5 Solid-liquid transition for the system germanium-silicon in the diagram temperature versus mole fraction x of germanium, at the pressure of 1 atm. The *circles* represent experimental data obtained by Stöhr and Klemm [132]

substitutional. The continuous line over the experimental data of Fig. 10.5 were obtained by (10.101) and (10.102) using the simple regular forms for the chemical potentials, that is,

$$\mu_1^{\mathrm{EL}}(x_{\mathrm{L}}) = w_{\mathrm{L}} x_{\mathrm{L}}^2 \qquad \text{and} \qquad \mu_2^{\mathrm{EL}}(x_{\mathrm{L}}) = w_{\mathrm{L}}(1 - x_{\mathrm{L}})^2, \tag{10.103}$$

for the liquid solution, and

$$\mu_1^{\mathrm{ES}}(x_{\mathrm{S}}) = w_{\mathrm{S}} x_{\mathrm{S}}^2 \qquad \text{and} \qquad \mu_2^{\mathrm{ES}}(x_{\mathrm{S}}) = w_{\mathrm{S}}(1 - x_{\mathrm{S}})^2, \tag{10.104}$$

for the solid solution. The parameters used were $w_{\mathrm{L}} = 450\,\mathrm{K}$ and $w_{\mathrm{L}} = 800\,\mathrm{K}$, in addition to the latent heats of the germanium and silicon, given by $\ell_1/R = 1212\,\mathrm{K}$ and $\ell_2/R = 1681\,\mathrm{K}$, respectively.

10.5 Partially Miscible Substances

Coexistence of Two Liquid Phases

When phenol (C_6H_6O) is added to water at room temperature one obtains a homogeneous solution, that is, a single thermodynamic phase, as long as the mass of phenol is not above 8 %. Above this value, phenol ceases to dissolve completely in water causing the appearance of another liquid phase, richer in phenol. Since the densities of the phases are distinct, the phase with lower density stays above that of higher density, separated by a meniscus. Adding more phenol, the two phase remain in coexistence until the mass of phenol in the mixture reaches 72 %. Above this value, the system returns to have single phase. If the mass of phenol is between 8 and 72 %, the system water-phenol will be, therefore, heterogeneous and composed by two liquid phases in coexistence. Changing the temperature, the range of phase coexistence changes, as shown in Fig. 10.6. For any concentration of phenol in the mixture, the two liquid phases in coexistence have each one a well defined mass fraction of phenol which, for the room temperature, is 0.08 for the phenol poor phase and 0.72 for the phenol rich phase.

To characterize the quantity of phenol in the mixture we will use the mole fraction x of phenol in the place of the mass fraction y, considered as the component 2 of the mixture. The relation between these two fractions is

$$\frac{y}{1-y} = \frac{M_1}{M_2}\frac{x}{1-x} \tag{10.105}$$

where M_1 and M_2 are the molar mass of components 1 and 2, respectively. Let us denote by x_{A} and x_{B} the mole fractions of component 2 in the phenol poor and phenol rich phases, respectively. Thus, given a mole fraction x of phenol, it is related to the fractions of phenol in number of moles in the phenol poor phase, r_{A}, and in the phenol rich phase, $r_B = 1 - r_A$, by the lever rule

$$x = r_A x_A + r_B x_B, \tag{10.106}$$

from which we get

$$r_A = \frac{x_B - x}{x_B - x_A} \quad \text{and} \quad r_B = \frac{x - x_A}{x_B - x_A}. \tag{10.107}$$

The mole fractions x_A and x_B of the two solutions in coexistence vary with temperature, describing the miscibility curve, or solubility, as seen in Fig. 10.6. Raising the temperature of the system water-phenol, from 20 °C, x_A increases and x_B decreases, so that the compositions of the two solutions become closer and, at a certain temperature, become equal. Above this temperature, called critical solution temperature, of consolute temperature, the two solutions become identical and the system becomes homogeneous. In other words, the two components of the mixture becomes completely miscible. At the critical temperature, the two mole fractions x_A and x_B have the same value x_c called critical mole fraction.

When the partial miscibility occurs at temperatures lower than the complete miscibility, as happens to the system water-phenol, the critical temperature is called upper consolute temperature. When the opposite occurs, it is called lower consolute temperature, as is the case of the system methylamine-water which has complete miscibility below 18.5 °C and partial miscibility above this temperature. Some binary mixtures, such as the system water-nicotine, have closed miscibility curve showing upper and lower consolute temperatures. Between 61 and 208 °C the miscibility of the system water-nicotine is partial. Out of this range the miscibility is total.

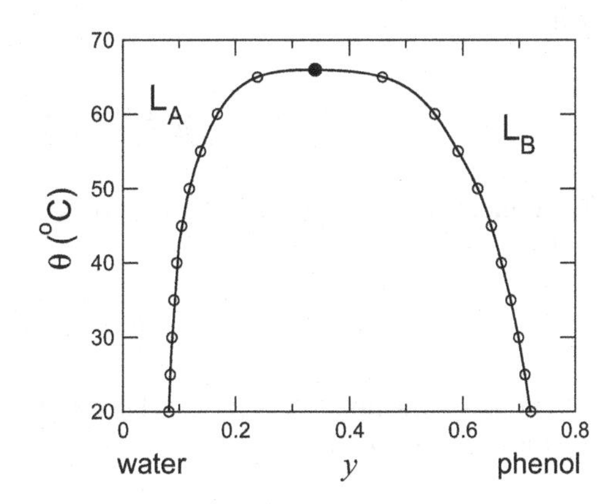

Fig. 10.6 Miscibility curve for the system water-phenol, at the pressure of 1 atm, where y is the mass fraction of phenol. In the region inside the miscibility curve, the phenol poor phase (L_A) and the phenol rich phase (L_B) coexist. Above the critical point, which occurs at $y = 0.34$ and $\theta = 66\,°C$, water and phenol are completely miscible. *Source*: LB

Hildebrand-Heitler Theory

The coexistence of two liquid phases and the critical point in systems of the type water-phenol can be described qualitatively by the theory of regular solutions. To this end, we use the expression for the Gibbs free energy $g(T, p, x)$ given by (10.39), that is,

$$g = wx(1-x) + (1-x)g_1 + xg_2 +$$
$$+ RT(1-x)\ln(1-x) + RTx\ln x, \tag{10.108}$$

where $g_1(T, p)$ and $g_2(T, P)$ depend on T and p and $w(p)$ depends only on pressure. We suppose that the pure substances do not undergo any transition, so that, unlike of what we have seen before, the functions g_1 and g_2 do not describe any transition. From the expression for g, we determine the chemical potential $\mu = \partial g/\partial x$, given by

$$\mu = w(1-2x) + g_2 - g_1 + RT\ln\frac{x}{1-x}. \tag{10.109}$$

At high temperatures, g is a convex function of x. However at low temperatures, g ceases to be convex and, in this case, we should consider the convex hull of g, obtained from a Maxwell construction. Initially, we determine the critical point. To this end, we calculate the derivatives of μ with respect to x, given by

$$\frac{\partial\mu}{\partial x} = -2w + \frac{RT}{x(1-x)} \tag{10.110}$$

and

$$\frac{\partial^2\mu}{\partial x^2} = RT\frac{2x-1}{x^2(1-x)^2}. \tag{10.111}$$

The critical point is such that the two derivatives vanish which occur when $x_c = 1/2$ and

$$RT_c = \frac{w}{2}. \tag{10.112}$$

The mole fractions x_A and x_B, related to the two phases, are determined by the equilibrium conditions

$$g_A - \mu_A x_A = g_B - \mu_B x_B \tag{10.113}$$

and

$$\mu_A = \mu_B, \tag{10.114}$$

which are equivalent to a Maxwell construction. In an explicit form

$$wx_A^2 + RT\ln(1 - x_A) = wx_B^2 + RT\ln(1 - x_B) \tag{10.115}$$

and

$$-2wx_A + RT\ln\frac{x_A}{1 - x_A} = -2wx_B + RT\ln\frac{x_B}{1 - x_B}, \tag{10.116}$$

equations that determine x_A and x_B as functions of T. It is easy to see that $x_B = 1-x_A$ because this condition makes the two equations above equivalent and equivalent to the equation

$$w(1 - 2x_A) + RT\ln\frac{x_A}{1 - x_A} = 0. \tag{10.117}$$

Defining the auxiliary quantity Δx by $\Delta x = x_B - x_A = 1 - 2x_A$ and, taking into account that $w = 2RT_c$, we have

$$\Delta x - \frac{T}{2T_c}\ln\frac{1 + \Delta x}{1 - \Delta x} = 0. \tag{10.118}$$

This equation can be solved numerically for Δx, from which we obtain the mole fractions $x_B = (1 + \Delta x)/2$ and $x_A = (1 - \Delta x)/2$, which describe the miscibility curve. Around the critical point, Δx is small, what allows the expansion of the left hand side of (10.118) in powers of Δx. Up to cubic terms,

$$(1 - \frac{T}{T_c})\Delta x - \frac{1}{3}(\Delta x)^3 = 0, \tag{10.119}$$

from which we get

$$\Delta x = \sqrt{3\frac{T_c - T}{T_c}}, \tag{10.120}$$

which gives the behavior of the miscibility curve around the critical point.

It is worth noticing that the quantity $(\partial x/\partial\mu)_{Tp}$, which we call susceptibility, diverges at the critical point along $x = x_c$. Replacing $x = x_c = 1/2$ in expression (10.110), we see that, for $T > T_c$,

$$\frac{\partial x}{\partial\mu} = \frac{1}{4R(T - T_c)}. \tag{10.121}$$

Using the result (10.120), we get the following expression for the susceptibility, for $T < T_c$,

$$\frac{\partial x}{\partial \mu} = \frac{1}{8R(T_c - T)}. \tag{10.122}$$

Critical Point

Around the critical point the thermodynamic quantities behave singularly. We assume that the singularities can be represented by power laws. According to this assumption, the miscibility gap $x_B - x_A$ vanishes at the critical point as

$$x_B - x_A \sim |T_c - T|^{\beta}. \tag{10.123}$$

Experimentally, one finds that the critical exponent β has values that are close to each other as seen in Table 10.1. The susceptibility $(\partial x/\partial \mu)_{Tp}$ diverges at the critical point as

$$\frac{\partial x}{\partial \mu} \sim |T - T_c|^{-\gamma}, \tag{10.124}$$

Table 10.1 Critical exponents related to the critical point of the liquid-liquid transition of incomplete miscible substances. The data refer to the pressure of 1 atm

Sistema	α	β	γ	θ_c (°C)
Isobutyric acid + water	0.12	0.33	1.24	26.14
Nitroethane + isooctane	0.11	0.32		30.03
Polystyreno + cyclohexane	0.14	0.33		30.50
Nitroethane + 3-methylpentane	0.14	0.34		26.47
Methanol + cyclohexane	0.10	0.33		49.1
Phenol + water		0.33	1.32	65.87
Triethylamina + water	0.11			18.5
Carbon tetrachloride + perfluoro methyl cyclohexane		0.33		28.64
Aniline + cyclohexane		0.33		31
Nitrobenzene + heptane		0.33		18
Carbon disulphide + nitromethane		0.32		63.5
Cyclohexane + anhydrous acetic		0.32		52.3
Gallium + mercury		0.34		203.3
Nitroethane + hexane			1.19	

along $x = x_c$. The value of the exponents β and γ obtained from the Hildebrand-Heitler theory, $\beta = 1/2$ and $\gamma = 1$, are distinct from the experimental values shown in Table 10.1, what means that this theory does not give a proper quantitative description if we are very close to the critical point although it gives a qualitative description.

A quantity that can be directly measured in binary liquid mixture is the molar heat capacity c_{px}, at x and p constant, defined by $c_{px} = T(\partial s/\partial T)_{px}$. Around the critical point this quantity behaves according to

$$c_{px} \sim |T - T_c|^{-\alpha}, \tag{10.125}$$

along $x = x_c$. Some experimental values of α are presented in Table 10.1. The isothermal compressibility at x constant and the coefficient of thermal expansion at x constant also diverge with the same exponent α, along $x = x_c$, that is,

$$-\frac{1}{v}(\frac{\partial v}{\partial p})_{Tx} \sim |T - T_c|^{-\alpha}, \tag{10.126}$$

$$\frac{1}{v}(\frac{\partial v}{\partial T})_{px} \sim |T - T_c|^{-\alpha}. \tag{10.127}$$

Coexistence of Two Solid Phases

The mixture of solid substances may result in both homogeneous or heterogeneous mixtures. Crystalline solids with the same structure can be miscible in all proportions, as happens with the system germanium-silicon, but may also give rise to solid mixtures with partial solubility, as happens to the system gold-platinum below 1260 °C or with the system silver-copper. If the structures are distinct, the solids will have partial solubility or will be immiscible. The alloy copper-zinc (brass) is homogeneous if the mole fraction of zinc is lower than 0.32. Above this value and until 0.48, the alloy consists of two solutions in coexistence. Copper and tin are practically immiscible at room temperature, comprising an alloy, the bronze, with two solid solutions in coexistence.

Below 1260 °C, the system gold-platinum may present coexistence of two solid solutions, as seen in Fig. 10.7. One rich in platinum (phase S_A), with mole fraction x_A, and the other rich in gold (phase S_B), with mole fraction $x_B > x_A$. The mole fractions x_A and x_B as functions of the temperature constitute the solubility curve (solvus) that delimits the coexistence region from the two solid solutions. If a fraction r_A in number of moles of the solid solution S_A is in equilibrium with a fraction r_B in number of moles of the solution S_B, then the system is represented by a point (x, T) inside the region of coexistence such that

$$x = r_A x_A + r_B x_B. \tag{10.128}$$

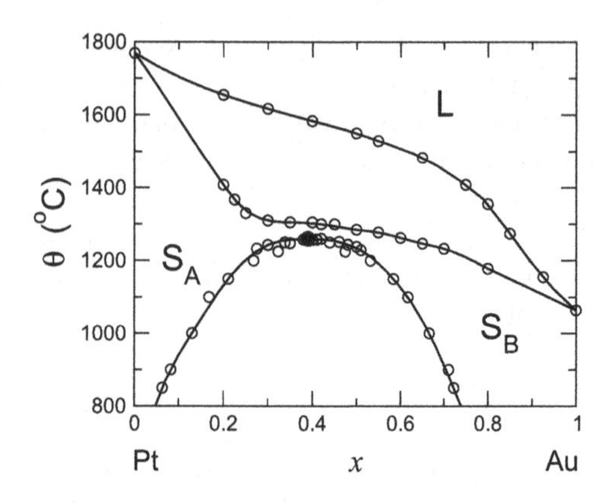

Fig. 10.7 Diagram temperature versus mole fraction of gold for the system gold-platinum. The mixture presents a liquid phase (L) and two solid phases (S_A and S_B). The critical point between the solid phases occur at $x = 0.39$ and $\theta = 1260\,°C$. The *circles* represent experimental data obtained by Darling et al. [129]

Equivalently

$$r_A = \frac{x - x_B}{x_A - x_B} \qquad \text{and} \qquad r_B = \frac{x_A - x}{x_A - x_B}. \tag{10.129}$$

Raising the temperature of the system gold-platinum with fixed composition and equal to $x = x_c = 0.39$, the two solutions in coexistence become identical at the critical temperature $T_c = 1260\,°C$, that is, $x_A - x_B \to 0$ when $T \to T_c$. Above this temperature, gold and platinum become miscible in all proportions.

Eutectic Point

Suppose that the system gold-platinum, with fixed composition, is cooled slowly from a state corresponding to a point just above the solubility curve until a final state corresponding to a point below this curve. In the initial state, the system is a single solid solution. Crossing the solubility curve, the system decomposes into two solid solutions forming an heterogeneous system. The process of cooling, represented in the T versus x diagram by a vertical straight line that crosses the solubility line, is called spinodal decomposition.

If we make the inverse process, that is, we start from a state inside the coexistence region and introduce heat, at fixed composition, the system will become homogeneous as we pass the solubility curve. Continuing the process, the solid solution begins to melt when it reaches the solid line, becoming a liquid solution when it reaches the liquid line. We see that the system gold-platinum turns into a single solid solution before it melts. However, there are binary systems for which the two solid solutions in coexistence melt without becoming a single solution. In this case we face a coexistence of three phase: the two solid solutions and the liquid solution.

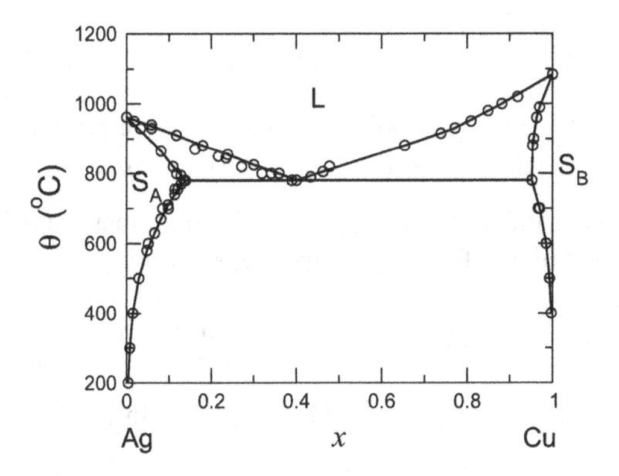

Fig. 10.8 Diagram temperature versus mole fraction of copper for the system silver-copper. The system presents the liquid phase (L) and two solid phases (S_A and S_B). The coexistence of the three phases occurs at $\theta = 780\,°C$ and the mole fractions of copper are 0.14 in phase S_A, 0.40 in the liquid phase and 0.95 in phase S_B. The *circle* represent experimental data obtained by Murray [131]

The system silver-copper, whose diagram is presented in Fig. 10.8, shows the coexistence of three phases at the temperature of $T_t = 780\,°C$, as long as the mole fraction of copper is between $x_A = 0.14$ and $x_B = 0.95$. Below this temperature the system displays coexistence between two solid solutions: one rich in silver (phase S_A) and the other rich in copper (phase S_B). Above this temperature it presents the coexistence of phase S_A and the liquid phase, if $x < x_E$, or the coexistence of phase S_B and the liquid phase, if $x > x_E$, where $x_E = 0.40$. The point (x_E, T_t) is called eutectic point.

Suppose that heat is introduced in the system silver-copper from a state in which the two solid phases are in coexistence and such that $x_A < x < x_E$, that is, such that the mole fraction is smaller that the mole fraction of the eutectic point. The temperature will increase and when reaching T_t the two solid phase begin to melt. The temperature will remain invariant until the whole phase S_B disappear, remaining only the phase S_A and the liquid phase. Just before the beginning of melting, the fractions r_A and r_B in number of moles of the phases S_A and S_B are given by (10.129). Just after the disappearing of the phase S_B, the fractions r'_A and r'_L in number of moles of phase S_A and the liquid phase are given by

$$r'_A = \frac{x - x_E}{x_A - x_E} \qquad \text{and} \qquad r'_L = \frac{x_A - x}{x_A - x_E}. \tag{10.130}$$

Notice that $r'_A < r_A$. Similarly, if the initial state is such that $x_E < x < x_B$ then phase S_A will disappear remaining only phase S_B and the liquid phase. If on the other hand the system is prepared so that it has exactly the composition of the eutectic point, then both solid phases will disappear simultaneously remaining only the liquid phase, that is, the system will pass directly to the liquid state, as shown in Fig. 10.8. In this sense, the heterogeneous alloy with the composition of the eutectic point behaves as a pure substance, melting entirely at a single temperature.

Problems

10.1 Show that at the azeotropic point, the derivatives of pressure along the boiling and condensation line vanish, that is, $\partial p/\partial x_L = 0$ and $\partial p/\partial x_G = 0$.

10.2 Determine the critical point of the liquid-liquid transition of partially miscible substances described by the following Gibbs free energy

$$g = wx(1-x) + cx^2(1-x) + (1-x)g_1 + xg_2 +$$

$$+RT(1-x)\ln(1-x) + RTx\ln x,$$

were c is a constant that depends only on pressure. In this case the critical point occurs at $x = 1/2$.

10.3 Show that, if the concentrations x_A and x_B solve (10.56) and (10.57), then they also solve the same equations with $g(x)$ replaced by $g(x) + ax$, for any value of a. The chemical potential changes but not the concentrations.

Chapter 11
Phase Diagrams

11.1 Gibbs Phase Rule

Fields and Densities

We study in this chapter the phase diagrams of multicomponent systems. The representation that provides the simplest diagrams is that composed only by thermodynamic fields. A system of c components is described by $c+2$ thermodynamic fields: the temperature T, the pressure p, and the chemical potentials $\mu_1, \mu_2, \ldots, \mu_c$ of the c components. However, these fields are not independent but are connected by the Gibbs-Duhem equation

$$-SdT + Vdp - N_1 d\mu_1 - N_2 d\mu_2 - \cdots - N_c d\mu_c = 0, \tag{11.1}$$

where S is the entropy, V is the volume, and $N_1, N_2, \ldots, N_c$ are the number of moles of the c components. One of the fields should be chosen as dependent and will work as a thermodynamic potential. The remaining $c+1$ fields, considered independent, comprises the thermodynamic space of dimension $c+1$.

Any of the $c+2$ fields can be chosen as the thermodynamic potential. If the Gibbs-Duhem equation is written in the form

$$dp = \bar{s}dT + \bar{\rho}_1 d\mu_1 + \bar{\rho}_2 d\mu_2 + \cdots + \bar{\rho}_c d\mu_c, \tag{11.2}$$

where $\bar{s} = S/V$ is entropy per unit volume and $\bar{\rho}_i = N_i/V$ is the number of moles per unit volume, it becomes clear that the pressure can act as a thermodynamic potential and the fields $T, \mu_1, \mu_2, \ldots, \mu_c$ as independent variables.

As a second example, we show how μ_1 can be chosen as the thermodynamic potential. To this end we write the Gibbs-Duhem equation in the form

$$-sdT + vdp - x_1 d\mu_1 - x_2 d\mu_2 - \cdots - x_c d\mu_c = 0, \tag{11.3}$$

M.J. de Oliveira, *Equilibrium Thermodynamics*, Graduate Texts in Physics,
DOI 10.1007/978-3-662-53207-2_11

where s denotes the molar entropy, v the molar volume, and x_i the mole fractions of the c components, which are related by

$$x_1 + x_2 + \cdots + x_c = 1. \quad (11.4)$$

Using (11.4) to eliminate x_1 from (11.3), then

$$d\mu_1 = -sdT + vdp - x_2 d\bar{\mu}_2 - x_3 d\bar{\mu}_3 - \cdots - x_c d\bar{\mu}_c, \quad (11.5)$$

where $\bar{\mu}_i = \mu_i - \mu_1$. The thermodynamic space associated to μ_1, which works as a thermodynamic potential, are composed by the variables $T, p, \bar{\mu}_2, \bar{\mu}_3, \ldots, \bar{\mu}_c$.

It is worth mentioning that this representation can be obtained by successive Legendre transformations from the molar Gibbs free energy $g(T, p, x_2, \ldots, x_c)$, which we used in the study of mixtures of pure substances and whose differential is given by

$$dg = -sdT + vdp + \bar{\mu}_2 dx_2 + \bar{\mu}_3 dx_3 + \cdots + \bar{\mu}_c dx_c. \quad (11.6)$$

The molar Gibbs free energy of g in its turn is obtained by Legendre transformation from the molar energy $u(s, v, x_2, \ldots, x_c)$, whose differential is given by

$$du = Tds - pdv + \bar{\mu}_2 dx_2 + \bar{\mu}_3 dx_3 + \cdots + \bar{\mu}_c dx_c. \quad (11.7)$$

Similarly, we can choose other representations involving only thermodynamic fields. Whatever the choice, the thermodynamic field space of a system of c components is composed by $c + 1$ variables.

Let us denote, generally, the $c + 1$ thermodynamic fields chosen to compose the thermodynamic space by $h_0, h_1, h_2, \ldots, h_c$ and the respective thermodynamic densities by $\rho_0, \rho_1, \rho_2, \ldots, \rho_c$. These two spaces are called space-h and space-ρ, respectively. The fields are understood as the components of the vector $\vec{h} = (h_0, h_1, \ldots, h_c)$, belonging to space-$h$, and the densities as components of the vector $\vec{\rho} = (\rho_0, \rho_1, \ldots, \rho_c)$, belonging to space-$\rho$. Denoting the thermodynamic associated to the fields by ϕ, then

$$d\phi = -\sum_i \rho_i dh_i, \quad (11.8)$$

that is,

$$\rho_i = -\frac{\partial \phi}{\partial h_i}. \quad (11.9)$$

Incidentally, by comparing (11.2) with (11.8), we see that the potential associated to the independent variables $T, \mu_1, \mu_2, \ldots, \mu_c$ is $-p$ and not p. The potential

$\phi(h_0, h_1, h_2, \ldots, h_c)$ is linked, by successive Legendre transformation, to the energy density $u(\rho_0, \rho_1, \rho_2, \ldots, \rho_c)$ whose differential is

$$du = \sum_i h_i d\rho_i, \tag{11.10}$$

that is,

$$h_i = \frac{\partial u}{\partial \rho_i}. \tag{11.11}$$

We remark that $u(\rho_0, \rho_1, \rho_2, \ldots, \rho_c)$ is a continuous and convex function of the densities altogether while $\phi(h_0, h_1, h_2, \ldots, h_c)$ is a continuous and concave function of the fields altogether. Although ϕ is a continuous function, its derivatives, the densities, may not be continuous. And indeed they are not continuous in the points of space-h corresponding to a phase coexistence. A point of discontinuity in the densities in this space indicates the occurrence of a phase coexistence.

Manifolds

The Gibbs phase rule concerns the number of phases that may coexist in a thermodynamic system with c components. The phase rule is a law that can not be derived from the laws of thermodynamics alone. It should be understood as an independent law, compatible with the laws of thermodynamics.

In a system composed by a single pure substance (one component) we have seen that two phases, for example, liquid and vapor coexist along a line in the phase diagram T versus p. We have also seen that the coexistence of three phases, solid, liquid and vapor occurs only at a single point of the two-dimensional diagram. A binary mixture (two components) on the other hand has a three-dimensional diagram. The usual thermodynamic space includes the temperature T the pressure p and a third field which we choose as the difference between the chemical potentials of the two components. In this three-dimensional phase diagram, the coexistence of two phases occurs on a surface, the coexistence of three phases takes place along a line and four phases can only coexist in a point of the diagram. These results and the generalization of them comprises the Gibbs phase rule.

To state the phase rule it is convenient to introduce what is known as the number of degrees of freedom f for the coexistence of a certain number of phases. If the coexistence occurs in a single point in space-h, then $f = 0$; if it occurs along a line, $f = 1$; if it occurs on a surface, $f = 2$, etc. That is, f is identified as the dimension of the manifold (a point, a line, a surface, etc.) in space-h corresponding to the phase coexistence. The Gibbs phase rule states that the number of degrees of freedom f for the occurrence of m thermodynamic phases in a system with c components is given by

$$f = c + 2 - m. \tag{11.12}$$

That is, the dimension of the manifold in space-h corresponding to the coexistence of m phases is equal to $c + 2 - m$.

The rule may also be expressed in terms of the manifold codimension, defined as the difference between the dimension of the space in which the manifold is immersed and the manifold dimension. As the dimension of space-h is $c + 1$ then a codimension κ of the manifold of phase coexistence is $\kappa = (c + 1) - f$. Thus, the Gibbs phase rule is equivalent to say that the codimension of the manifold of the coexistence of m phases in space-h is $m - 1$, or

$$\kappa = m - 1. \tag{11.13}$$

Simplices

Suppose that at a given point in space-h, occurs the coexistence of m phases with densities $\vec{\rho}^{\,(1)}, \vec{\rho}^{\,(2)}, \ldots, \vec{\rho}^{\,(m)}$. In space-$\rho$, these densities are a set of m distinct points whose spatial distribution is a simplex of dimension $m - 1$ (Fig. 11.1). A zero-dimensional simplex is a point. A one-dimensional simplex is a line segment, a two-dimensional simplex is a triangle, a three-dimensional simplex is a tetrahedron, etc. A $(m - 1)$-dimensional simplex comprises a set of m generic vertices located at arbitrary positions of a space with dimension greater than or equal $m - 1$. The generality of the position means that two vertices of a simplex never coincide, three vertices are never collinear, four vertices are never at the same plane, and so on.

The proposition that the coexistence of m phases in a system of several components is represented in space-ρ by a $(m - 1)$-dimensional simplex constitutes the Griffiths-Wheeler postulate. In the following we show that the Gibbs phase rule can be derived from this postulate.

Denoting by $\phi^{(1)}, \phi^{(2)}, \ldots, \phi^{(m)}$ the thermodynamic potentials that describe each of the m single phases in the neighborhood of the manifold of the phase coexistence, then, for each single-phase region, we write the following equation

$$d\phi^{(j)} = -\sum_{i=0}^{c} \rho_i^{(j)} dh_i. \tag{11.14}$$

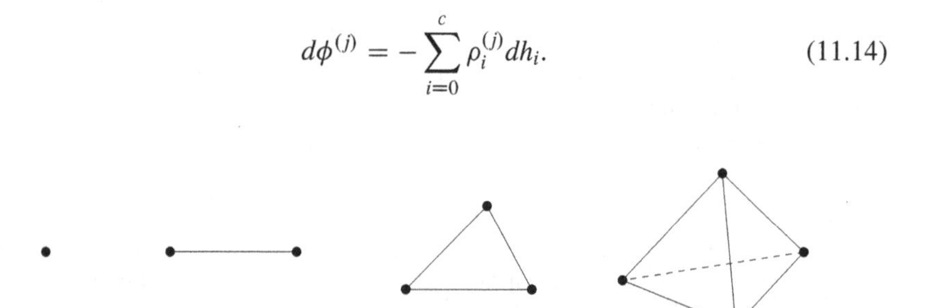

Fig. 11.1 Simplices: zero-dimensional (*point*), one-dimensional (*line segment*), two-dimensional (*triangle*) and three-dimensional (*tetrahedron*)

Subtracting the equations corresponding to $j = 2, \ldots, m$ from that corresponding to $j = 1$ and taking into account the equality of the thermodynamic potentials $\phi^{(j)}$ at the points of the manifold of phase coexistence, we reach the Clausius-Clapeyron equation for a system of c components and m phases in coexistence,

$$\sum_{i=0}^{c} [\rho_i^{(j)} - \rho_i^{(1)}] dh_i = 0, \tag{11.15}$$

valid for $j = 2, \ldots, m$. We may also write the Clausius-Clapeyron equation as

$$\sum_{i=0}^{c} \sigma_i^{(j)} dh_i = 0 \tag{11.16}$$

valid for $j = 2, \ldots, m$, where $\sigma_i^{(j)}$ are the components of the vector $\vec{\sigma}^{(j)}$ defined by

$$\vec{\sigma}^{(j)} = \vec{\rho}^{(j)} - \vec{\rho}^{(1)} \tag{11.17}$$

for $j = 2, 3, \ldots, m$.

The set of equations (11.16) can be written as a scalar product

$$\vec{\sigma}^{(j)} \cdot d\vec{h} = 0, \tag{11.18}$$

valid for $j = 2, 3, \ldots, m$. This equation describes locally the manifold in space-h corresponding the coexistence of m phases. Taking into account that the densities $\vec{\rho}^{(1)}, \vec{\rho}^{(2)}, \ldots, \vec{\rho}^{(m)}$ constitute a simplex $(m-1)$-dimensional, it easy to see that the vectors $\vec{\sigma}^{(2)}, \vec{\sigma}^{(3)}, \ldots, \vec{\sigma}^{(m)}$ comprises a set of $m - 1$ vectors linearly independent. On the other hand, each one of the equations given by (11.18) describes a hyperplane in space-h perpendicular to the vector $\vec{\sigma}^{(j)}$. Therefore, the $m - 1$ (11.18) describe the intersection of $m - 1$ hyperplanes. Since no hyperplane coincides with another because the vectors $\vec{\sigma}^{(j)}$ are linearly independent, then the intersection of the hyperplanes comprises a manifold of dimension $f = (c+1) - (m-1) = c+2-m$, which is the Gibbs phase rule.

The Gibbs phase rule is applicable to systems of several components that are devoid of symmetry. It can not be used indiscriminately when symmetry has an important role. In some cases, however, we can restore the Gibbs rule if thermodynamic space of symmetric systems is extended properly. For example, suppose that two or more phases in coexistence are connected by symmetrical operations. The Gibbs phase rule may be applied provided that the thermodynamic fields, that completely break the symmetry between the phases, are added to the space-h.

11.2 Structure of Phase Diagrams

Ordinary Manifold

The phase diagram in space-h of a system of c components comprises the manifolds (points, lines, surfaces, etc.) of phase coexistence which are connected to each other forming a spatial structure. With the exception of zero-dimensional manifold, which is a point, other manifolds extend continuously in space-h and can reach the boundaries of space-h. However, it is possible, and is actually more common, for it to finish by meeting or by transforming into other manifolds. To examine these possibilities, we must take into account that each point in space-h that belongs to a manifold of m coexisting phases is associated with a $(m-1)$-dimensional simplex in space-ρ.

A simplex corresponding to m coexisting phases is composed by m vertices in space-ρ. A property relevant to our analysis is that any set of $m-1$ vertices of this simplex is also a simplex but of dimension smaller by one unit. Therefore, we can imagine that a simplex of m vertices has as neighbors m simplices of $m-1$ vertices. This implies that in space-h, from a manifold of coexistence of m phases emerge m manifolds of coexistence of $m-1$ phases. It is noteworthy that the latter has dimension one unit smaller.

Based on this property, we can build possible structures of the manifolds of phase coexistence, which we call ordinary manifolds in opposition to critical manifolds which we will see later. In a space of dimension $c+1$, we start from a zero-dimensional manifold (a point) which corresponds to the coexistence of $c+2$ phases. From this point emerge $c+2$ lines of coexistence of $c+1$ phases. From each of these coexistence lines, in turn, spring $c+1$ surfaces of coexistence of c phases. From these surfaces can sprout manifolds of dimension three, and so on.

According to this scheme, we examine the phase structure of a system of one component. In this case, the space-h is two-dimensional. The zero-dimensional manifold (a point) corresponds to the coexistence of three phases (triple point). Three lines of two phase coexistence converge to this point, as shown in Fig. 7.6a. For a two component system, the space-h is three dimensional. The zero-dimensional manifold (a point) corresponds to the coexistence of four phases. To this point, four lines of coexistence of three phases converge, as shown in Fig. 11.2. To each of these lines, surfaces of two phases coexistence converge.

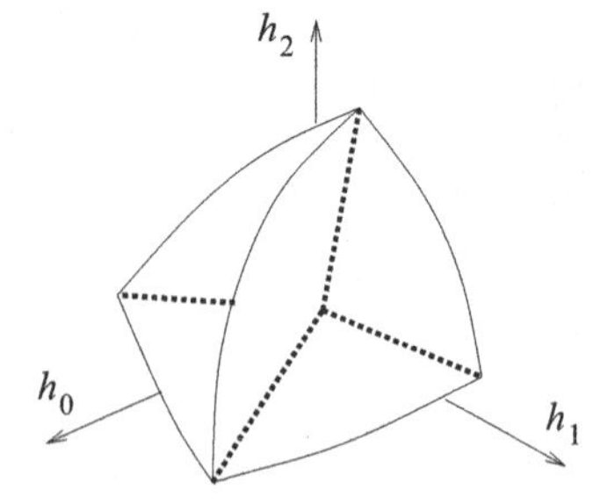

Fig. 11.2 Phase diagram in field space (h_0, h_1, h_2) of a system with two components. The surfaces of coexistence of two phases meet at lines of coexistence of three phases (*dashed line*). These converge to a point of coexistence of four phases (*quadruple point*)

A phase diagram comprises one or more structures of the type shown in Figs. 7.6a and 11.2. For example, a pure substance can have multiple triple points. The most notable of these is the point of coexistence solid-liquid-vapor. However, a pure substance can show various crystal forms that can coexist with each other or together with the liquid forming several point of three-phase coexistence.

We should also note that cuts performed in a phase diagram in space-h of a system of c components may represent a possible phase diagram of a system of $c-1$ components. However, this cut should be generic, that is, it should avoid passing through peculiar points. For example, a cut in the diagram of Fig. 11.2, represented by $h_2 = \text{const}$ is a possible diagram of a system of one component, since the cutting is not done on the quadruple point or tangentially to the line of triple points.

Critical Manifolds

In the study of the possible structures of a phase diagram in space-h, it is convenient to think that an ordinary manifold of dimension f is described in parametric form by a set of f parameters. The continuous variation of these parameters allows us to walk on the manifold. Recall that to each point belonging to the ordinary manifold of coexistence of m phases corresponds a simplex in the space-ρ whose number of vertices is equal to m. Therefore, walking along the points of an ordinary manifold, the number of vertices of the corresponding simplices is invariant. However, the distances between the vertices may change when walk along the manifold. Thus, one possible mechanism for the disappearance of a manifold is that for which two vertices of a simplex collapse into a single vertex, which means that two phases become identical. The corresponding points in space-h define a manifold that we call critical manifold, which has dimension equal to that corresponding to three ordinary phases in coexistence.

One can imagine critical manifolds where more than two phases become equal. For example, we may suppose that three phases become identical. In this case the critical manifold has dimension equal to that corresponding to the manifold of five ordinary phases in coexistence. In general, when i ordinary phases become identical, the critical manifold has dimension equal to that of the manifold corresponding to $2i-1$ ordinary phases in coexistence, which is the Zernike rule.

Let us imagine now a configuration of phases of a system in which there are the coexistence of m_1 ordinary phases, m_2 critical phases (two ordinary phases becoming identical), m_3 tricritical phases (three ordinary phases becoming identical), etc. According to the Zernike rule this configuration of phases correspond, as regards to the Gibbs phase rule, to a number of ordinary phases equal to

$$m = m_1 + 3m_2 + 5m_3 + \cdots . \quad (11.19)$$

According to this rule, a system comprising a pure substance, $c = 1$, may display only a critical zero-dimensional manifold in which two phases become equal. This manifold is the critical point, the terminal point of a line of coexistence of two phases, which was the object of our study in previous chapters. The phase diagram in the vicinity of this point are shown in Fig. 7.5, both in space fields and space of densities.

Critical Line and Critical End Point

The thermodynamic space of fields, corresponding to a binary mixture, consists of three variables. Usually, they include temperature, pressure, and a third thermodynamic field that we choose as the difference between the chemical potentials of the two components. In this space, the coexistence of two phases takes place on a surface, which can extend to the boundaries of the thermodynamic space or may end in a line of critical points, which is identified as the edge of the surface of coexistence, as shown in Fig. 11.3.

The phase diagram of a binary system may also contain lines of coexistence of three phases, as can be seen in Figs. 11.2 and 11.4. Walking along the line of coexistence of three phases, it is possible that it terminates at a quadruple point as seen in Fig. 11.2. It is also possible that at some point two phases become identical in the presence of third, as illustrated in Fig. 11.4. The corresponding point is called critical end point because in addition to being the terminal point of a line of triple points it is the terminal point of a line of critical points. As shown in Fig. 11.4, the critical line, which is the edge of the surface of coexistence, ends on another surface of the coexistence of two phases.

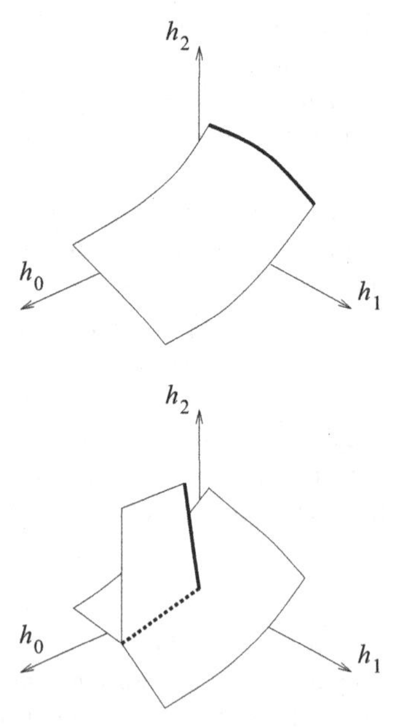

Fig. 11.3 Phase diagram in space (h_0, h_1, h_2) of a system of two components. The edge (*thick line*) of the surface of two-phase coexistence is a critical line

Fig. 11.4 Phase diagram in space (h_0, h_1, h_2) of a system of two components. The surfaces of two-phase coexistence meet at the line (*dashed line*) of three-phase coexistence. The end point of the line of critical points (*thick line*) is a critical end point

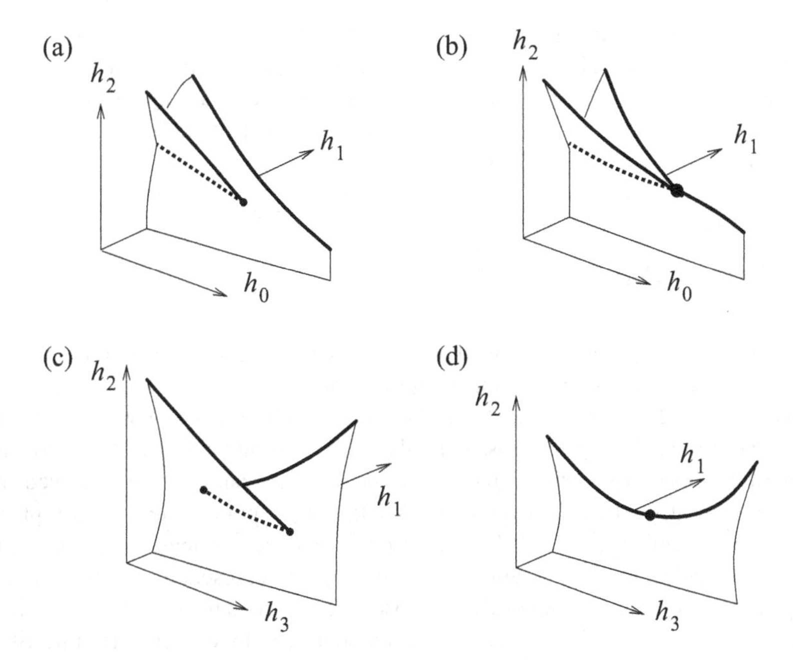

Fig. 11.5 Phase diagram of a ternary system corresponding to cuts in space (h_0, h_1, h_2, h_3) such that one of the four fields is kept constant. The surfaces correspond to the coexistence of two phases. The *continuous lines* are critical lines. The *dashed line* is a line of triple points. *Small filled circles* are critical end points. The *full large circle* is the tricritical point

Tricritical Point

A tricritical point is the one in which three phases become identical. It should be distinguished from an ordinary critical in which only two become identical. According to the Gibbs and Zernike rules three phase can only become identical in a system that has at least three components. In a three-component system, whose space-h is four-dimensional, three phases become identical in a zero-dimensional manifold (a point). As the viewing of a space of more than three dimensions is impractical, it is expedient to examine projections or cuts of this space. We can also use the fact that a phase diagram in a given space can be seen as a possible cut of a phase diagram in a space constituted by an additional field. This additional field can then be continuously varied to change the phase diagram.

Consider the diagram of Fig. 11.5a, which can be considered a phase diagram of a binary system and also as a cutting of the four-dimensional space (h_0, h_1, h_2, h_3) of a ternary system. In this diagram, we are assuming that the variable h_3 has a fixed value. We then vary h_3 appropriately so that the critical end point moves toward the other critical line. The tricritical point will occur when the critical end point touches the other critical line, as can be seen in Fig. 11.5b. We see that the tricritical point is the meeting point of three critical lines and is also the terminus of the line of triple points. It is also possible to imagine other cuts, like those shown in Fig. 11.5c, d. Figure 11.5c presents a line of triple points whose ends are critical end points and is a possible phase diagram of a binary system. Appropriately varying one of the fields, we can make the two critical end points to approach each other. The tricritical point occurs when they coalesce, as shown in Fig. 11.5d.

Multicritical Points

In addition to the points mentioned above, we can imagine other points in which several ordinary phases become identical in the presence or absence of other phases. However, according to Gibbs and Zernike rules, the observation of such points on a mixture of pure substances is possible only above a certain number of components. A bicritical point (two critical phases in coexistence) can only be observed in a mixture with at least four components. A tetracritical point (four ordinary phases that become identical), can only be observed in a mixture of at least five components.

It is convenient to use a notation of product type to represent the configuration of the phases in coexistence. An ordinary phase is represented by the letter A; a critical phase, in which two ordinary phases become identical, by the letter B; a tricritical phase, in which three phases become identical, by the letter C, etc. A state composed of two ordinary phases in coexistence is represented by A^2, a triple point by A^3. A critical end point by AB, a bicritical point by B^2. In general, a multicritical point is represented by $A^{m_1} B^{m_2} C^{m_3} D^{m_4} \ldots$. In accordance with the Gibbs and Zernike laws, the minimum number of components necessary to observe a multicritical point is given by

$$c = m_1 + 3m_2 + 5m_3 + 7m_4 + \cdots - 2. \qquad (11.20)$$

Other Spaces

So far, the phase diagrams were constructed in space-h, composed only by thermodynamic fields. From the experimental point of view, it is sometimes more convenient to use thermodynamic densities instead of thermodynamic fields as independent variables. In a mixture of several components it is common the use of the mole fractions, in addition to temperature and pressure, as control variables. It is useful, therefore, to know how to construct a diagram in the space of densities from a diagram in the space of fields. The fundamental rule is the Clausius-Clapeyron equation (11.16) which tells us that the tie lines should be perpendicular to the lines of coexistence, provided that the axes of space and ρ are parallel to the axes of space-h.

Figure 11.6a, c, e present diagrams corresponding to three cuts of the phase diagram of Fig. 11.4, made at constant pressure. We are identifying the fields h_0, h_1 and h_2 as the temperature T, the activity ζ and the pressure p, respectively.

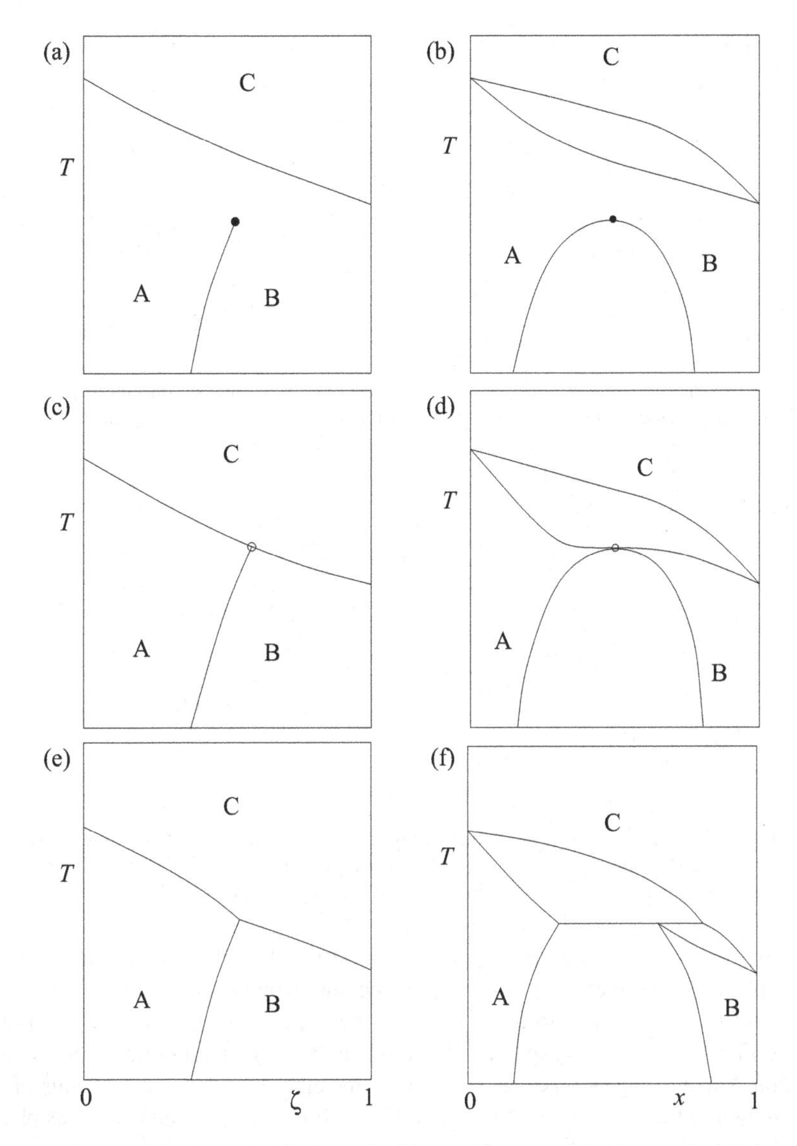

Fig. 11.6 Figures (**a**), (**c**) and (**e**) show the phase diagrams of a binary mixture in space temperature T versus activity ζ, at constant pressure. They correspond to cuts in the diagram shown in Fig. 11.4. Figures (**b**), (**d**) and (**f**) are the respective phase diagrams in space T versus mole fraction x, at constant pressure. The *full circle* represents an ordinary critical point and an *empty circle* represents a critical end point

Figure 11.6a corresponds to a cut made at a pressure just above the pressure of the critical end point. In this diagram, we see that, above the critical temperature, the substances become completely miscible, passing to phase C at higher temperatures. Figure 11.6e corresponds to a cut made at a pressure just below the pressure of the critical end point. An increase in temperature of the phases A and B

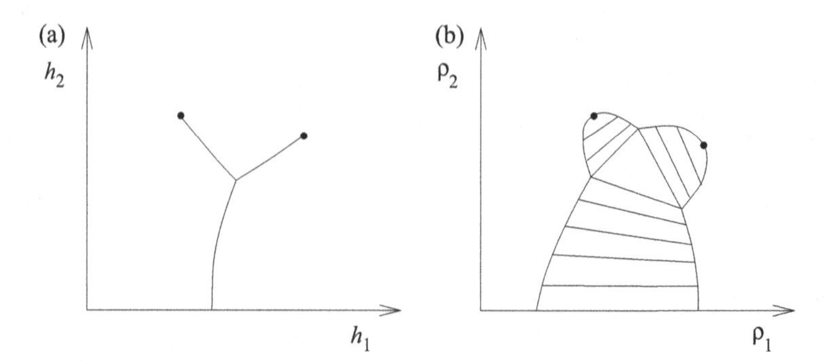

Fig. 11.7 (**a**) Phase diagram corresponding to a cut in Fig. 11.5c done at constant h_3 and passing between the two critical end points. (**b**) Analogous diagram in the space of conjugated densities

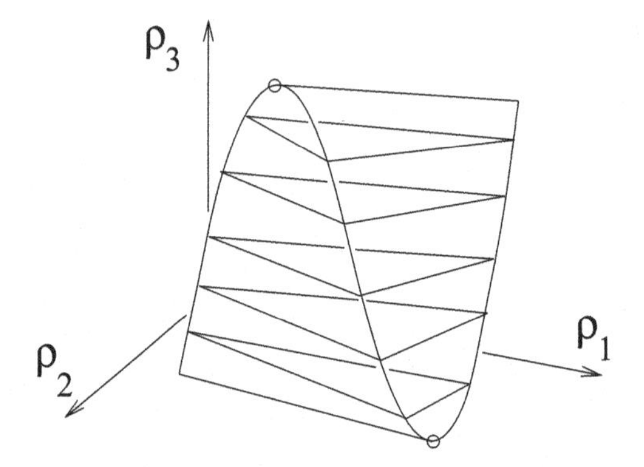

Fig. 11.8 Phase diagram analogous to that of Fig. 11.5c. The *triangles* represent the coexistence of three phases. The two-phase regions and the ordinary critical point are not shown in the figure but are similar to that shown in Fig. 11.7b. The *open circles* represent critical end points

in coexistence causes the appearance of phase C when the temperature reaches the triple point temperature, whereupon the three phases remain in coexisting. Figure 11.6c corresponds to a cut made at the pressure of the critical end point. Figure 11.6b, d, f are the respective diagrams in the space temperature versus mole fraction. Various binary mixtures of partially miscible liquids have diagrams of the type shown in Fig. 11.6. In this case A and B are liquid phases and C is a gas phase. We also find diagrams of this type in binary mixtures of partially miscible solids, such as gold-platinum system whose diagram is shown in Fig. 10.7. In this system, A and B are solid phases and C is a liquid phase.

Figure 11.7a presents a phase diagram corresponding to a cut in Fig. 11.5c made between the two critical end points. The diagram analogous in the space of densities is shown in Fig. 11.7b. We see that the tie lines are perpendicular to the lines of two-phase coexistence. The triangle represents the coexistence of three phases.

Figure 11.8 shows the analogous of Fig. 11.5c in the space of densities. The coexistence of three phases, which is a line in the space of fields, is represented by a stack of triangles in the space of densities.

11.3 Griffiths-Landau Theory

Landau Function

The Griffiths-Landau theory allows us to determine the topology of phase diagrams in the vicinity of critical and multicritical points. It is still able to provide quantitative results, but its importance lies in the simplicity in obtaining the topology of phase diagrams.

The Griffiths-Landau theory takes as its starting point a thermodynamic potential f which has a mixed dependency on fields and densities. The potential f is a polynomial function of densities. The polynomial coefficients depend only on the fields. These two properties are sufficiently generic in such a way that any thermodynamic potential that is analytic function can be approximated by a potential of Griffiths-Landau type by means of a Taylor series expansion in the densities. These include, for example, those relating to the theory of van der Waals and regular solution theory.

Next we develop the theory for the case in which the thermodynamic potential f is a function of only one density, which we denote by ρ. The dominant term of the polynomial is chosen as being convex which means that the highest power must be even. However, this is not sufficient to guarantee the convexity of f since a polynomial function do not have in general the property of convexity. The thermodynamic potential is therefore taken as the convex hull of f, obtained from a Legendre transformation

$$\phi = \min_{\rho}\{\psi(\rho)\}, \tag{11.21}$$

where

$$\psi = f(\rho) - h\rho \tag{11.22}$$

is called Landau function. Being $f(\rho)$ a polynomial then $\psi(\rho)$ is also a polynomial of the same degree. We remark that the Landau function depend on all fields relating to the potential ϕ and not only on those relating to the potential f. The dependence on the fields, however, is found only on the coefficients of the polynomial of $\psi(\rho)$.

Next we introduce a change of variables to make the analysis simpler. Defining the new density $x = \rho - \bar{\rho}$, we see that $\psi(x)$ will also be a polynomial and of the same degree of $\psi(\rho)$. In addition, the potential ϕ will be given by

$$\phi = \min_{x}\{\psi(x)\}. \tag{11.23}$$

The choice of $\bar{\rho}$ will be done so that one of the coefficients of the polynomial in x, other than the dominant, vanishes. We point out that, although the coefficients of the new polynomial are distinct from the older ones, they are functions of the fields only since $\bar{\rho}$ will depend only on the older coefficients and thus only on the fields.

The description of the topology around the critical and tricritical points, as we will see below, is obtained from the Landau polynomial

$$\psi(x) = a_0 + a_1 x + a_2 x^2 + \cdots + a_{n-2} x^{n-2} + x^n, \tag{11.24}$$

where n is the degree of the polynomial, that must be even. The coefficient of the dominant term was taken as being equal to unity. This expedient is justified by dividing both sides of (11.21) by the coefficient of the dominant term, which is strictly positive. The choice of the variable x was done so that the penultimate term is absent, that is, so that the coefficient of the term x^{n-1} vanishes.

As we said above, the coefficients a_i of the terms of the Landau polynomial depend only on the thermodynamic fields. According to the Griffiths-Landau theory, this dependence is smooth so that it is always possible to assume that it is linear. Thus we consider that the coefficients a_i are themselves fields, obtained from the original fields by the linear dependence. This implies that the topology of the diagram is the same whether we use the original fields or the coefficients a_i, identified as fields.

Adopting the coefficients a_i as fields, it is convenient to determine the density conjugated to them, which we denote by ξ_i and are given by

$$\xi_\ell = -\frac{\partial \phi}{\partial a_\ell}, \tag{11.25}$$

from which we obtain

$$\xi_\ell = -x^\ell, \tag{11.26}$$

where x, in this equation, represents an absolute minimum of $\psi(x)$.

Factorized Representation

It is convenient in our analysis to consider the polynomial $\tilde{\psi}$ defined by $\tilde{\psi}(x) = \psi(x) - \phi$. This polynomial has the following fundamental properties. If x^* is an absolute minimum of the Landau function $\psi(x)$, then x^* is identified as the root of $\tilde{\psi}(x)$ with even multiplicity (double, quadruple, etc.). In fact, according to (11.23), $\phi = \psi(x^*)$ and therefore $\tilde{\psi}(x^*) = 0$. In addition, according to (11.23), $\psi(x) \geq \phi$ and therefore $\tilde{\psi}(x) \geq 0$ for any x. As $\tilde{\psi}(x)$ is analytic then in the neighborhood of $x = x^*$ this function behaves as $(x - x^*)^k$, where k is even.

Based on these results, we can construct the specific Landau functions for each one of the manifolds of the phase diagram. To this end we use the factorized representation

$$\tilde{\psi}(x) = (x - r_1)(x - r_2) \cdots (x - r_n). \tag{11.27}$$

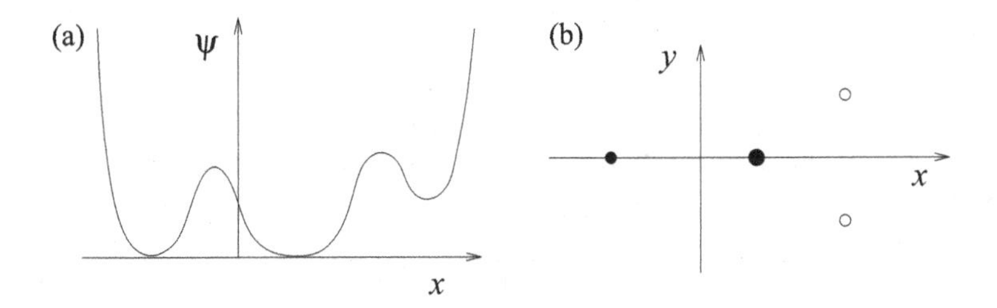

Fig. 11.9 (**a**) Landau function corresponding to the coexistence of an ordinary phase (quadratic minimum, double root) and a critical phase (quartic minimum, quadruple root). (**b**) Roots of $\tilde{\psi}$ in the complex plane: *small full circle*, double root; *big full circle*, quadruple root; and *empty circles*, simple roots

where r_i are the roots of the polynomial $\tilde{\psi}$ which are chosen according to the following rules.

(a) To each ordinary phase in coexistence there corresponds a double real root.
(b) To each critical phase (two ordinary phase becoming identical) in coexistence there corresponds a real quadruple root.
(c) To each tricritical phase (three ordinary phases becoming identical) in coexistence there corresponds a real sixfold root.
(d) In general, to each multicritical phase such that p phases becomes identical there corresponds a real root of multiplicity $2p$.

The remaining roots cannot be real because otherwise $\tilde{\psi}(x)$ would vanish since x takes on real values. These non real roots must appear in pairs of complex conjugated numbers because the coefficients of the polynomial are real. The real roots constitute the absolute minima of $\tilde{\psi}(x)$ and, because they are also minima of $\psi(x)$, they are identified as the densities of the phases in coexistence.

As an example of a Landau function, Fig. 11.9 shows a polynomial of degree eight corresponding to a critical end point (coexistence of an ordinary phase and a critical phase). The polynomial has one double real root, one quadruple real root and a pair of complex conjugated roots.

Finally, the absence of the term x^{n-1} in the development of the product (11.27) means that the sum of the roots vanishes, that is,

$$r_1 + r_2 + \cdots + r_n = 0. \tag{11.28}$$

If all roots are equal then they should vanish.

The determination of the equation that describe a certain manifold in the space of the fields is reduced to the comparison of the development of the product (11.27) with (11.24). This comparison allow the determination of the relations between the coefficients a_i and the roots r_i. Understanding the roots r_i as parameters, these relations constitute the equations that describe the manifold in a parametric way.

Critical Point

To describe the region of the phase diagram around the critical point it suffices to consider a polynomial of fourth order in x, Thus, the expression which we use for the Landau function is

$$\psi = a_0 + a_1 x + a_2 x^2 + x^4. \tag{11.29}$$

The coefficients of the Landau function define the two-dimensional space (a_1, a_2).

To determined the occurrence of two phases in coexistence, the polynomial $\tilde{\psi}(x)$ should have two distinct double roots what allows us to write

$$\tilde{\psi} = (x - r_1)^2 (x - r_2)^2. \tag{11.30}$$

Since the sum of the roots vanishes, then $r_2 = -r_1$, so that

$$\tilde{\psi} = (x^2 - r_1^2)^2 = x^4 - 2r_1^2 x^2 + r_1^4. \tag{11.31}$$

Therefore,

$$a_2 = -2r_1^2 \qquad \text{and} \qquad a_1 = 0, \tag{11.32}$$

and the conditions for the coexistence of two phases are

$$a_1 = 0 \qquad \text{and} \qquad a_2 < 0, \tag{11.33}$$

which define the line of coexistence. The densities of the two phases en coexistence are given by

$$r_1 = \frac{1}{\sqrt{2}} |a_2|^{1/2} \qquad \text{and} \qquad r_2 = -\frac{1}{\sqrt{2}} |a_2|^{1/2}. \tag{11.34}$$

When the two phases become identical, the double roots become identical. Therefore, a critical point is characterized by the occurrence of quadruple root, that is, $\tilde{\psi} = x^4$, which gives the result

$$a_1 = 0 \qquad \text{and} \qquad a_2 = 0, \tag{11.35}$$

which are the conditions that determine the critical point. The topology around the critical point can be seen in the diagram of a_1 versus a_2 as shown in Fig. 11.10. We show also the diagram in the variables ξ_1 and a_2, where ξ_1 is the density conjugated to the field a_1 and given by $\xi_1 = -r_1$ or $\xi_1 = -r_2$.

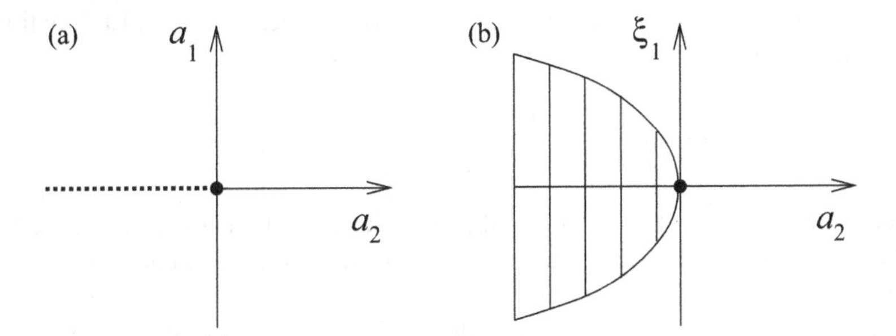

Fig. 11.10 (**a**) Phase diagram around the critical point in the space of the coefficients a_1 and a_2. The *dashed line* is a two-phase coexistence line. The *full circle* is the critical point. (**b**) Same diagram in the space density ξ_1 versus a_2, where ξ_1 is the density conjugated to the field a_1. The region of coexistence of two phases is indicated by *tie lines*

Tricritical Point

The topology of the region around the tricritical point is obtained from a polynomial of sixth degree, that is, from the Landau function

$$\psi = a_0 + a_1 x + a_2 x^2 + a_3 x^3 + a_4 x^4 + x^6. \tag{11.36}$$

Notice that the term x^5 is absent. The coefficients of the Landau function define a four-dimensional space (a_1, a_2, a_3, a_4).

To describe the coexistence of three phases (triple point) it is necessary that $\tilde{\psi}(x)$ has three distinct double roots, that is,

$$\tilde{\psi} = (x - r_1)^2 (x - r_2)^2 (x - r_3)^2, \tag{11.37}$$

where r_1, r_2 and r_3 are distinct and are the densities of each phase. Since the sum of the roots vanishes then $r_3 = -r_1 - r_2$, what allows to write

$$\tilde{\psi} = (x - r_1)^2 (x - r_2)^2 (x + r_1 + r_2)^2. \tag{11.38}$$

Developing the product, we get

$$\tilde{\psi} = (x^3 - px + q)^2 = x^6 - 2px^4 + 2qx^3 + p^2 x^2 - 2pqx + q^2 \tag{11.39}$$

where

$$p = r_1^2 + r_2^2 + r_1 r_2 \qquad \text{and} \qquad q = (r_1 + r_2) r_1 r_2, \tag{11.40}$$

Comparing expression (11.39) with (11.36), we get the relations

$$a_4 = -2p, \qquad a_3 = 2q, \qquad a_2 = p^2, \qquad \text{and} \qquad a_1 = -2pq, \tag{11.41}$$

with $a_4 < 0$ because $p > 0$ for any values of the densities r_1 and r_2. Elimination p and q, we obtain the following relations among the parameters a_i

$$a_2 = \frac{1}{4}a_4^2, \qquad a_1 = \frac{1}{2}a_3 a_4 \qquad \text{and} \qquad a_4 < 0, \tag{11.42}$$

which define surfaces of coexistence of three phases in a four-dimensional space.

Let us set up now Landau functions corresponding to critical end points. These points correspond to the occurrence of two coexistence phases, one of them an ordinary phase and the other a critical phase. Therefore, the critical end points are characterized by one double root and one quadruple root, that is,

$$\tilde{\psi} = (x - r_1)^4 (x - r_2)^2, \tag{11.43}$$

where r_1 is the density of the critical phase and r_2 of the ordinary phase. Since the sum of the roots vanishes, then $r_2 = -2r_1$ yielding the result

$$\tilde{\psi} = (x - r_1)^4 (x + 2r_1)^2. \tag{11.44}$$

Developing the product and comparing it with the expression (11.36), we get the relations

$$a_4 = -6r_1^2, \qquad a_3 = 4r_1^3, \qquad a_2 = 9r_1^4, \qquad \text{and} \qquad a_1 = -12r_1^5. \tag{11.45}$$

Eliminating r_1 we get the following relations among the parameters a_i

$$a_1 = \pm\frac{1}{3\sqrt{6}}|a_4|^{5/2}, \qquad a_3 = \pm\frac{1}{3\sqrt{6}}|a_4|^{3/2},$$

$$a_2 = \frac{1}{4}a_4^2 \qquad \text{and} \qquad a_4 < 0. \tag{11.46}$$

These relations determine lines of critical end points in a four-dimensional space.

The Landau function corresponding to an ordinary critical point is characterized by the occurrence of quadruple root. Since the critical phase is not in coexistence with other phases then the other roots must be non real. Therefore

$$\tilde{\psi} = (x - r_1)^4 (x - r_2)(x - r_3), \tag{11.47}$$

where r_1 is the density of the critical phase and r_2 and r_3 are complex conjugate. As the sum of the roots vanishes then $r_2 + r_3 = -4r_1$ what permits us to write $\tilde{\psi}$ in the form

$$\tilde{\psi} = (x - r_1)^4 (x^2 - 4r_1 x + r^2), \tag{11.48}$$

where $r^2 = r_2 r_3$ is real and positive. Developing the product and comparing it with expression (11.36), we get the following relations

$$a_4 = -10r_1^2 + r^2, \qquad a_3 = 20r_1^3 - 4r_1 r^2,$$

$$a_2 = -15r_1^4 + 6r_1^2 r^2 \qquad \text{and} \qquad a_1 = 4r_1^5 - 4r_1^3 r^2. \tag{11.49}$$

These relation determine in a parametric form a critical surface in a four-dimensional space. They in fact define three critical surfaces. One of them is obtained when $r_1 = 0$, which results in

$$a_1 = a_2 = a_3 = 0 \qquad \text{and} \qquad a_4 > 0. \tag{11.50}$$

The other two correspond to $r_1 \neq 0$ and are given implicitly by (11.49).

Finally, a tricritical point is described by a Landau function $\tilde{\psi} = x^6$ from which we get

$$a_1 = a_2 = a_3 = a_4 = 0, \tag{11.51}$$

which determined the tricritical point in a four-dimensional space.

Diagrams

To visualize the topology of the phase diagram around the critical and tricritical points, we use as the space of thermodynamic fields the space of the coefficients a_i. For the critical point, the space is two-dimensional and is composed by the variables a_1 and a_2. Figure 11.10 shows the critical line and the critical points traced according to the conditions (11.33) and (11.35), respectively.

For the tricritical point the space is four-dimensional and is compose by the variables a_1, a_2, a_3 and a_4. Due to the impossibility of visualizing such a space we examine cuts made at constant a_3 and at constant a_4. The various manifolds defined by the conditions obtained above can be seen in Fig. 11.5 considering that a_1, a_2, a_3 and a_4 are, respectively, parallel to h_1, h_2, h_3 and h_0. In these three-dimensional diagrams, the triple points and the critical points are lines. The critical end points and the tricritical point are isolated points.

Figure 11.5a, b correspond to cuttings made at constant a_3. In the first a_3 is nonzero and in the second $a_3 = 0$. Only in this second diagram we can observe the tricritical point since this point occurs when $a_1 = a_2 = a_3 = a_4 = 0$. For $a_3 = 0$, the critical lines are determined explicitly from (11.49). For this it suffices to set $r^2 = 5r_1^2$ to get

$$a_4 = -5r_1^2, \qquad a_2 = 15r_1^4 \qquad \text{and} \qquad a_1 = -16r_1^5, \tag{11.52}$$

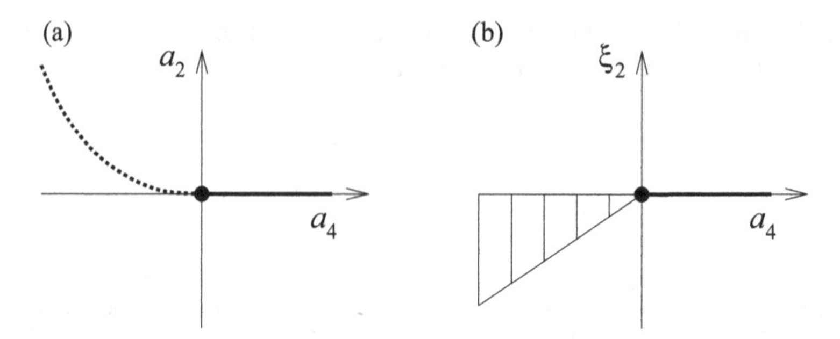

Fig. 11.11 (**a**) Phase diagram around the tricritical point in the space of the coefficients a_2 and a_4, corresponding to $a_1 = a_3 = 0$. The *dashed curve* is a three-phase coexistence line. The *full circle* is the tricritical point. The *thick line* is a critical line. The region below the *two lines* correspond to points of two-phase coexistence. (**b**) Same diagram in space ξ_2 versus a_4, where ξ_2 is the density conjugated to the field a_2. The region of three-phase coexistence is indicated by *tie lines*

from which we get the equations for both critical lines

$$a_2 = \frac{3}{5}a_4^2, \qquad a_1 = \pm\frac{16}{25\sqrt{5}}|a_4|^{5/2} \qquad \text{and} \qquad a_4 < 0 \tag{11.53}$$

The other critical line is described by (11.50).

If in the diagram corresponding to $a_3 = 0$ we consider a cutting defined by $a_1 = 0$, we obtain the two-dimensional diagram in the variables a_2 and a_4, as shown in Fig. 11.11. In this diagram the critical line is given by (11.50) and the line of triple points by $a_2 = a_4^2/4$ and $a_4 < 0$, obtained from (11.42). In this figure we show also the diagram in the variables ξ_2 versus a_4, where ξ_2, the density conjugated to the field a_2, is given by $\xi_2 = a_4/2$ and $a_4 < 0$ or $\xi_2 = 0$ and $a_4 < 0$.

We can also examine a cut made at constant a_4 which are shown in Fig. 11.5c, d, and correspond respectively to $a_4 < 0$ and $a_4 = 0$. In the diagram of Fig. 11.5c, the line of triple points finish in two critical end points. The location of these points are given by (11.46). From them, we see that the distance between these two points vanishes when $a_4 \to 0$. Equivalently, the line of triple points shrinks and disappears in this limit, as seen in Fig. 11.5d.

A diagram analogous to that of Fig. 11.5c can be seen if Fig. 11.8. The line of triple points correspond to a stack of triangles. At the bottom and at the top of the stack one finds line segments (degenerated triangles) that represent the coexistence of a critical phase and an ordinary phase (critical end point). When $a_4 \to 0$, the stack of triangles shrinks to a single point which is the tricritical point. However, the stack does not shrink uniformly in the three directions. The height of the stack ℓ_3, which is the distance between the critical end points, decreases as

$$\ell_3 = |a_4|^{\beta_3}, \qquad \beta_3 = 3/2, \tag{11.54}$$

a result that is obtained from (11.46). The length of the base of the triangles ℓ_2 is equal to the length of the tie lines shown in Fig. 11.11b, which behave linearly with a_4, that is,

$$\ell_2 = |a_4|^{\beta_2}, \qquad \beta_2 = 1. \tag{11.55}$$

The height of the triangles ℓ_1 is associated to the variable ξ_1, conjugated to a_1. Therefore,

$$\ell_1 = |a_4|^{\beta_1}, \qquad \beta_1 = 1/2. \tag{11.56}$$

The exponents β_1, β_2 and β_3 characterize the behavior of the densities around the tricritical point. According to the Griffiths-Landau theory the values of these exponents are 1/2, 1 and 3/2, respectively. The experimental results obtained in ternary and quaternary mixtures are in agreement with these results.

Chapter 12
Order-Disorder Transition

12.1 Binary Alloys

Ordered Alloys

A crystalline solid has as a fundamental property the ordered structure of its atoms. This structure consists of an array of sites forming a regular three-dimensional periodic lattice. The ordered structure, strictly speaking, does not mean that the atoms are located exactly on the sites of the lattice as they are in constant motion due to thermal agitation. In fact, the average positions of the atoms are the points in space that should be considered as the sites of the ordered lattice. Only for simplicity we say that atoms themselves form the crystal lattice.

The possible crystalline structures are classified according to the symmetry groups. The crystalline solids are ordered according to one or other structure that defined the thermodynamic state. Many metals are ordered according to the structure called face-centered cubic (fcc). In this structure the atoms of a simple solid are located at the vertices and centers of cubes that make up the structure. Among the metals that are arranged according to the fcc structure are copper, gold, silver, platinum, nickel, lead and aluminum. Binary alloys of any of these metals form substitutional disordered alloys with the same structure.

Gold and silver form continuous solid solution for these two metals are miscible in all proportions. A solution of gold and silver has the fcc spatial structure where the gold and silver atoms are distributed randomly among the lattice sites, forming an alloy called disordered alloy. The gold and platinum also form alloys whose atoms are distributed at random on the sites of a fcc lattice. However, gold and platinum are not completely miscible. For certain compositions, the gold-platinum system separates into two phases which are also disordered alloys. These two phases in coexistence differ in the composition one being gold-rich and the other platinum-rich as shown in the diagram of Fig. 10.7. Raising the temperature the two

M.J. de Oliveira, *Equilibrium Thermodynamics*, Graduate Texts in Physics,
DOI 10.1007/978-3-662-53207-2_12

disordered alloys become identical at a critical point. Above the critical temperature, gold and platinum are miscible in all proportions.

The alloy consisting of gold and copper, on the other hand, offers the opportunity to illustrate an alloy where the atoms are not randomly distributed on the sites of a fcc lattice but are preferably located at certain sites of the crystal lattice. For a better understanding, we must consider that a fcc lattice can be understood as composed of four interpenetrating cubic sublattices. For a mole fraction of copper around 0.25, the copper atoms are located preferentially in a sublattice and the gold atoms preferentially in the other three sublattices, thereby forming the ordered alloy Au_3Cu. Heating this alloy it passes from the ordered state to the disordered state.

The binary alloy consisting of magnesium and cadmium also presents order of their atoms. Figure 12.1 presents the phase diagram temperature versus mole fraction of magnesium where one can observe the regions corresponding to the ordered alloys Cd_3Mg, CdMg and $CdMg_3$ as well as the disordered alloy (Cd,Mg). The transitions between all these phases are first order including that which occurs at $x = 0.5$ and at a temperature of 253 °C. This transition is congruent, that is, it occurs with no change in composition.

A binary system particularly interesting is the one consisting of copper and zinc, and generically named brass. This system has several phases according to the composition. For a mole fraction of zinc from 0.45 to 0.482 and temperature around 450 °C, the brass is in a phase called β'-brass. At this phase the alloy is in a cubic ordered structure in which the zinc atoms are preferably at the vertices of the cube and the copper atoms preferably in the centers of the cubes. Heating the alloy, it passes to a disordered phase called β-brass, where the atoms of zinc and copper are distributed indiscriminately at the corners and center of the cubes of the lattice, comprising a lattice called body centered cubic (bcc). The transition occurs at a

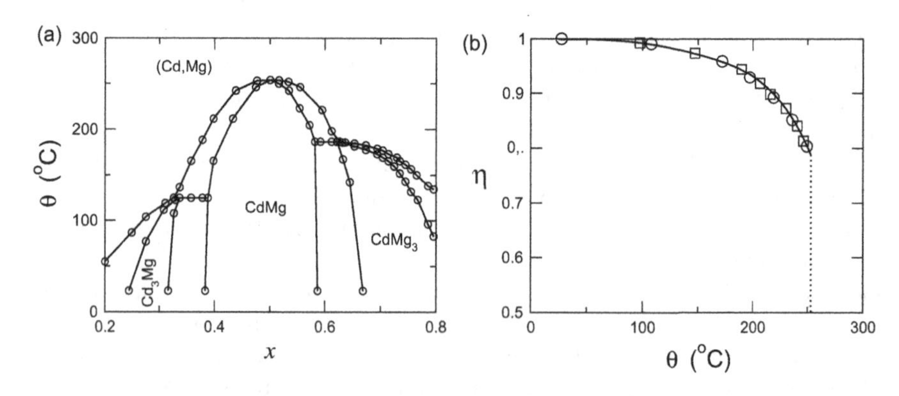

Fig. 12.1 Phase diagram and order parameter of the binary alloy cadmium-magnesium obtained experimentally by Franz and Gantois [136]. (**a**) Diagram temperature θ versus mole fraction x of magnesium. (**b**) Order parameter η as a function of temperature θ along $x = 0.5$. It displays a jump at the transition to the ordered phase that occurs at a 253 °C

temperature of 454 °C for a mole fraction of zinc of 0.448 and at a temperature of 468 °C for a mole fraction of zinc of 0.482. This transition is a continuous transition.

Order Parameter

Consider an ordered alloy of type brass-β' formed by two kinds of atoms A and B, which are arranged in a cubic lattice in which the atoms are located either at the vertices or at the centers of the cubes. The vertices and the centers in turn form two sublattices that are intertwined. Suppose that the fraction x of atoms of type A is exactly 0.5. The order does not necessarily mean that all the atoms A are found in one sublattice and all atoms B are found in the other sublattice. This only occurs when the order is complete, which should occur at low temperatures. The ordering means that the atoms of one type are found preferably in a sublattice and the atoms of the other type preferably in the other sublattice. It is convenient therefore to define the fractions x_1 and x_2 of atoms A in the sublattice formed by the vertices and centers of the cubes, respectively. The ordered alloy does not only mean that $x_1 = 1$ and $x_2 = 0$ or that $x_1 = 0$ and $x_2 = 1$ but that $x_1 \neq x_2$. The alloy becomes disordered when $x_1 = x_2$.

Suppose now that the fraction x of atoms of type A is different from 0.5. The disordered alloy implies that the fraction of atoms of type A is the same in any of the sublattices and hence $x_1 = x_2 = x$. If the alloy is ordered, then $x_1 \neq x_2$. The difference

$$\eta = x_1 - x_2 \tag{12.1}$$

is therefore a measure of the degree of ordering of the alloy. The quantity η, called the order parameter, vanishes in the disordered phase and is nonzero in the ordered phase. Notice that $(x_1 + x_2)/2 = x$ and therefore

$$x_1 = x + \frac{\eta}{2} \quad \text{and} \quad x_2 = x - \frac{\eta}{2}. \tag{12.2}$$

Figures 12.1b and 12.2b show the order parameter of two binary alloys as a function of temperature. In the cadmium-magnesium alloy the order parameter decreases and shows a jump at the transition temperature to the disordered phase, characterizing a discontinuous transition. The ordered and disordered alloys coexist at the transition. In the copper-zinc alloy on the other hand, the order parameter decreases continuously and vanishes at the critical temperature, characterizing a continuous transition.

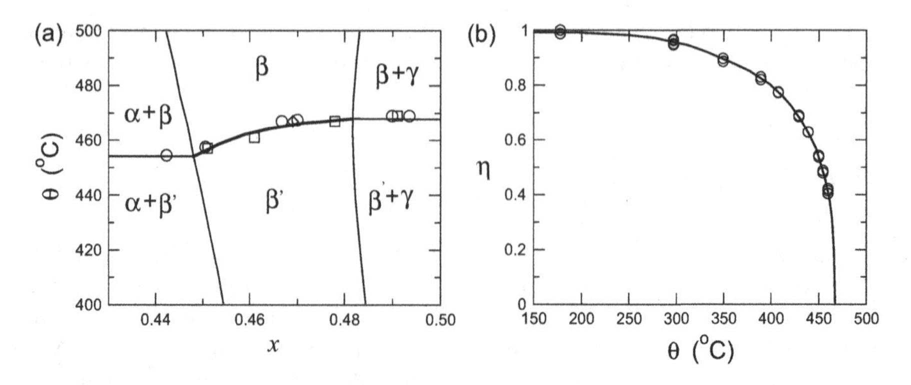

Fig. 12.2 Binary alloy copper-zinc. (**a**) Experimental phase diagram temperature θ versus mole fraction x of zinc around the order-disorder transition, represented by the *thick line*, between the ordered phase β' and the disordered phase β. The *symbols* represent the experimental data. (**b**) Order parameter η as a function of temperature θ along $x = 0.469$. The order parameter vanishes continuously at the critical temperature, $\theta_c = 467\,^\circ\mathrm{C}$

Gibbs Free Energy

The thermodynamic space related to the molar Gibbs free energy g of a two component system consists of the temperature T, the pressure p and the mole fraction x of the second component. The variables conjugated to these quantities are respectively the molar entropy s, the molar volume v and chemical potential μ which is the difference between the chemical potentials of the components.

However, to treat appropriately the ordered alloys, it is desirable to extend the thermodynamic space. To this end, let us examine the simplest case of a binary alloy A-B whose atoms are arranged in two sublattices that we call 1 and 2. A possible ordered structure is that in which the atoms of type A is preferably located in the sublattice 1. On the other hand, an equally possible structure is that in which the A atoms are located preferably in the sublattice 2. Therefore, the system is capable of displaying two different thermodynamic states, that has to be understood as two ordered phases in coexistence. We remark that the two ordered phases in coexistence have the same composition. Therefore, unlike what happens with disordered phases in coexistence, the ordered phases in coexistence can not be distinguished by the composition.

To properly describe the coexistence of two ordered phases it is necessary to use another variable since the composition, described by the mole fraction x of one of the component, does not distinguish the two ordered phases. The quantity that can distinguish these phases is the order parameter defined above. Thus we add the order parameter η to the thermodynamic space which earns an extra dimension. The molar Gibbs free energy then depends also on η so that

$$dg = -Tds + vdp + \mu dx + \mu_* d\eta, \tag{12.3}$$

where μ_* is the variable conjugate to the order parameter. The chemical potential μ_* is a thermodynamic field that breaks the symmetry between the sublattices.

An equivalent description is obtained by using the variables x_1 and x_2, which are the fractions of atoms of type A in sublattice 1 and 2, respectively, in place of x and η. In this representation we have

$$dg = -Tds + vdp + \mu_1 dx_1 + \mu_2 dx_2, \tag{12.4}$$

where μ_1 and μ_2 are conjugated to x_1 and x_2, respectively. Using (12.2) in (12.4) we obtain the following relations between the chemical potentials

$$\mu = \mu_1 + \mu_2 \qquad \text{and} \qquad \mu_* = \frac{1}{2}(\mu_1 - \mu_2). \tag{12.5}$$

We notice that the properties of the ordered alloy are invariant by the permutation of the labels of the sublattice. As a consequence, the Gibbs free energy $g(T, p, x_1, x_2)$ must be invariant by the permutation of x_1 and x_2. Therefore, the phase diagram is invariant by the permutation of μ_1 by μ_2, so that the coexistence of the two ordered phases occurs when $\mu_1 = \mu_2$, that is, when $\mu_* = 0$.

12.2 Bragg-Williams Theory

Gibbs Free Energy

The orderly arrangement of atoms in an alloy is a direct result of the forces of attraction and repulsion between atoms. The Bragg-Williams theory considers that the ordering of an alloy is established due to the greater attractive force between atoms of different kind than that between atoms of the same type. To understand the consequences of this idea, let us examine the simple binary alloy in which the atoms are located on a bcc lattice, composed of two interpenetrating sublattices, one formed by the vertices of the cube and the other by the centers of the cubes. Each site belonging to a sublattice has as neighbors only sites belonging to the other sublattice. Assuming that the dominant interaction is the one that occurs between atoms located at neighboring sites, then the relevant interaction occurs between atoms located in different sublattice. But, in accordance with Bragg-Williams, the energy of interaction between atoms of different kind must be less than that between atoms of the same type. Therefore, the local configuration of lowest energy is the one in which one atom of one type are surrounded only by atoms of distinct type, which are on another sublattice. This favors the segregation of atoms of one type of atoms in a sublattice and another type in another sublattice, constituting an ordered alloy.

According to the Bragg-Williams theory the molar energy of a binary alloy has the following form

$$u = u_{AA}x_1x_2 + u_{AB}(x_1y_2 + y_1x_2) + u_{BB}y_1y_2, \tag{12.6}$$

where x_1 and x_2 are the mole fraction of component A of the alloy in sublattice 1 and 2, respectively; and $y_1 = 1 - x_1$ and $y_2 = 1 - x_2$ are the mole fraction of component B of the alloy in sublattice 1 and 2, respectively. The parameters u_{AA}, u_{BB} and u_{AB} are related, respectively, to the interaction between two atoms of type A, two atoms of type B and two atoms of distinct types. Suppose for a moment that $x_1 + x_2 = 1$. If the atoms of the same type are all in the same sublattice (complete ordering) then $x_1 = 1$ and $x_2 = 0$ or $x_1 = 0$ and $x_2 = 1$. In both cases $u = u_{AB}$. If half the atoms of one type is in one sublattice and the other half is in the other sublattice (disorder) then $x_1 = x_2 = 1/2$ and therefore $u = (u_{AA} + 2u_{AB} + u_{BB})/4$. Therefore, the ordering will be favored if $u_{AB} < (u_{AA} + u_{BB})/2$, that is, when the quantity, defined by

$$w = u_{AA} + u_{BB} - 2u_{AB}, \tag{12.7}$$

is positive, that is, $w > 0$.

The variation of energy that occurs when an alloy is made up from pure substances is given by $\Delta u = u - (u_1 + u_2)$ where u_1 and u_2 are the molar energies of the pure substances given by

$$u_1 = \frac{1}{2}u_{AA}(x_1 + x_2) \qquad \text{and} \qquad u_2 = \frac{1}{2}u_{BB}(y_1 + y_2). \tag{12.8}$$

Therefore, using the results (12.6) and (12.8), we get

$$\Delta u = -\frac{1}{2}w(x_1y_2 + y_1x_2), \tag{12.9}$$

where w is given by (12.7).

The second hypothesis of the Bragg-Williams theory concerns the entropy. We assume that the change in entropy that occurs when we mix two metals A and B to compose each one of the sublattices of the alloy is the same as that of an ideal solution. Therefore, the change in molar entropy molar Δs of an alloy which orders according to the two sublattices is given by

$$\Delta s = -\frac{R}{2}\{x_1 \ln x_1 + y_1 \ln y_1 + x_2 \ln x_2 + y_2 \ln y_2\}. \tag{12.10}$$

The variation in the molar Gibbs free energy is $\Delta g = \Delta h - T\Delta s$. On the other hand, the variation in enthalpy is $\Delta h = \Delta u + p\Delta v$. Under ordinary pressures, the term $p\Delta v$ can be neglected so that $\Delta g = \Delta u - T\Delta s$.

The Gibbs free energy g of the alloy is set up from

$$g = \frac{1}{2} g_A(x_1 + x_2) + \frac{1}{2} g_B(y_1 + y_2) + \Delta g, \qquad (12.11)$$

where the first two terms are the molar Gibbs free energies of the pure substances. Using the results (12.9) and (12.10) and taking into account that $\Delta g = \Delta u - T\Delta s$, we get

$$g = \frac{1}{2} g_A(x_1 + x_2) + \frac{1}{2} g_B(y_1 + y_2) - \frac{1}{2} w(x_1 y_2 + y_1 x_2)$$

$$+ \frac{RT}{2}\{x_1 \ln x_1 + y_1 \ln y_1 + x_2 \ln x_2 + y_2 \ln y_2\}. \qquad (12.12)$$

The function g can also be written in the form

$$g = w x_1 x_2 + \frac{b}{2}(x_1 + x_2) + a$$

$$+ \frac{RT}{2}\{x_1 \ln x_1 + (1 - x_1)\ln(1 - x_1) + x_2 \ln x_2 + (1 - x_2)\ln(1 - x_2)\}, \qquad (12.13)$$

where $b = g_A - g_B - w$ and $a = g_B$.

The chemical potential $\mu_1 = \partial g/\partial x_1$ and $\mu_2 = \partial g/\partial x_2$ are given by

$$\mu_1 = w x_2 + \frac{b}{2} + \frac{RT}{2} \ln \frac{x_1}{1 - x_1} \qquad (12.14)$$

and

$$\mu_2 = w x_1 + \frac{b}{2} + \frac{RT}{2} \ln \frac{x_2}{1 - x_2}. \qquad (12.15)$$

Two Sublattices

The two symmetric ordered phases occur when the chemical potential $\mu_* = \mu_1 - \mu_2$, conjugated to the order parameter $\eta = x_1 - x_2$, vanishes. In this case $\mu_1 = \mu_2 = \mu/2$ and using the relations (12.14) and (12.15), we obtain

$$\mu = 2w(x - \frac{\eta}{2}) + b + RT \ln \frac{x + \frac{\eta}{2}}{1 - x - \frac{\eta}{2}} \qquad (12.16)$$

and

$$\mu = 2w(x + \frac{\eta}{2}) + b + RT \ln \frac{x - \frac{\eta}{2}}{1 - x + \frac{\eta}{2}}. \tag{12.17}$$

These two equations determine x and η as functions of T and μ.

Eliminating μ from these two equation we get

$$2w\eta = RT \ln \frac{(x + \frac{\eta}{2})(1 - x + \frac{\eta}{2})}{(x - \frac{\eta}{2})(1 - x - \frac{\eta}{2})}. \tag{12.18}$$

The coexistence of the ordered phases corresponds to a nonzero solution for η. This solution is monotonically decreasing with temperature and vanishes continuously. The critical line is defined by the limit $\eta \to 0$ of this solution. Dividing both sides of (12.18) by η and taking this limit we obtain

$$2w = \frac{RT}{x(1-x)}, \tag{12.19}$$

or

$$RT = 2wx(1-x), \tag{12.20}$$

which is the equation that describes the critical line in the diagram temperature versus mole fraction, shown in Fig. 12.3. Therefore, the Bragg-Williams theory predicts a continuous transition occurring along a line of critical points in the diagram temperature versus mole fraction. Such a line is observed experimentally in the copper-zinc alloy in the transition between the ordered phase β'-brass and the disordered phase β-brass, as shown in Fig. 12.2a.

Next we determine the behavior of the order parameter near the critical line. For simplicity we will consider the case of an alloy with the same concentrations of the

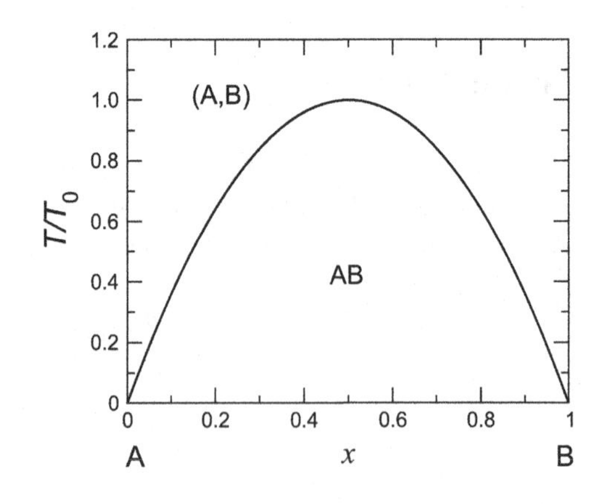

Fig. 12.3 Phase diagram of the binary alloy A-B in a bcc lattice according to the Bragg-Williams theory. The transition between the ordered and disordered phases is continuous. The temperature T_0 is defined by $T_0 = w/2R$

two components, that is, for $x = 1/2$. In this case, (12.18) reduces to

$$w\eta = RT \ln \frac{1+\eta}{1-\eta}, \tag{12.21}$$

equation that yields η as a function of temperature, which was obtained originally by Gorsky.

The behavior of η in the neighborhood of the critical point is obtained by the expansion of the right hand side of (12.21) up to cubic terms in η,

$$\frac{3}{2RT}(w - 2RT)\eta = \eta^3. \tag{12.22}$$

If $T > T_c = w/(2R)$ there is only the solution $\eta = 0$. For $T < T_c$ the nonzero solutions are

$$\eta = \pm\sqrt{3\frac{T_c - T}{T_c}}. \tag{12.23}$$

Near the critical point, the order parameter behaves therefore as

$$\eta \sim (T_c - T)^\beta. \tag{12.24}$$

with $\beta = 1/2$. It is possible to show from (12.18) that this behavior, with $\beta = 1/2$, is valid not only at $x = 1/2$ but for any value of x between 0 and 1. That is, the critical behavior predicted by the Bragg-Williams theory is the same around any point of the critical line.

Experimentally, one verifies that the exponent β has values distinct from 1/2. Table 12.1 shows the values of this exponent and other critical exponents for the order-disorder transition between the β'-brass and β-brass.

Four Sublattices

The Gibbs free energy built for two sublattices can be generalized to four sublattices. In this case, in addition to temperature and pressure, g depends on the mole fractions x_1, x_2, x_3 and x_4 of component A in each of four sublattices. The molar Gibbs free

Table 12.1 Critical exponents related to the order-disorder transition of binary alloys obtained experimentally

Liga	α	β	γ
CuZn	0.11	0.31	1.25
Fe_3Al		0.31	
FeCo		0.30	

energy is given by

$$g = w(x_1x_2 + x_1x_3 + x_1x_4 + x_2x_3 + x_2x_4 + x_3x_4) + b(x_1 + x_2 + x_3 + x_4) + a+$$
$$+\frac{RT}{4}\{x_1 \ln x_1 + (1 - x_1)\ln(1 - x_1) + x_2 \ln x_2 + (1 - x_2)\ln(1 - x_2)+$$
$$+ x_3 \ln x_3 + (1 - x_3)\ln(1 - x_3) + x_4 \ln x_4 + (1 - x_4)\ln(1 - x_4)\}. \tag{12.25}$$

The relation of the mole fractions of the sublattices with the mole fraction x is given by

$$x = \frac{x_1 + x_2 + x_3 + x_4}{4}. \tag{12.26}$$

From g we get the chemical potentials

$$\mu_1 = w(x_2 + x_3 + x_4) + b + \frac{RT}{4}\ln\frac{x_1}{1 - x_1}, \tag{12.27}$$

$$\mu_2 = w(x_1 + x_3 + x_4) + b + \frac{RT}{4}\ln\frac{x_2}{1 - x_2}, \tag{12.28}$$

$$\mu_3 = w(x_1 + x_2 + x_4) + b + \frac{RT}{4}\ln\frac{x_3}{1 - x_3}, \tag{12.29}$$

$$\mu_4 = w(x_1 + x_2 + x_3) + b + \frac{RT}{4}\ln\frac{x_4}{1 - x_4}. \tag{12.30}$$

Let us now consider the case where one of the components preferably lies in a sublattice and the other component, preferably in the other three sublattices. Assuming that these three sublattices are 2, 3 and 4, then $x_2 = x_3 = x_4$, so that $\mu_2 = \mu_3 = \mu_4$. Therefore

$$\mu_1 = 3wx_2 + b + \frac{RT}{4}\ln\frac{x_1}{1 - x_1}, \tag{12.31}$$

$$\mu_2 = w(x_1 + 2x_2) + b + \frac{RT}{4}\ln\frac{x_2}{1 - x_2}. \tag{12.32}$$

The condition for coexistence is $\mu_1 = \mu_2 = \mu/4$. On the other hand

$$x = \frac{x_1 + 3x_2}{4}, \tag{12.33}$$

which combined with the relation that defines the order parameter

$$\eta = x_1 - x_2, \tag{12.34}$$

yields

$$x_1 = x + \frac{3\eta}{4} \qquad \text{and} \qquad x_2 = x - \frac{\eta}{4}. \tag{12.35}$$

Therefore

$$\mu = 4w(3x - \frac{3\eta}{4}) + 4b + RT \ln \frac{x + \frac{3\eta}{4}}{1 - x - \frac{3\eta}{4}}, \tag{12.36}$$

$$\mu = 4w(3x + \frac{\eta}{4}) + 4b + RT \ln \frac{x - \frac{\eta}{4}}{1 - x + \frac{\eta}{4}}. \tag{12.37}$$

Eliminating μ from these two equations we get

$$4w\eta = RT \ln \frac{(x + \frac{3\eta}{4})(1 - x + \frac{\eta}{4})}{(x - \frac{\eta}{4})(1 - x - \frac{3\eta}{4})}. \tag{12.38}$$

Let us now consider the case where one of the components is found preferably and equally in two of the sublattices and the other component preferably and equally in the other two sublattices. Assuming that the first two sublattices are 1 and 3 and the other two are 2 and 4, then $x_1 = x_3$ and $x_2 = x_4$ so that $\mu_1 = \mu_3$ and $\mu_2 = \mu_4$. Therefore,

$$\mu_1 = w(2x_2 + x_1) + b + \frac{RT}{4} \ln \frac{x_1}{1 - x_1}, \tag{12.39}$$

$$\mu_2 = w(2x_1 + x_2) + b + \frac{RT}{4} \ln \frac{x_2}{1 - x_2}. \tag{12.40}$$

When there is coexistence, then $\mu_1 = \mu_2 = \mu/4$. Using relations (12.1) and (12.2), we get

$$\mu = 4w(3x - \frac{\eta}{2}) + 4b + RT \ln \frac{x + \frac{\eta}{2}}{1 - x - \frac{\eta}{2}}, \tag{12.41}$$

$$\mu = 4w(3x + \frac{\eta}{2}) + 4b + RT \ln \frac{x - \frac{\eta}{2}}{1 - x + \frac{\eta}{2}}. \tag{12.42}$$

Eliminating μ from these two equations we get

$$4w\eta = RT \ln \frac{(x + \frac{\eta}{2})(1 - x + \frac{\eta}{2})}{(x - \frac{\eta}{2})(1 - x - \frac{\eta}{2})}. \tag{12.43}$$

The solutions of (12.38) and (12.43) allow the construction of the phase diagram shown in Fig. 12.4. The transitions between the phases are all discontinuous except

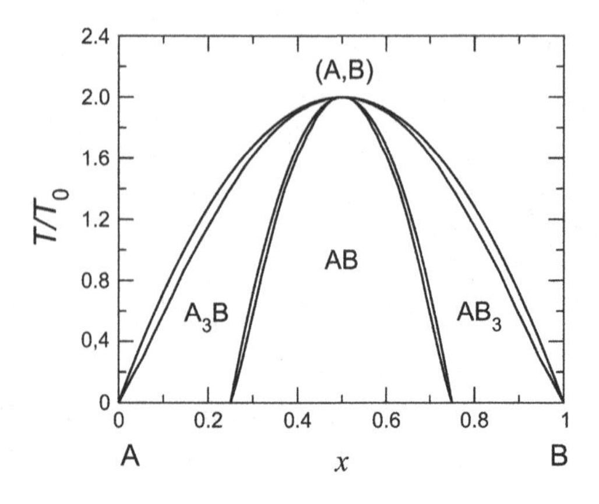

Fig. 12.4 Phase diagram of a binary alloy A-B in a fcc lattice according to the Bragg-Williams theory, obtained originally by Shockley [139]. The transitions between phases are discontinuous except that occurring along the $x = 0.5$ which is continuous. The temperature T_0 is defined by $T_0 = w/2R$

that occurring along the $x = 0.5$ which is a continuous transition. The phase diagram predicted by the Bragg-Williams theory is distinct from those observed experimentally for binary alloys that are ordered into four sublattices, as that shown in Fig. 12.1. Results closer to the experimental data, however, can be obtained from more elaborate theories.

12.3 Landau Theory

Spontaneous Symmetry Breaking

Landau theory deals with the transition between phases which have different symmetries. More specifically between a phase of higher symmetry, usually occurring at high temperatures, called disordered phase to a lower symmetry one called ordered phase. The transition to the ordered phase corresponds to a symmetry breaking as some symmetries existing in the disordered phase will be absent in the ordered phase. Since the symmetry breaking is not induced by any external field, the transition is associated with a spontaneous breaking of symmetry.

The ordered phase is actually made up of several distinct ordered phases in coexistence which have similar structures. The structures are not arbitrary but are connected to each other by certain symmetry operations which comprises a group of symmetry, which we denote by $\mathscr{G}$. The symmetry operations of the group $\mathscr{G}$ transform an ordered state into another, thus generating all ordered states in coexistence from one of them. The disordered phase, on the other hand, is invariant under the transformations of the group $\mathscr{G}$. According to Landau theory all ordered states have the same Gibbs free energy which results in the following rule: the Gibbs free energy is invariant under the symmetry operations of the group $\mathscr{G}$.

To describe the ordered phase we introduce an order parameter: a thermodynamic variable that takes nonzero values in the ordered phase and vanishes in the disordered phase. To address the possibility of the occurrence of distinct ordered states, we assume that the order parameter has several components which define a vector space whose vectors we denote by $\vec{\eta}$. Thus, each of the coexisting ordered states is represented by a point in that space. The positions of representative points of the ordered states form a geometric figure that is invariant by the symmetry operations of the group $\mathscr{G}$. The disordered phase is represented by $\vec{\eta} = 0$. The space vector is constructed so that its dimension is the same as that of the geometric figure representing the ordered states. The thermodynamic state associated with the molar Gibbs free energy is thus defined by the components η_i of the vector $\vec{\eta}$ in addition to the usual variables such as temperature, pressure and mole fractions.

The thermodynamic field h_i conjugated to the component η_i of the order parameter is given by $h_i = \partial g/\partial \eta_i$, where $g(\vec{\eta})$ is the molar Gibbs free energy. This field, when nonzero, breaks the symmetry of the disordered phase and destroys the coexistence of ordered phases by favoring one of them. In general this field, which breaks the symmetry, is not accessible experimentally. The order-disorder transition described by the Landau theory, wherein the symmetry breaking is spontaneous, corresponds to the case where this field is identically zero. Thus, the molar Gibbs free energy is such that

$$\frac{\partial g}{\partial \eta_i} = 0. \tag{12.44}$$

Since g must be a convex function of η_i then condition (12.44) means that the ordered and disordered phases correspond to the minima of g with respect to the order parameter.

The Landau theory for the order-disorder transition assumes that $g(\vec{\eta})$ is a polynomial in η_i. In many applications it is enough to assume that the polynomial is of degree four, that is,

$$g(\vec{\eta}) = a_0 + a_2 I_2(\vec{\eta}) + a_3 I_3(\vec{\eta}) + a_4 I_4(\vec{\eta}), \tag{12.45}$$

where $I_n(\vec{\eta})$ is a monomial or a sum of monomials of degree n in η_i and the coefficients a_n are dependent of other thermodynamic variables with the exception of η_i. Since $g(\vec{\eta})$ is invariant under the symmetry operations of group $\mathscr{G}$, the same occurs with $I_n(\vec{\eta})$ which for this reason is called invariant of degree n.

The linear term is absent because the invariant of the first degree is nonexistent. The invariant of second order, on the other hand, always exists and is given by

$$I_2 = \sum_i \eta_i^2, \tag{12.46}$$

ensuring that $\vec{\eta} = 0$ is a minimum of g for $a_2 > 0$. Assuming that a_2 varies with temperature, other minima of g arise leading to the ordered states. We assume that a_2 varies linearly with temperature and vanishes at a temperature T_0. That is, we assume that $a_2 = A(T - T_0)$ with $A > 0$ so that the disordered phase occurs at high temperatures.

Continuous Transition

Let us first examine a binary alloy A-B whose atoms can be ordered in a crystalline structure composed by two sublattices. The ordered alloy is such that the mole fractions x_1 and x_2 of component A of the alloy related to sublattices 1 and 2, respectively, are distinct. An ordered state is such that the atoms of type A are preferably found in sublattice 1 and is characterized by $x_1 > x_2$. The other ordered state is the one in which the atoms of type A are preferably found in the sublattice 2. This state is characterized by $x_2 > x_1$. One state can be generated from the other by a spatial translation which in this case is equivalent to the exchange of atoms between the sublattices 1 and 2. Therefore, the Gibbs free energy is invariant by the exchange of x_1 by x_2.

Using the order parameter η defined by

$$\eta = x_1 - x_2 \tag{12.47}$$

we see that $\eta > 0$ characterizes an ordered state and $\eta < 0$ characterizes the other ordered state. The permutation of x_1 by x_2 is equivalent therefore to the permutation of η by $-\eta$ and therefore $g(\eta)$ must be invariant under this permutation, that is,

$$g(-\eta) = g(\eta). \tag{12.48}$$

According to Landau theory we postulate that g is a polynomial of fourth degree in η. Using the property (12.48) we conclude that the invariants of odd degree are absent, which allows us to write

$$g = a_0 + a_2\eta^2 + a_4\eta^4. \tag{12.49}$$

Furthermore, still according to Landau theory, the coefficient a_2 is positive in the disordered phase and negative in the ordered phase, implying

$$a_2 = A(T - T_c), \tag{12.50}$$

where $A > 0$. The coefficient a_4 is considered to be positive.

The consequences of the result (12.50) are illustrated in Fig. 12.5. For temperatures above T_c, the function $g(\eta)$ is convex and has a single minimum at $\eta = 0$. For temperatures below T_c, the function $g(\eta)$ is no longer convex and has two symmetric minima. In this case we consider the convex hull of $g(\eta)$. The appearance of two minima is a manifestation of spontaneous symmetry breaking. It is important to

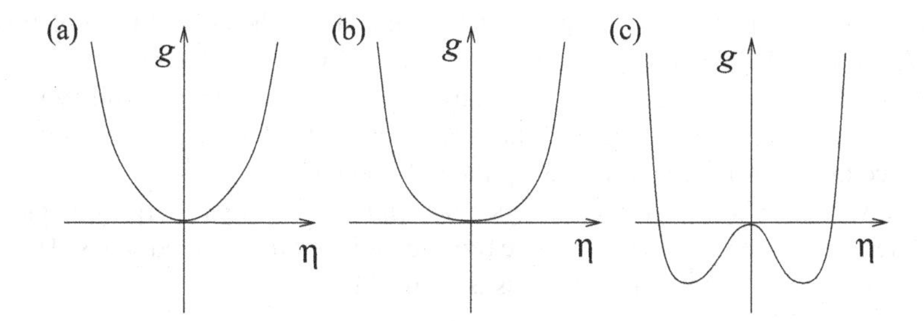

Fig. 12.5 Molar Gibbs free energy g versus order parameter η for a binary alloy that is ordered in two sublattices according to the Landau theory for a temperature above the critical temperature (**a**), equal to the critical temperature (**b**), and below the critical temperature (**c**)

notice that as the temperature approaches the critical temperature, the minimum continuously approaches zero.

The coexistence of two ordered phases occurs when the field $h = \partial g/\partial\eta$ vanishes which is equivalent to saying that the order parameter η corresponds to the minimum of g. In fact, the convex hull has a horizontal line segment (corresponding to $h = 0$) that connects the two minima. From

$$\frac{\partial g}{\partial \eta} = 2a_2\eta + 4a_4\eta^3 = 0 \tag{12.51}$$

we obtain the two minima

$$\eta = \pm\sqrt{\frac{|a_2|}{2a_4}}, \tag{12.52}$$

result valid for $T < T_c$. For $T > T_c$ we get $\eta = 0$. Therefore, the order parameter vanishes continuously and behaves in accordance with

$$\eta \sim |T - T_c|^{1/2}. \tag{12.53}$$

Discontinuous Transition

Next we study an example where the phase transition is discontinuous, that is, the order parameter undergoes a jump in the passage from the order to the disorder phase. We will see that this is a consequence of the existence of an invariant of third order.

Let us examine a binary alloy A-B that orders in a structure composed by three sublattices. We denote by x_1, x_2 and x_3 the mole fractions of component A in each of the three sublattices. An ordered phase is that in which the atoms of type A are found preferably in sublattice 1 and indiscriminately in the other two sublattices. Let us suppose that in this ordered state $x_1 = a$ and $x_2 = x_3 = b$ with $a > b$. The other two ordered states are characterized by $x_2 = a$ and $x_3 = x_1 = b$ and by $x_3 = a$ and $x_1 = x_2 = b$. The three ordered states are represented by three distinct points in the space (x_1, x_2, x_3) which form an equilateral triangle.

Next we define a new coordinate system such that one axis is perpendicular and the other two are parallel to the triangle representative of the ordered states. The one corresponding to the perpendicular axis is defined by

$$x = \frac{1}{3}(x_1 + x_2 + x_3), \tag{12.54}$$

which is the mole fraction of component A. The other two are defined by

$$\eta = x_1 - \frac{1}{2}(x_2 + x_3) \qquad \text{and} \qquad \xi = \frac{\sqrt{3}}{2}(x_2 - x_3) \tag{12.55}$$

and work as components of the vector order parameter (η, ξ), since they vanish when $x_1 = x_2 = x_3$. One of the ordered phase is defined by $\xi = 0$ and $\eta > 0$. The other are obtained by symmetry operations that leave the equilateral triangle invariant. Using this symmetry operation, we can determine the invariants up to fourth order, that are given by

$$I_2 = \eta^2 + \xi^2, \qquad I_3 = \eta^3 - 3\eta\xi^2, \qquad I_4 = (\eta^2 + \xi^2)^2. \tag{12.56}$$

For temperatures high enough, $g(\eta, \xi)$ is a convex function with a minimum at $\eta = \xi = 0$. For sufficient low temperatures, $g(\eta, \xi)$ develops three minima that form an equilateral triangle. In this case the function $g(\eta, \xi)$ ceases to have the convexity properties and we should consider the convex hull. The three minima correspond to the three phases in coexistence. One of the minima occurs along $\xi = 0$. Along $\xi = 0$ the molar Gibbs free energy is given by

$$g = a_0 + a_2\eta^2 + a_3\eta^3 + a_4\eta^4. \tag{12.57}$$

Figure 12.6 shows the graph of $g(\eta)$ for several values of temperature. We recall that $a_2 = A(T - T_0)$, where $A > 0$ and that $a_4 > 0$. For simplicity, the other coefficients are considered constants.

For low temperatures, the absolute minimum corresponds to a nonzero value of η, characterizing an ordered phase. Raising the temperature, this minimum ceases to be absolute. The local minimum at $\eta = 0$, corresponding to the disordered phase, becomes then the absolute minimum. The transition, which is discontinuous, occurs at temperature T_1 which is determined by the condition $g(\eta) = g(0)$, as seen in Fig. 12.6b. Therefore, at this temperature $g(\eta)$ has the form

$$g(\eta) = a_0 + a_4\eta^2(\eta - b)^2, \tag{12.58}$$

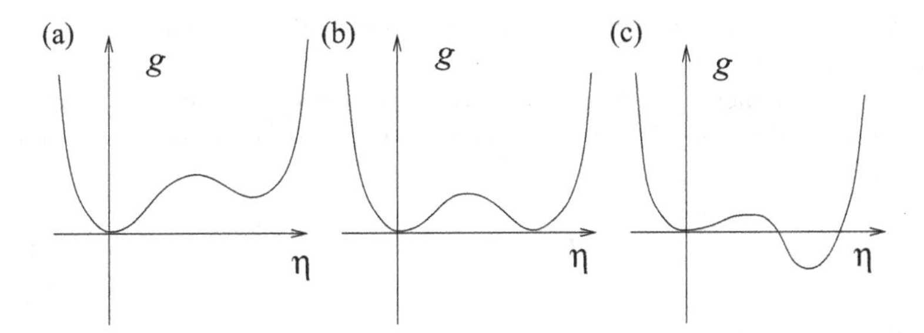

Fig. 12.6 Molar Gibbs free energy g versus the order parameter η for a binary alloy that orders in three sublattices according to the Landau theory for a temperature above the critical temperature (**a**), equal to the critical temperature (**b**), and below the critical temperature (**c**). The phase transition is discontinuous

which, compared to (12.57), gives us $a_3 = -2ba_4$ and $a_2 = b^2 a_4$. Eliminating b, we get

$$a_2 = \frac{a_3^2}{4a_4}. \tag{12.59}$$

Therefore the transition temperature occurs at $T_1 = T_0 + a_3^2/4Aa_4$.

Four Sublattices

Let us study now a binary alloy A-B that orders in four sublattices. We denote by x_1, x_2, x_3 and x_4 the mole fraction of component A in each of the four sublattices. The ordered state in which the atoms A are found preferable in sublattice 1 is characterized by $x_1 = a$ and $x_2 = x_3 = x_4 = b$ with $a > b$. The other three states are characterized by $x_2 = a$ and $x_3 = x_4 = x_1 = b$, by $x_3 = a$ and $x_4 = x_1 = x_2 = b$, and by $x_4 = a$ and $x_1 = x_2 = x_3 = b$. The representative points of the ordered state make up a tetrahedron in space (x_1, x_2, x_3, x_4). Next we use a new coordinate system defined by

$$x = \frac{1}{4}(x_1 + x_2 + x_3 + x_4), \tag{12.60}$$

which is the mole fraction of component A, and by

$$\eta = x_1 + x_2 - x_3 - x_4, \tag{12.61}$$

$$\xi = x_1 - x_2 + x_3 - x_4, \tag{12.62}$$

$$\theta = x_1 - x_2 - x_3 + x_4. \tag{12.63}$$

The variables η, ξ and θ works as components of the order parameter since they vanish when $x_1 = x_2 = x_3 = x_4$.

Using the symmetry operation that leaves the tetrahedron invariant we can obtain the invariants of second, third and fourth order. The invariant of second order is given by

$$I_2 = \eta^2 + \xi^2 + \theta^2 \tag{12.64}$$

and that of third order by

$$I_3 = \eta\xi\theta. \tag{12.65}$$

There are two invariants of fourth order:

$$I_{41} = (\eta^2 + \xi^2 + \theta^2)^2 \tag{12.66}$$

and

$$I_{42} = \eta^4 + \xi^4 + \theta^4. \tag{12.67}$$

The molar Gibbs free energy is given by

$$g = a_0 + a_2 I_2 + a_3 I_3 + a_{41} I_{41} + a_{42} I_{42}. \tag{12.68}$$

For temperatures sufficiently small $g(\eta, \xi, \theta)$ has four minima corresponding to the four ordered states. One of them occurs along $\theta = \xi = \eta$. Along this line the molar Gibbs free energy is given by

$$g = a_0 + 3a_2\eta^2 + a_3\eta^3 + (9a_{41} + 3a_{42})\eta^4. \tag{12.69}$$

The presence of the third order invariant implies a discontinuous phase transition.

Chapter 13
Magnetic Systems

13.1 Magnetic Materials

Introduction

The most evident magnetic materials are iron (Fe), cobalt (Co) and nickel (Ni). At room temperature they exhibit a natural magnetism or spontaneous magnetization and therefore are used as permanent magnets. If heated, however, such materials lose their natural magnetization at a given temperature, called critical temperature T_c or Curie temperature, and become paramagnetic. The iron loses its natural magnetism at 770 °C, cobalt at 1122 °C and nickel at 358 °C. Many compounds of the iron group metals, for example, MnSb, CrTe and CrO_2 are also ferromagnetic at room temperature.

Certain materials, such as gadolinium (Gd) and the compounds $CrBr_3$, EuO and EuS, are paramagnetic already at room temperature. To acquire spontaneous magnetization, it is necessary to lower the temperature. Gadolinium becomes ferromagnetic below 20 °C. The compounds $CrBr_3$, EuO and EuS become ferromagnetic at low temperatures.

Magnetite (Fe_3O_4), mineral known since ancient times by the magnetic properties as well as other iron oxides with the natural spinel crystal structure, such as $MgFe_2O_4$, $MnFe_2O_4$, $NiFe_2O_4$, $CoFe_2O_4$ and $CuFe_2O_4$, also exhibit a natural magnetization. Although presenting a spontaneous magnetization, these oxides, called ferrites, are not exactly ferromagnetic but ferrimagnetic. In any event, at sufficiently high temperatures they lose their natural magnetism becoming paramagnetic. Magnetite loses magnetization at 585 °C. Another class of ferrimagnetic materials is that formed by the iron oxides from the crystal structure of the garnet. An important example of this category is the garnet of iron and yttrium, $Y_3Fe_5O_{12}$, whose Curie temperature is 287 °C

Other magnetic materials such as manganese (Mn) cease to be paramagnetic but do not become ferromagnetic. They acquire magnetic states known generically as

M.J. de Oliveira, *Equilibrium Thermodynamics*, Graduate Texts in Physics,
DOI 10.1007/978-3-662-53207-2_13

antiferromagnetic states, which are characterized by the absence of spontaneous magnetization. Manganese (Mn) becomes antiferromagnetic below −173 °C. Likewise, the oxides FeO, MnO and Mn_2O_3 and the fluorine compounds FeF_2, MnF_2 and $RbMnF_3$ are paramagnetic at room temperature, becoming antiferromagnetic at low temperatures. On the other hand, chromium (Cr) and hematite (Fe_2O_3) and the oxides CoO and Cr_2O_3 are antiferromagnetic at room temperature becoming paramagnetic at higher temperature. Chromium becomes paramagnetic above 39 °C. The temperature below which the material becomes antiferromagnetic is called Néel temperature, T_N. Although the antiferromagnetic state do not present spontaneous magnetization it should not be confused with the paramagnetic state.

Microscopic State

From the microscopic viewpoint each of the compounds mentioned above should be understood as a large collection of microscopic magnetic dipoles which, in the case of ionic compounds, are located at the magnetic ions. The magnetic ions are the ions of transition metals, especially those of the iron group of iron, or the rare earth ions. The interactions between the magnetic dipoles are the cause of several types of magnetic ordering. These interactions are not properly magnetic, as one might imagine at first, but originate from the Pauli exclusion principle of quantum mechanics.

In the ferromagnetic state, the interactions between the dipoles favor the parallel alignment of them. In the simplest antiferromagnetic state the interactions cause an antiparallel alignment of dipoles. At high temperatures or if the interactions between the dipoles are negligible, the thermal agitation causes the dipole moments of ions to point in arbitrary directions in space giving rise to a magnetically disordered state, with a zero total dipole moment. This state magnetically disordered is called paramagnetic state. Thus, when increasing the temperature of the magnetic substances mentioned above, they undergo a transition from a magnetically ordered thermodynamic phase, which may be ferromagnetic, antiferromagnetic or other more complex, to a magnetically disorderly thermodynamic phase, the paramagnetic phase.

Paramagnetism

The paramagnetic state is characterized from the macroscopic point of view by the linear response to an applied magnetic field. In absence of the field, a sample of paramagnetic material does not exhibit magnetization. Applying a field the sample acquires a magnetization that increases linearly with the field. If, on the other hand, the field is reduced and vanishes, then the magnetization is also reduced and vanishes. For small values of the applied field H, the magnetization m of a sample

Table 13.1 Paramagnetic substances and their respective Curie constants C, in K A m^2/Tmol, obtained experimentally. The adimensional quantity p is defined by $p = \sqrt{3RC}/\mu_B$ where μ_B is the molar Bohr magneton. *Source*: AIP

Substances	C	p
$Ce(NO_3)_3{\cdot}5H_2O$	7.17	2.39
$Pr_2(SO_4)_3$	16.4	3.62
$Nd_2(SO_4)_3{\cdot}8H_2O$	18.2	3.81
$Gd_2(SO_4)_3{\cdot}8H_2O$	78.0	7.90
$Tb_2(SO_4)_3{\cdot}8H_2O$	118.6	9.74
$Dy_2(SO_4)_3{\cdot}8H_2O$	137.4	10.5
$Ho_2(SO_4)_3{\cdot}8H_2O$	136	10.4
$Er_2(SO_4)_3{\cdot}8H_2O$	111.8	9.46
$Tm_2(SO_4)_3$	63.3	7.11
$Yb_2(SO_4)_3{\cdot}8H_2O$	29.2	4.83
$CrK(SO_4)_2{\cdot}12H_2O$	18.4	3.84
MnF_3	30.1	4.91
$MnSO_4$	43.4	5.89
$Fe(NH_4)(SO_4)_2{\cdot}12H_2O$	43.9	5.92
$Co(NO_3)_2{\cdot}6H_2O$	25.8	4.54
$NiSiF_6{\cdot}6H_2O$	13.1	3.24
$CuSO_4{\cdot}5H_2O$	4.6	1.92

in the paramagnetic state is proportional to the field, that is,

$$m = \chi_0 H, \tag{13.1}$$

where χ_0, the magnetic susceptibility, is positive. Increasing the field, the behavior of m with H is no longer linear and at sufficiently high values of the field, the magnetization saturates reaching a maximum value.

The susceptibility χ_0 depends on temperature T. For materials which are paramagnetic at all temperatures, which we call ideal paramagnetism, it behaves in accordance with the Curie law

$$\chi_0 = \frac{C}{T}, \tag{13.2}$$

where C is a positive constant. Equation (13.1) together with (13.2) define thus from the thermodynamic point of view, an ideal paramagnet. From the microscopic point of view, an ideal paramagnetic material corresponds to a system consisting of non interacting permanent magnetic microscopic dipoles. Examples of such materials include the paramagnetic salts of the elements of the iron group and rare earths such as those presented in Table 13.1. In these compounds the magnetic ions are far from each another so that the interaction between them can be considered negligible.

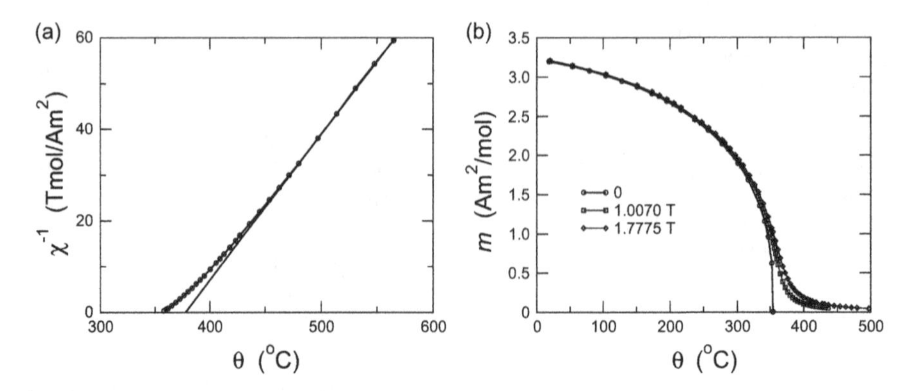

Fig. 13.1 Experimental data for nickel obtained by Weiss and Forrer [148]. (**a**) Susceptibility at zero field in the paramagnetic phase. The graph shows the inverse of the susceptibility χ as a function of temperature. The asymptote intercepts the axis of temperatures at 378 °C, above the Curie temperature that occurs at 357 °C. (**b**) Molar magnetization m as a function of temperature for various values of the applied magnetic field

For materials that have a paramagnetic phase, but are not ideal paramagnets, the quantity χ_0 behaves, at sufficiently high temperatures according to the Curie-Weiss law

$$\chi_0 = \frac{C}{T - \Theta}. \tag{13.3}$$

The constant Θ is positive for materials that undergo the transition to a ferromagnetic state and is negative for those that undergo transition to an antiferromagnetic state. The two constants C and Θ can be determined by fitting an asymptote to the experimental data of $1/\chi_0$ versus T, as shown in Fig. 13.1. The asymptote intersects the axis of temperatures at $T = \Theta$ and the slope provides $1/C$. The constant Θ should not be confused with the transition temperature.

Ferromagnetism

Suppose that a ferromagnetic substance is subjected to a magnetic field. If the substance is heated while the field is maintained constant, then the magnetization of the substance varies continuously with increasing temperature, as shown in Fig. 13.1b. For any temperature below or above the critical temperature, the substance exhibits a magnetization due to the presence of the field, as shown in Fig. 13.2. Then assume that the field is turned off slowly at a given temperature, kept unchanged. If the temperature exceeds the Curie temperature, $T > T_c$, the magnetization vanishes linearly with the field, $m = \chi_0 H$, characterizing the paramagnetic state. If the temperature is lower, $T < T_c$, the magnetization does not disappear and the

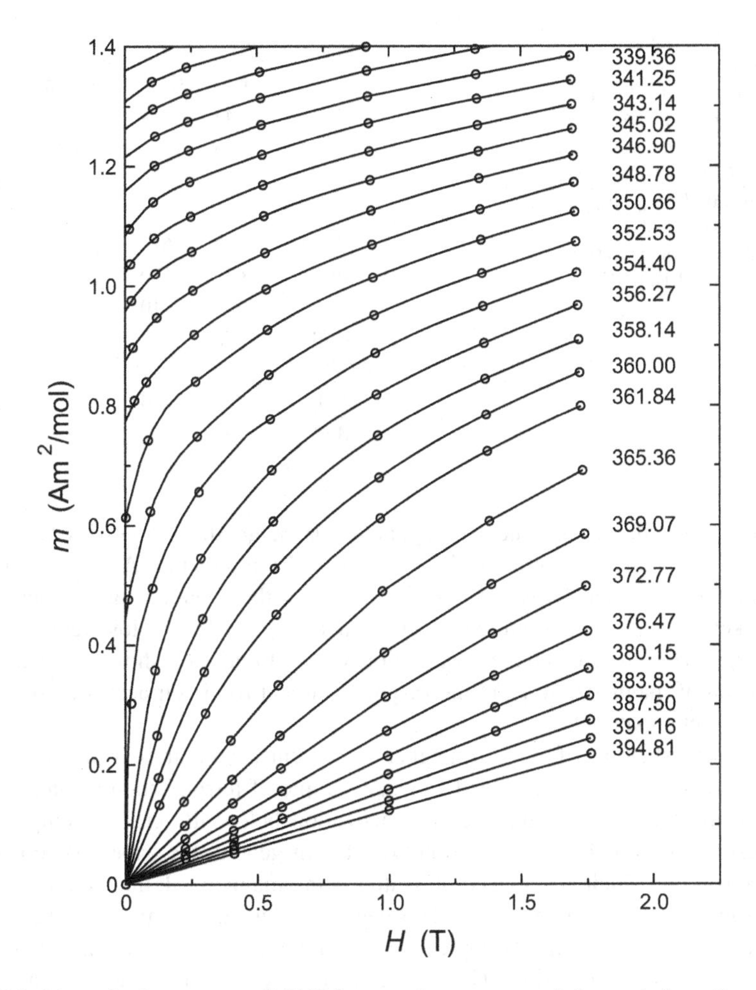

Fig. 13.2 Magnetization m versus field H for several temperatures below and above the critical temperature obtained experimentally for nickel by Weiss and Forrer [148]. The temperatures are given in °C and the critical temperature occurs at 357.6 °C. Above the critical temperature, $m \to 0$ when $H \to 0$. Below, m approaches the spontaneous magnetization when $H \to 0$

substance remains showing a magnetization even after the field has been turned off, which for that reason is called the remanent magnetization. For small values of the field the magnetization depends on the field according to

$$m = m^* + \chi_0 H \tag{13.4}$$

where the spontaneous magnetization m^* and the susceptibility χ_0 depends only on temperature. The different behaviors, above and below the critical temperature can be seen in Fig. 13.2, which shows isotherms obtained experimentally for nickel both above and below the Curie temperature, occurring at 357 °C.

Table 13.2 Critical temperature T_c, volumetric magnetization of saturation, $\mathscr{M}_s$, and at 20 °C, $\mathscr{M}$, of ferromagnetic materials. The number b of Bohr magnetons is defined by $b = \mu/\mu_B$ where μ is the molar magnetization of saturation. *Sources*: AIP, LB, CRC

Material	T_c (K)	$\mathscr{M}$ (kA/m)	$\mathscr{M}_s$ (kA/m)	b
Fe	1043	1714	1735	2.22
Co	1395	1422	1445	1.71
Ni	631	484	509	0.604
Gd	289		1980	7.12
MnAs	318	670		3.40
MnBi	633	620		3.52
MnSb	587	710		3.53
MnB	578	152		1.92
CrTe	339	247		2.39
CrO_2	393	515		2.07
MnP	298			1.20

At zero field, the remanent or spontaneous magnetization, varies with temperature and vanishes continuously at the critical temperature. Above the critical temperature it is identically zero. The behavior of the spontaneous magnetization of nickel with the temperature can be seen in Fig. 13.1b. At low temperatures, the magnetization m reaches its saturation value. Table 13.2 shows the values of saturation of magnetization and the magnetization at room temperature for several ferromagnetic materials.

The understanding of the ferromagnetism presents, sometimes, a situation which seems paradoxical. As mentioned above, a sample of ferromagnetic material such as iron must have a spontaneous magnetization below the critical temperature. However, any sample of iron does not display in general the expected permanent magnetization, which is achieved only after being magnetized, that is, after having been subjected to a magnetic field. This can be explained assuming that the sample is composed of several macroscopic magnetic domains, each possessing a magnetic dipole in a certain direction. The dipoles of different domains point to arbitrary directions which results in a vanishing total dipole. To restore the natural magnetism of the sample it is necessary to subject it to a magnetic field, that is, to magnetize it.

13.2 Magnetic Thermodynamic Potentials

Magnetic Work

To study the effects of a magnetic field on a sample of magnetic material we use a solenoid inside which we place the sample. The solenoid is traversed by an electric current supplied by a battery, thus producing a magnetic field in its interior. The magnetic induction field **B** inside the solenoid is associated with the magnetic field **H** and the magnetic dipole moment per unit volume or volumetric magnetization, $\mathscr{M}$ of the sample by the relation

$$\mathbf{B} = \mu_0(\mathbf{H} + \mathscr{M}), \tag{13.5}$$

where μ_0 is the permeability of vacuum. In the absence of the sample, the magnetic induction field within the solenoid is given by

$$\mathbf{B}_\mathrm{e} = \mu_0 \mathbf{H}_\mathrm{e}. \tag{13.6}$$

The local field **H** and the external field $\mathbf{H}_\mathrm{e}$ are produced by the same free currents that run through the solenoid. However, they may not be identical. For a uniform external field, we can show that the infinitesimal magnetic work dW_mag performed on the magnetic sample is given by

$$dW_\mathrm{mag} = \mathbf{B}_\mathrm{e} \cdot d\mathbf{M}, \tag{13.7}$$

where **M** is the total magnetic dipole moment of the sample.

To show the result (13.7) we proceed as follows. For simplicity, we consider a cylindrical solenoid of length ℓ and cross-sectional area equal to A and that the sample fills the entire space within the solenoid. We also assume that **H** and $\mathscr{M}$ are uniform and parallel to the axis of the solenoid and that edge effects are neglected. The local field H inside the solenoid, which in this case coincides with the external field, is given by

$$H = \frac{NI}{\ell}, \tag{13.8}$$

where N is the number of spiral of the solenoid and I is the electric current through the solenoid. The magnetic induction field B inside the solenoid varies in time so that the potential difference $\mathscr{V}$ at the ends of the solenoid is equal to the time variation of the flux of magnetic induction field. Since the total flow is NAB, then

$$\mathscr{V} = NA\frac{dB}{dt}. \tag{13.9}$$

The power P transferred to the system solenoid-sample is $P = \mathscr{V}I$. Using (13.8) and (13.9) we reach the result

$$P = H\ell A\frac{dB}{dt} = VH\frac{dB}{dt}, \tag{13.10}$$

where $V = \ell A$ is the volume of the solenoid and is also the volume of the sample.

Let us now determine the power transferred P_0 to the solenoid when it is empty. Using the same arguments, we obtain

$$P_0 = VH\mu_0\frac{dH}{dt}. \tag{13.11}$$

Therefore, the magnetic work rate $dW_{\text{mag}}/dt = P - P_0$ performed only on the sample will be

$$\frac{dW_{\text{mag}}}{dt} = VH\mu_0\frac{d\mathscr{M}}{dt}, \tag{13.12}$$

since $B - \mu_0 H = \mu_0\mathscr{M}$. Taking into account that the total magnetic dipole of the sample is $M = V\mathscr{M}$ then

$$\frac{dW_{\text{mag}}}{dt} = \mu_0 H\frac{dM}{dt}. \tag{13.13}$$

Since in the present case the local field H and the external field H_{e} coincide then $B_{\text{e}} = \mu_0 H_{\text{e}} = \mu_0 H$ and we arrive at the following result for the infinitesimal magnetic work performed on the magnetic sample:

$$dW_{\text{mag}} = B_{\text{e}}dM. \tag{13.14}$$

In the International System of Units, the magnetic dipole moment M is measured in ampère square meter (A m^2) and the magnetic induction fields B and B_{e} are measured in tesla (T). The magnetic fields H and H_{e} as well as the volumetric magnetization $\mathscr{M}$ are measured in ampère per meter (A/m). The magnetic dipole moment per mole, or molar magnetization, m is given in A m^2/mol.

A standard dipole moment widely used is the Bohr magneton defined as $e\hbar/2m_{\text{e}}$, where e is the charge of the electron, $\hbar$ is the Planck constant and m_{e} is the mass of the electron. The Bohr magneton is almost identical to the dipole moment of the free electron and is $9.2740154 \times 10^{-24}$ A m^2. Here it is useful to define the molar Bohr magneton by $\mu_{\text{B}} = (e\hbar/2m_e)N_{\text{A}}$, where N_{A} is the Avogadro constant. In the International System of Units

$$\mu_{\text{B}} = 5.5849388 \qquad \text{A m}^2/\text{mol}. \tag{13.15}$$

Thermodynamic Potentials

The infinitesimal change of internal energy dU of a magnetic system is equal to the sum of heat introduced $dQ = TdS$ and the work done on the magnetic system $dW_{\text{mag}} = HdM$, that is,

$$dU = TdS + HdM. \tag{13.16}$$

To simplify notation we denote by H the external magnetic induction field B_{e}, which we call external field or the applied field. Importantly, the total dipole moment M must be understood as an extensive thermodynamic quantity and H

as a thermodynamic field. Thus we understand that internal energy U of a simple magnetic system is a function of the extensive variables S and M. More precisely we consider that the internal energy is also a function of the number of moles N so that $U(S,M,N)$ is the fundamental relationship of a simple magnetic system in the representation of internal energy. The entropy $S(U,M,N)$ is also a fundamental relationship for which we have

$$dS = \frac{1}{T}dU - \frac{H}{T}dM, \tag{13.17}$$

which we get directly from (13.16).

From $U(S,M,N)$, we obtain other representations by Legendre transformations. The magnetic enthalpy $\mathscr{H}(S,H,N)$, the magnetic Helmholtz free energy $F(T,M,N)$ and the magnetic Gibbs free energy $G(T,H,N)$ are defined by the Legendre transformations

$$\mathscr{H} = \min_M (U - HM), \tag{13.18}$$

$$F = \min_S (U - TS), \tag{13.19}$$

$$G = \min_M (F - HM). \tag{13.20}$$

From them we get the results

$$d\mathscr{H} = TdS - MdH, \tag{13.21}$$

$$dF = -SdT + HdM, \tag{13.22}$$

$$dG = -SdT - MdH. \tag{13.23}$$

The thermodynamic potentials have the usual properties of convexity. In particular U is a convex function of the extensive variables S and M, while G is concave function of the thermodynamic fields T and H.

Thermodynamic Coefficients

From (13.23), it follows

$$S = -\left(\frac{\partial G}{\partial T}\right)_H \quad \text{and} \quad M = -\left(\frac{\partial G}{\partial H}\right)_T. \tag{13.24}$$

Two thermodynamic coefficients are particularly important: the magnetic isothermal susceptibility X_T, defined by

$$X_T = \left(\frac{\partial M}{\partial H}\right)_T, \tag{13.25}$$

and the heat capacity at constant field C_H, defined by

$$C_H = T\left(\frac{\partial S}{\partial T}\right)_H = \left(\frac{\partial \mathscr{H}}{\partial T}\right)_H. \tag{13.26}$$

The convexity of G implies the results $X_T \geq 0$ and $C_H \geq 0$. Other thermodynamic coefficients are: the heat capacity at constant magnetization

$$C_M = T\left(\frac{\partial S}{\partial T}\right)_M = \left(\frac{\partial U}{\partial T}\right)_M, \tag{13.27}$$

and the magnetocaloric coefficient

$$A = \left(\frac{\partial M}{\partial T}\right)_H. \tag{13.28}$$

The convexity of U implies $C_M \geq 0$.

Molar Quantities

The thermodynamic properties of a simple magnetic system can be obtained from the fundamental relation $u(s,m)$, where $u = U/N$ is the molar internal energy, $s = S/N$ is the molar entropy and $m = M/N$ is the molar dipole moment or simply molar magnetization. For this system, we have

$$du = Tds + Hdm. \tag{13.29}$$

The molar entropy $s(u,m)$ as a function of u and m is also a fundamental relation and

$$ds = \frac{1}{T}du - \frac{H}{T}dm. \tag{13.30}$$

Other molar quantities can also be defined that include the molar potentials $h = \mathscr{H}/N, f = F/N$ and $g = G/N$, for which

$$dh = Tds - mdH, \tag{13.31}$$

$$df = -sdT + Hdm, \tag{13.32}$$

$$dg = -sdT - mdH. \tag{13.33}$$

We also highlight the isothermal molar susceptibility $\chi = X_T/N$, given by

$$\chi = \left(\frac{\partial m}{\partial H}\right)_T \tag{13.34}$$

and the molar heat capacities $c_m = C_M/N$ and $c_H = C_H/N$ at constant magnetization and field, respectively, given by

$$c_m = T\left(\frac{\partial s}{\partial T}\right)_m = \left(\frac{\partial u}{\partial T}\right)_m \tag{13.35}$$

and

$$c_H = T\left(\frac{\partial s}{\partial T}\right)_H = \left(\frac{\partial h}{\partial T}\right)_H \tag{13.36}$$

13.3 Ideal Paramagnetic Systems

Equation of State

In analogy with the properties of an ideal gas we postulate that for an ideal paramagnetic system, the molar energy $u(T)$ is a function of temperature only. Considering that $1/T = (\partial s/\partial u)$ and $H/T = -(\partial s/\partial m)$ we see that

$$\frac{\partial}{\partial m}\left(\frac{1}{T}\right) = -\frac{\partial}{\partial u}\left(\frac{H}{T}\right). \tag{13.37}$$

The left hand side of this equation vanishes identically because, inverting $u(T)$, we see that T is a function of u but not of m. As the right hand side must also be identically zero, we conclude that H/T is independent of u, depending only on m. We reached then the following result due to Langevin: the magnetization m of a paramagnetic ideal system depends on temperature T and field H through the combination H/T.

The result of Langevin leads directly to the Curie law. It suffices to recall that for a paramagnetic system, the magnetization m is proportional to the field H for small values of the field, that is, $m = \chi_0 H$, where χ_0 depends on T. The dependence χ_0 with temperature must be such that m is a function of H/T. Therefore, it is necessary that $\chi_0 = C/T$ so that

$$m = C\frac{H}{T}. \tag{13.38}$$

Equation (13.38) defines the equation of state $m(H,T)$ of an ideal paramagnetic system, which is valid in the regime of small values of H. An equation of state valid for any values of H is the one introduced by Langevin

$$m = \mu\left(\coth\frac{\mu H}{RT} - \frac{RT}{\mu H}\right), \tag{13.39}$$

where μ is the saturation magnetization and R is the universal gas constant. The Langevin equation of state, however, is based on a classical description of microscopic magnetic dipoles. A more appropriate description, which takes into account the quantum aspects, is that given by the equation of state used by Brillouin.

Brillouin Theory

From the microscopic point of view an ideal paramagnetic solid comprises a set of non-interacting magnetic dipoles located at the magnetic atoms. As the origin of the microscopic magnetic dipole moment is the angular momentum of the electrons and that it is quantized, we must presume that the same occurs with the dipole moment of magnetic atoms. Assuming that the component of the moment of each dipole in the direction of the applied field H takes discrete values and equally spaced, then in accordance with Brillouin, the magnetization m is given by

$$m = \mu \left(\frac{2J+1}{2J} \coth \frac{2J+1}{2J} \frac{\mu H}{RT} - \frac{1}{2J} \coth \frac{1}{2J} \frac{\mu H}{RT} \right). \tag{13.40}$$

The number of values that the component of the magnetic dipole can take is $2J + 1$ and J can take one of the following values: $1/2, 1, 3/2, 2, 5/2, \ldots$. For small values of the field, the magnetization behaves linearly with the field in accordance with

$$m = \frac{J+1}{3J} \frac{\mu^2 H}{RT}. \tag{13.41}$$

Therefore the molar magnetic susceptibility at zero field behaves according to the Curie law $\chi_0 = C/T$ where

$$C = \frac{J+1}{3J} \frac{\mu^2}{R}. \tag{13.42}$$

It is usual to present the saturation magnetization μ in terms of the Bohr magneton μ_{B} as follows $\mu = b\mu_{\mathrm{B}}$. It is also customary to present the Curie constant as follows

$$C = \frac{p^2 \mu_{\mathrm{B}}^2}{3R}. \tag{13.43}$$

Therefore, the saturation magnetization and Curie constant can be equivalently be given by the dimensionless quantities b and p. If the magnetization of a paramagnetic substance follows the Brillouin equation of state then these two quantities are related by $p = b\sqrt{(J+1)/J}$.

Table 13.1 shows the experimental data of C obtained by the Curie law for various paramagnetic substances and the respective values of dimensionless

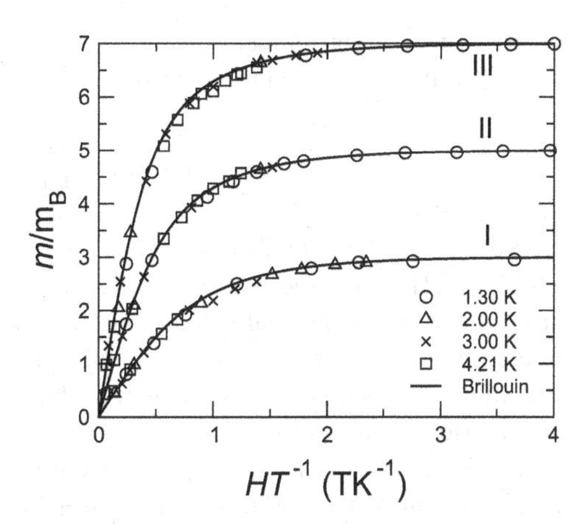

Fig. 13.3 Magnetization m versus H/T obtained experimentally by Henry [145] for three paramagnetic salts: (I) $CrK(SO_4)_2{\cdot}12H_2O$, (II) $Fe(NH_4)(SO_4)_2{\cdot}12H_2O$ and (III) $Gd_2(SO_4)_3{\cdot}8H_2O$

quantity p. Figure 13.3 shows the magnetization of three paramagnetic salts as functions of H/T. They are consistent with (13.40) corresponding to $J = 3/2$ for chromium salt, $J = 5/2$ for iron salt and $J = 7/2$ for the gadolinium salt.

Case J = 1/2

Let us consider here the simplest case of an ideal paramagnetic system, which is the one corresponding to $J = 1/2$. In this case the expression (13.40) is reduced to

$$m = \mu \tanh \frac{\mu H}{RT}. \tag{13.44}$$

For small fields the magnetization behaves linearly with the field according to $m = \mu^2 H/RT$ and the susceptibility at zero field is in agreement with the Curie Law $\chi_0 = C/T$ with $C = \mu^2/R$. From the magnetization we get by differentiation the molar susceptibility $\chi = (\partial m/\partial H)_T$,

$$\chi = \frac{\mu^2}{RT}\left(\operatorname{sech}\frac{\mu H}{RT}\right)^2, \tag{13.45}$$

and the molar magnetocaloric coefficient $\alpha = (\partial m/\partial T)_H$,

$$\alpha = -\frac{\mu^2 H}{RT^2}\left(\operatorname{sech}\frac{\mu H}{RT}\right)^2. \tag{13.46}$$

Integrating $m = -(\partial g/\partial H)_T$ in H, we get the molar Gibbs free energy

$$g = g_0 - RT \ln(2\cosh \frac{\mu H}{RT}), \tag{13.47}$$

where g_0 depends on temperature only. The derivative of this expression with respect to temperature gives the molar entropy

$$s = s_0 + R\ln(2\cosh\frac{\mu H}{RT}) - \frac{\mu H}{RT}\tanh\frac{\mu H}{RT}, \tag{13.48}$$

where $s_0 = -dg_0/dT$. To determine s_0 we notice on one hand that at zero field, $s = s_0 + R\ln 2$. On the other hand, at zero field the entropy must be the same for any temperature and its value for one mole of dipoles, according to the microscopic interpretation of entropy, must be equal to $R\ln 2$ since each dipole moment takes two values. From this result we conclude that $s_0 = 0$ and that g_0 is a constant which we choose as being zero.

From the above results for g and s we note that the molar enthalpy $h = g + Ts$ is $h = -Hm$ which replaced in $u = h + Hm$ gives $u = 0$. This result leads us to conclude that the molar heat capacity at constant magnetization $c_m = (\partial u/\partial T)_m$ vanishes identically. The molar heat capacity at constant field $c_H = (\partial h/\partial T)_m$, however, is nonzero and is given by

$$c_H = R\left(\frac{\mu H}{RT}\text{sech}\frac{\mu H}{RT}\right)^2. \tag{13.49}$$

Inverting expression (13.44), we get H as a function of m,

$$H = \frac{RT}{2\mu}\ln\frac{\mu + m}{\mu - m}, \tag{13.50}$$

from which we reach the molar Helmholtz free energy f by integration of $H = (\partial f/\partial m)_T$. Performing the integration, we reach the result

$$f = RT\left(\frac{\mu + m}{2\mu}\ln\frac{\mu + m}{2\mu} + \frac{\mu - m}{2\mu}\ln\frac{\mu - m}{2\mu}\right). \tag{13.51}$$

where the constant of integration was fitted so that f calculated at $m = 0$ is equal to g calculated at $H = 0$ since these two quantities are related by $g = f - Hm$.

The molar entropy is obtained from $s = -(\partial f/\partial T)_m$, which gives

$$s = -R\left(\frac{\mu + m}{2\mu}\ln\frac{\mu + m}{2\mu} + \frac{\mu - m}{2\mu}\ln\frac{\mu - m}{2\mu}\right). \tag{13.52}$$

From these two last results we see that $f = -Ts$ from which follows $u = 0$, as seen before.

13.4 Weiss Theory

Spontaneous Magnetization

To explain the emergence of spontaneous magnetization in ferromagnetic materials, Weiss imagined that each microscopic dipole magnetic material is subjected to the action of the field produced by neighboring dipoles. Weiss assumed that this field should be proportional to the magnetization m so that the total field on a dipole would equal $H + \lambda m$, where the parameter $\lambda > 0$. Thus, even in the absence of applied field H, the dipole could be subjected to a local field, called by Weiss the molecular field, which would orient the dipole.

If in the equation of state of an ideal paramagnetic system we replace H by $H + \lambda m$, we reach the equation of state of a system that undergoes a ferromagnetic-paramagnetic transition. Using (13.44), we get

$$m = \mu \tanh(\frac{\mu}{RT}(H + \lambda m)). \tag{13.53}$$

For zero applied field, $H = 0$, this equation becomes

$$m = \mu \tanh(\frac{\mu\lambda}{RT} m). \tag{13.54}$$

To see that this equation has a nonzero solution, corresponding to a spontaneous magnetization, let us resort to a graphic method. Using an auxiliary variable $x = \lambda\mu m/RT$, equation (13.54) can be written in the form

$$\frac{RT}{\lambda\mu^2} x = \tanh x. \tag{13.55}$$

The function $\tanh x$ has the aspect shown in Fig. 13.4a. On the graph we draw the straight line $(RT/\lambda\mu^2)x$. One solution is $x = 0$ and exists for any value of temperature. It is the sole solution if the slope of the straight line is greater or equal to the slope of $\tanh x$ at $x = 0$, that is, as long as $RT/\lambda\mu^2 \geq 1$. This condition is equivalent to $T \geq T_c$, where T_c is given by

$$T_c = \frac{\lambda\mu^2}{R}. \tag{13.56}$$

For $T < T_c$, there are two nontrivial solutions corresponding to the interception of the straight line with the curve, as seen in Fig. 13.4a. The positive solutions are presented in Fig. 13.4b for each value of temperature.

The value of the constant λ can be determine from (13.56) as long as the saturation magnetization μ is known. It is more interesting however to determine the value of the Weiss molecular field $H_W = \lambda\mu = RT_c/\mu$. To estimate H_W we assume that μ is of the order of magnitude of the Bohr magneton μ_B. Therefore

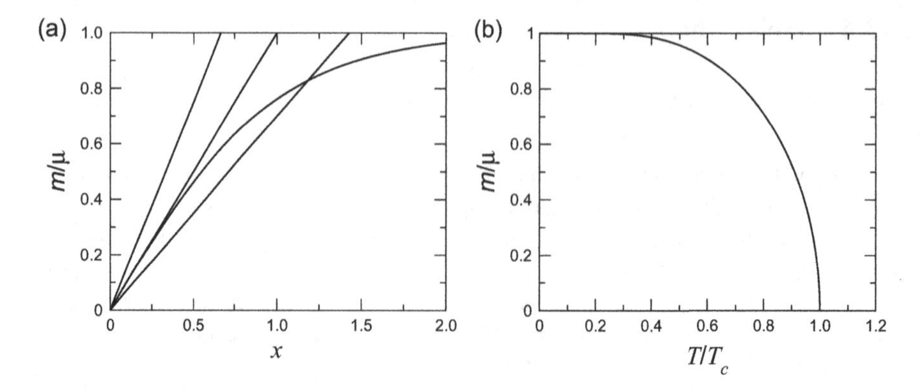

Fig. 13.4 (**a**) Graphic solution of (13.55). The *curve* represents the function tanh x. The *straight lines* correspond to the cases in which the temperature is greater, equal to or smaller than the critical temperature T_c. (**b**) Magnetization corresponding to the graphic solution of (**a**) as a function of temperature

$H_W \approx RT_c/\mu_B$ and using the value of the critical temperature of iron, we get the estimate $H_W \approx 10^3$ T.

Let us compare now H_W with the field H_d created by a magnetic dipole of the order of the Bohr magneton at a point situated at a distance a corresponding to the spacing between magnetic atoms in a solid. Such a field is $H_d = 10^{-7}\mu_B^2/N_A^2 a^3$. Using the value for a equal to $2 \cdot 10^{-10}$ m, we obtain the estimate $H_d \approx 10^{-1}$ T. Thus we see that H_d is much smaller than H_W, which means that in a ferromagnetic material the interaction between the magnetic dipoles cannot be of magnetic origin, as we should expect. We conclude that ferromagnetism is due to a nonmagnetic interaction between atoms. In fact, this interaction has quantum origin and is based on the Pauli exclusion principle.

Free Energy

Inverting (13.53), we get

$$H = -\lambda m + \frac{RT}{2\mu} \ln \frac{\mu + m}{\mu - m}, \tag{13.57}$$

which must be understood as an equation of state in the representation of the Helmholtz free energy $f(T, m)$. The graph of H versus m, for several temperatures is presented in Fig. 13.5. The spontaneous magnetization m^* obeys (13.57) for $H = 0$, that is,

$$\lambda m^* = \frac{RT}{2\mu} \ln \frac{\mu + m^*}{\mu - m^*}, \tag{13.58}$$

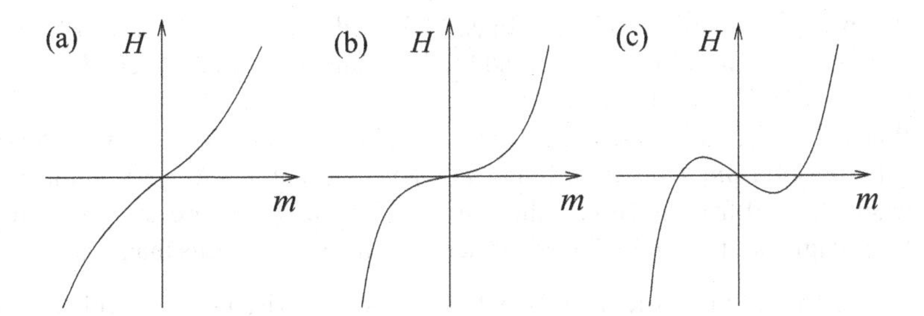

Fig. 13.5 External field H versus magnetization for a ferromagnetic system according to the Weiss theory for a temperature above the critical temperature (**a**), equal to the critical temperature (**b**), and below the critical temperature (**c**)

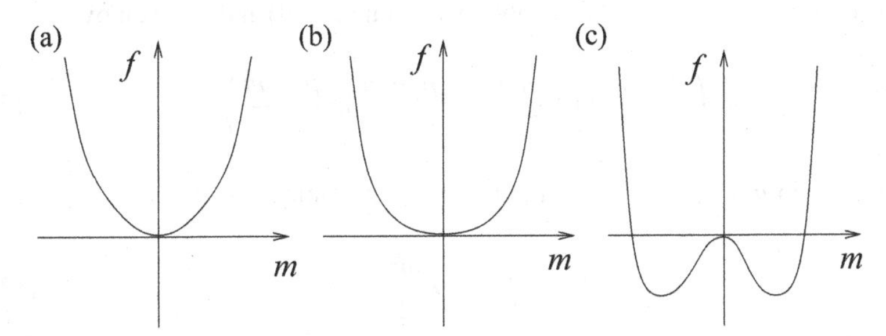

Fig. 13.6 Molar magnetic Helmholtz free energy f versus magnetization for a ferromagnetic system according the Curie theory. For a temperature above the critical temperature (**a**), equal to the critical temperature (**b**), and below the critical temperature (**c**)

which is equivalent to (13.54). The condition that gives the critical temperature T_c is obtained when the nonzero solution $m^* \to 0$. Dividing both sides by m^* and taking the limit, we get $\lambda = RT_c/\mu^2$, which is equivalent to (13.56).

Using $H = (\partial f/\partial m)_T$ and integrating (13.57), we get

$$f = -\lambda\frac{m^2}{2} + RT\left(\frac{\mu+m}{2\mu}\ln\frac{\mu+m}{2\mu} + \frac{\mu-m}{2\mu}\ln\frac{\mu-m}{2\mu}\right). \tag{13.59}$$

where we have chosen the integration constant in such a way that $f(T, m)$ reduces to the expression (13.51) when $\lambda = 0$.

Figure 13.6 shows f as a function of magnetization m for several temperatures. For temperatures below the critical temperature, given by (13.56), $f(m)$ ceases to have the convexity properties and in these cases we need to consider the convex hull. The convex hull is obtained by the construction of the double tangent which in the present case corresponds to a horizontal line segment. The double tangent connects the two minima of f that occur at $m = \pm m^*$, where m^* is the positive root of $H(m^*) = 0$. Thus we distinguish three regimes:

(a) When $T > T_c$, corresponding to Figs. 13.5a and 13.6a, (13.57) and (13.59) are valid for any values of m. Since m vanishes linearly with H when $H \to 0$ then we are facing a paramagnetic state.

(b) When $T = T_c$, corresponding to Figs. 13.5b and 13.6b, (13.57) and (13.59) are also valid for any values of m. In addition m vanishes with H when $H \to 0$, but not linearly. We call this state critical.
(c) When $T < T_c$, corresponding to Figs. 13.5c and 13.6c, (13.57) and (13.59) are valid only for $|m| \geq m^*$. For $|m| < m^*, f$ is constant and $H = 0$. In this case, when $H \to 0$ from positive values $m \to m^* > 0$. Therefore, we are facing a ferromagnetic state and m^* is identified with the spontaneous magnetization.

The diagram H versus temperature T displays a coexistence line which occurs along $H = 0$ and for $T < T_c$. On the line two thermodynamic phases coexist: one corresponding to the magnetization $m = m^*$ and the other corresponding to the magnetization $m = -m^*$.

The entropy $s = -(\partial f/\partial T)_m$ is obtained from (13.59) and is given by

$$s = -R\left(\frac{\mu+m}{2\mu}\ln\frac{\mu+m}{2\mu} + \frac{\mu-m}{2\mu}\ln\frac{\mu-m}{2\mu}\right). \tag{13.60}$$

Replacing s in $u = f + Ts$ and using (13.59), we obtain

$$u = -\lambda\frac{m^2}{2}. \tag{13.61}$$

Susceptibility and Heat Capacity

The susceptibility χ is obtained from its inverse, given by $1/\chi = (\partial H/\partial m)_T$. Deriving the expression (13.57) with respect to m, we get

$$\frac{1}{\chi} = -\lambda + \frac{RT}{\mu^2 - m^2}. \tag{13.62}$$

Along $H = 0$ and for $T > T_c$, the magnetization vanishes yielding the result

$$\frac{1}{\chi} = -\lambda + \frac{RT}{\mu^2}. \tag{13.63}$$

Using the result (13.56), $T_c = \lambda\mu^2/R$, we get

$$\chi = \frac{\mu^2}{R(T - T_c)}, \tag{13.64}$$

which is the Curie-Weiss law (13.3) with $\Theta = T_c$ and $C = \mu^2/R$.

Below the critical temperature and at zero field, the two phases are in coexistence and we should consider a susceptibility of each one of them. However, both phase have the same susceptibility. The susceptibility as a function of temperature is

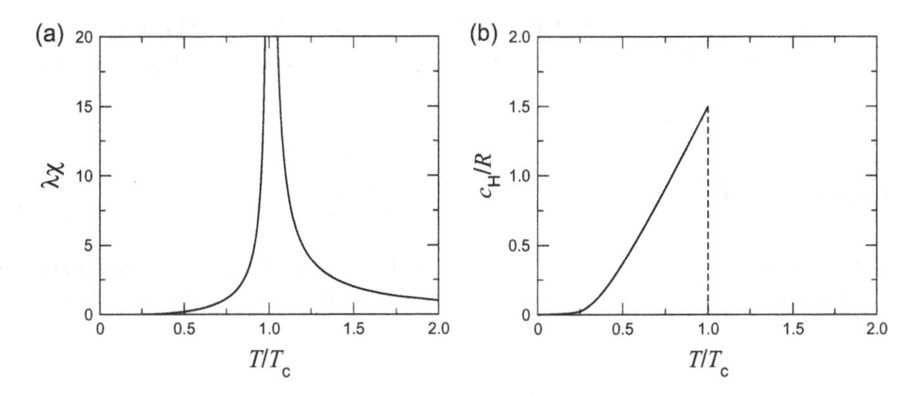

Fig. 13.7 Behavior of a ferromagnetic system at zero field according to the Weiss theory. (**a**) Molar susceptibility χ versus temperature T. (**b**) Molar heat capacity c_H versus temperature

obtained by the replacement of the nonzero solution m^* of (13.58) in (13.62). The result is shown in Fig. 13.7a.

To determine the molar heat capacity at constant field $c_H = (\partial h/\partial T)_H$, we need the molar enthalpy $h = u - Hm$, which is

$$h = -\lambda\frac{m^2}{2} - Hm. \tag{13.65}$$

At zero field and for $T > T_c$, the magnetization vanishes, so that $h = 0$ and $c_H = 0$. At zero field and for $T < T_c$, we should take into account that the phases are found in coexistence. However, both phase have the same value of h, given by $h = -\lambda(m^*)^2/2$, so that

$$c_H = -\frac{\lambda}{2}\frac{d}{dT}(m^*)^2. \tag{13.66}$$

Figure 13.7b shows the graph of c_H versus T.

Next we determine the molar heat capacity at constant magnetization, $c_m = (\partial u/\partial T)_m$, along $H = 0$. For $T \geq T_c$ the magnetization vanishes so that the result (13.61) gives $u = 0$, from which we conclude that $c_m = 0$. For $T < T_c$, we should consider the molar heat capacity of each one of the coexisting phases. However, they have the same molar energy given by $u = -\lambda(m^*)^2/2$, which becomes equal to the molar enthalpy h. Therefore, $c_m = c_H$ along $H = 0$.

The equality between the molar heat capacities along $H = 0$ can be obtained directly from the following relation between them

$$c_H = c_m + T\left(\frac{\partial H}{\partial T}\right)_m^2 \chi. \tag{13.67}$$

It suffices to take into account that $(\partial H/\partial T)_m = 0$ along $H = 0$. This behavior is distinct from that related to fluids for which c_p and c_v are different along the critical density.

Critical Point

Near the critical point the magnetization is small and therefore it can be determined by an expansion of the right hand side of (13.57) in powers of m. Up to cubic terms in m, we get

$$H = (\frac{T}{T_c} - 1)\lambda m + \frac{\lambda}{3\mu^2} m^3. \tag{13.68}$$

The nonzero root of $H(m) = 0$ is given by

$$(m^*)^2 = 3\mu^2(1 - \frac{T}{T_c}), \tag{13.69}$$

from which we get the positive root

$$m^* = \mu\sqrt{3(1 - \frac{T}{T_c})}. \tag{13.70}$$

To determine the behavior of χ just below the critical temperature, we use result (13.69) in (13.62) to obtain

$$\chi = \frac{\mu^2}{2R(T_c - T)}. \tag{13.71}$$

To determined the behavior of c_H just below the critical temperature, we replace result (13.69) in (13.66) to conclude that $c_H = 3R/2$ what means that c_H has a jump in T_c as seen in Fig. 13.7b. To obtain the linear behavior seen in the same figure we should use the expression of $(m^*)^2$ up to quadratic terms in $(T_c - T)$. Using such an expression we reach the result

$$c_H = \frac{3}{2}R + \frac{12}{5}R(\frac{T}{T_c} - 1). \tag{13.72}$$

Anisotropy

The results obtained so far are valid for isotropic magnetic systems and, therefore, such that the spontaneous magnetization can appear in any direction. The magnetic system, however, can be anisotropic so that the spontaneous magnetization may appear only in certain crystalline directions. In systems with uniaxial anisotropy, for example, it appears along the anisotropic axis. If the magnetic field is applied along this preferential direction, the free energy (13.59) and other results obtained above remain valid. But, if the field is applies in another direction, the free energy is no longer appropriated. In these cases, we adopt the following molar Helmholtz free energy

$$f = -\frac{\lambda\gamma}{2}(m_x^2 + m_y^2) - \frac{\lambda}{2}m_z^2 +$$

$$+ RT\left(\frac{\mu+m}{2\mu}\ln\frac{\mu+m}{2\mu} + \frac{\mu-m}{2\mu}\ln\frac{\mu-m}{2\mu}\right), \tag{13.73}$$

where m_x, m_y and m_z are the Cartesian components of the magnetization and $m = (m_x^2 + m_y^2 + m_z^2)^{1/2}$. This free energy describes a ferromagnetic system with uniaxial anisotropy, whose preferential axis extends along the z direction. The parameter γ is a measure of the anisotropy with respect to the preferential axis z. The value $\gamma = 1$ corresponds to the isotropic case and $\gamma = 0$ to the extreme anisotropy. We consider $\gamma < 1$, so that the spontaneous magnetization will occur along the preferential direction z.

In the present case the following relation is valid

$$df = -sdT + H_x dm_x + H_y dm_y + H_z dm_z, \tag{13.74}$$

where H_x, H_y and H_z are the components of the external field. From f we found the equations of state

$$H_x = -\lambda\gamma m_x + \frac{RT}{2\mu}\frac{m_x}{m}\ln(\frac{\mu+m}{\mu-m}), \tag{13.75}$$

$$H_y = -\lambda\gamma m_y + \frac{RT}{2\mu}\frac{m_y}{m}\ln(\frac{\mu+m}{\mu-m}) \tag{13.76}$$

$$H_z = -\lambda m_z + \frac{RT}{2\mu}\frac{m_z}{m}\ln(\frac{\mu+m}{\mu-m}) \tag{13.77}$$

If the external field is applies along the preferential direction z, that is, if $H_x = H_y = 0$, then $m_x = m_y = 0$ and equation (13.77) reduces to equation (13.57). The results obtained previously become valid for this case and, in particular, the susceptibility obtained previously becomes what we call parallel susceptibility.

Next we examine the case where the field is applied in a direction perpendicular to the preferential axis z, that is, such that $H_y = H_z = 0$ and $H_x \neq 0$. We wish to determine the perpendicular susceptibility $\chi_\perp = \partial m_x/\partial H_x$ at zero field. To this end we examine (13.75) for small values of H_x. In this case, m_x will also be small and will vanish when $H_x \to 0$ even in the ferromagnetic phase since the spontaneous magnetization emerges only in direction z. For small values of m (13.75) reduces to

$$H_x = -\lambda\gamma m_x + \frac{RT}{2\mu}\frac{m_x}{m^*}\ln(\frac{\mu + m^*}{\mu - m^*}), \tag{13.78}$$

in which we set $m_z = m^*$.

Above the critical temperature $m^* = 0$ and we get

$$H_x = -\lambda\gamma m_x + \frac{T}{T_c}\lambda m_x, \tag{13.79}$$

where we used the result (13.56). Below the critical temperature, we use the result (13.58) to obtain

$$H_x = -\lambda\gamma m_x + \lambda m_x, \tag{13.80}$$

From these equations we reach the expressions for the perpendicular susceptibility at zero field

$$\chi_\perp = \frac{T_c}{\lambda(T - \gamma T_c)}, \tag{13.81}$$

valid above the critical temperature and

$$\chi_\perp = \frac{1}{\lambda(1 - \gamma)}, \tag{13.82}$$

valid below the critical temperature. Therefore, the perpendicular susceptibility is finite and constant below the critical temperature.

13.5 Criticality

Critical Exponents

In the neighborhood of the critical point, the thermodynamic quantities have a singular behavior. The molar susceptibility χ has a divergent behavior and the spontaneous magnetization m^* vanishes $T \to T_c$, at zero field. The results of the previous section, obtained from the Weiss theory, describe the ferromagnetic-paramagnetic transition and in particular predict the occurrence of these singularities. The results predicted by the Weiss theory however show deviations from the experimental results as seen in Fig. 13.3a where χ^{-1} deviates from the linear behavior predicted by (13.64). The experimental data show that only at high temperature this linear behavior occurs, what constitutes the Curie-Weiss law.

Table 13.3 Critical exponents and critical temperature, in K, of ferromagnetic materials obtained experimentally

Ferromagnets	α	β	γ	δ	T_c
Fe	−0.12	0.39	1.34	4.2	1043
Ni	−0.10	0.38	1.34	4.6	631
Co		0.42	1.23		1395
$CrBr_3$		0.37	1.22	4.3	32.7
EuO	−0.04	0.37	1.40	4.5	69.4
EuS	−0.13	0.36	1.39		16.4
$Cu(NH_4)_2Cl_4{\cdot}2H_2O$			1.37		0.701
$CuK_2Cl_4{\cdot}2H_2O$			1.36		0.877
$Cu(NH_4)_2Br_4{\cdot}2H_2O$		0.38	1.33	4.3	1.83
$CuRb_2Br_4{\cdot}2H_2O$		0.37			1.87
$CuCs_2Cl_4{\cdot}2H_2O$			1.33		

To describe appropriately the behavior near the critical point we assume that the spontaneous magnetization vanishes according to

$$m^* \sim (T_c - T)^\beta, \tag{13.83}$$

the susceptibility at zero field, $H = 0$, diverges according to

$$\chi \sim |T - T_c|^{-\gamma}, \tag{13.84}$$

and along the critical isotherm, $T = T_c$, the magnetization and the field are related by

$$H \sim m^\delta, \tag{13.85}$$

where β, γ and δ are certain critical exponents. The results (13.70), (13.64), (13.71) and (13.68), obtained from the Weiss theory, predicts the following values for the critical exponents: $\beta = 1/2$, $\gamma = 1$ and $\delta = 3$. The experimental data on the other hand give distinct values for these exponents, as can be seen in Table 13.3 for several magnetic materials.

Experimentally, one also observes that the molar heat capacity c_H at zero field has singular behavior. This quantity may diverge according to

$$c_H \sim |T - T_c|^{-\alpha}, \tag{13.86}$$

where α is positive. It is possible that c_H does not diverge although has a singular behavior. In this case, to characterize appropriately the singularity we assume the following behavior

$$c_c - c_H \sim |T - T_c|^{-\alpha}, \tag{13.87}$$

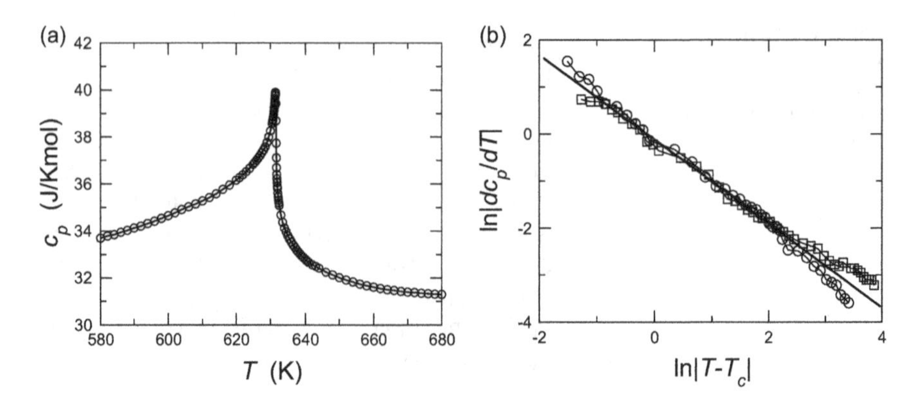

Fig. 13.8 (**a**) Molar isobaric heat capacity c_p of nickel obtained experimentally by Connelly et al. [142], as a function of temperature T. (**b**) Log-log plot of dc_p/dT versus T. The slope of the straight line fitted to the experimental data gives -0.90 which yields an exponent $\alpha = -0.10$

with α negative, where c_c is the finite value of c_H at the critical point. Figure 13.8 shows the molar heat capacity of nickel whose exponent α is negative. To determine the exponent, it is convenient to use the quantity dc_H/dT to eliminate the constant c_c. According to (13.87) this quantity behaves as $|T - T_c|^{-1-\alpha}$.

A singularity characterized by an exponent $\alpha = 0$ must be described by a more specific form. It can be a divergence of the logarithm type

$$c_H \sim \ln |T - T_c|, \tag{13.88}$$

but can also corresponds to a jump in c_H, as predicted by the Weiss theory and shown in Fig. 13.7b.

The critical exponents α, β, γ and δ are not independent but keep certain relations such as the Rushbrooke relation,

$$\alpha + 2\beta + \gamma = 2, \tag{13.89}$$

the Griffiths relation,

$$\alpha + \beta(\delta + 1) = 2, \tag{13.90}$$

and the Widom relation,

$$\gamma = \beta(\delta - 1). \tag{13.91}$$

The relations between the exponents can be demonstrated from the Widom scaling theory, as we will see below.

Singular Part

To analyze the critical behavior, it is convenient to separate the molar Gibbs free energy $g(T, H)$ in a regular part $g_r(T, H)$ and a singular part $g_s(T, H)$ that contains the singularities associated to the critical point as well as the discontinuities along the line of coexistence. Thus, we write

$$g(T, H) = g_r(T, H) + g_s(T, H). \tag{13.92}$$

Any quantity obtained from g will also have a regular and a singular part. The singular part of the nondiverging quantities is chosen so that it vanishes at the critical point, what always can be done by the addition of appropriate terms in g_r.

The regular part of the molar Gibbs free energy has an expansion around the critical point whose first terms are

$$g_r = g_c - s_c(T - T_c) - \frac{c_c}{2T_c}(T - T_c)^2, \tag{13.93}$$

where g_c, s_c and c_c are constant. The last term is absent if the molar heat capacity is divergent.

It is convenient to use variables related to the critical point which occurs at $H = 0$ and $T = T_c$. We define thus $\varepsilon = T - T_c$ and write

$$g_s(T, H) = \mathcal{G}(\varepsilon, H). \tag{13.94}$$

Notice that $g_s(T, H)$ is an even function of H and the same happens to $\mathcal{G}(\varepsilon, H)$, that is,

$$\mathcal{G}(\varepsilon, H) = \mathcal{G}(\varepsilon, -H). \tag{13.95}$$

The magnetization m and the molar entropy s are given by

$$m = -\frac{\partial g}{\partial H} = -\mathcal{G}_2(\varepsilon, H) \tag{13.96}$$

and

$$s = -\frac{\partial g}{\partial T} = s_c + \frac{c_c}{T_c}\varepsilon - \mathcal{G}_1(\varepsilon, H), \tag{13.97}$$

where $\mathcal{G}_1 = \partial\mathcal{G}/\partial\varepsilon$ and $\mathcal{G}_2 = \partial\mathcal{G}/\partial H$.

Deriving m and s with respect to H and to T, respectively, we get

$$\chi = \left(\frac{\partial m}{\partial H}\right)_T = -\mathcal{G}_{22}(\varepsilon, H) \tag{13.98}$$

and

$$\frac{1}{T}c_H = \left(\frac{\partial s}{\partial T}\right)_H = \frac{c_c}{T_c} - \mathcal{G}_{11}(\varepsilon, H), \tag{13.99}$$

where $\mathcal{G}_{11} = \partial^2\mathcal{G}/\partial\varepsilon^2$ and $\mathcal{G}_{22} = \partial^2\mathcal{G}/\partial H^2$.

Along the coexistence line, $H = 0$ and $\varepsilon < 0$, the spontaneous magnetization m^* is obtained by taking the limit $H \to 0$ in (13.96),

$$m^* = -\mathcal{G}_2(\varepsilon, 0^+). \tag{13.100}$$

Along $\varepsilon = 0$

$$m = -\mathcal{G}_2(0, H). \tag{13.101}$$

Along $H = 0$, the susceptibility χ is given by

$$\chi = -\mathcal{G}_{22}(\varepsilon, 0) \tag{13.102}$$

and the molar heat capacity at constant field is given by

$$c_H = c_c - T_c\mathcal{G}_{11}(\varepsilon, 0). \tag{13.103}$$

Scaling Theory

According to the Widom scaling theory, the singular part of the Gibbs free energy around the critical point is a generalized homogeneous function, that is,

$$\mathcal{G}(\varepsilon, H) = \lambda^{-a}\mathcal{G}(\lambda\varepsilon, \lambda^b H), \tag{13.104}$$

valid for any positive values of λ. The exponent a and b are related to the critical exponents introduced above, as we will see below.

The homogeneity of $\mathcal{G}$ implies the homogeneity of its derivatives. Indeed, deriving (13.104) with respect to ε and to H, we get

$$\mathcal{G}_1(\varepsilon, H) = \lambda^{-a+1}\mathcal{G}_1(\lambda\varepsilon, \lambda^b H) \tag{13.105}$$

and

$$\mathcal{G}_2(\varepsilon, H) = \lambda^{-a+b}\mathcal{G}_2(\lambda\varepsilon, \lambda^b H). \tag{13.106}$$

Deriving these expressions with respect to ε and to H, respectively, we obtain in addition

$$\mathscr{G}_{11}(\varepsilon, H) = \lambda^{-a+2}\mathscr{G}_{11}(\lambda\varepsilon, \lambda^b H), \tag{13.107}$$

and

$$\mathscr{G}_{22}(\varepsilon, H) = \lambda^{-a+2b}\mathscr{G}_{22}(\lambda\varepsilon, \lambda^b H). \tag{13.108}$$

Next we determine the behavior of these quantities along $H = 0$. For $\varepsilon < 0$, we choose $\lambda = |\varepsilon|^{-1}$ to obtain

$$\mathscr{G}_1(\varepsilon, 0) = |\varepsilon|^{a-1}\mathscr{G}_1(-1, 0), \tag{13.109}$$

$$\mathscr{G}_2(\varepsilon, 0^+) = |\varepsilon|^{a-b}\mathscr{G}_2(-1, 0^+), \tag{13.110}$$

$$\mathscr{G}_{11}(\varepsilon, 0) = |\varepsilon|^{a-2}\mathscr{G}_{11}(-1, 0), \tag{13.111}$$

$$\mathscr{G}_{22}(\varepsilon, 0) = |\varepsilon|^{a-2b}\mathscr{G}_{22}(-1, 0). \tag{13.112}$$

For $\varepsilon > 0$ and $H = 0$, we choose $\lambda = \varepsilon^{-1}$ to obtain

$$\mathscr{G}_{11}(\varepsilon, 0) = \varepsilon^{a-2}\mathscr{G}_{11}(1, 0), \tag{13.113}$$

$$\mathscr{G}_{22}(\varepsilon, 0) = \varepsilon^{a-2b}\mathscr{G}_{22}(1, 0). \tag{13.114}$$

Replacing these results into the expressions (13.100), (13.102) and (13.103), we get

$$m^* \sim |\varepsilon|^{a-b}, \tag{13.115}$$

$$\chi \sim |\varepsilon|^{a-2b}, \tag{13.116}$$

$$c_H \sim |\varepsilon|^{a-2}, \tag{13.117}$$

or

$$c_c - c_H \sim |\varepsilon|^{a-2}, \tag{13.118}$$

if c_H is finite. We see therefore that the Widom scaling theory predicts power laws for the critical behavior of these quantities.

Comparing these expressions with (13.83), (13.84), (13.86) and (13.87) we find the following relations between the critical exponents and the indices a and b:

$$\alpha = -a + 2, \tag{13.119}$$

$$\beta = a - b, \tag{13.120}$$

$$\gamma = -a + 2b. \tag{13.121}$$

For $\varepsilon = 0$ and $H > 0$, we choose $\lambda = H^{-1/b}$ to obtain

$$\mathscr{G}_2(0, H) = H^{(a-b)/b}\mathscr{G}_2(0, 1), \tag{13.122}$$

which, replaced in (13.101), gives

$$m \sim H^{(a-b)/b}. \tag{13.123}$$

Comparing with expression (13.85), we get the following relation between the exponent δ and the indices a and b:

$$\frac{1}{\delta} = \frac{a-b}{b}. \tag{13.124}$$

The four exponents α, β, γ and δ are not independent. As they are related to the indices a and b, then only two of them can be chosen as independent. Eliminating a and b, we find several relations that include the Rushbrooke relation (13.89), the Griffiths relation (13.90) and the Widom relation (13.91).

Universal Function

The homogeneity of the singular part of the free energy and its derivatives implies, as we have seen, certain relations between the critical exponents. It also give rise to certain geometric relations between the thermodynamic quantities, which we examine now. Let us consider initially the magnetization $m = -\mathscr{G}_2(\varepsilon, H)$. If, in the regime $\varepsilon > 0$, we choose $\lambda = \varepsilon^{-1}$ and substitute it in (13.106) we get the result

$$\mathscr{G}_2(\varepsilon, H) = \varepsilon^{\beta}\mathscr{G}_2(1, H\varepsilon^{-\Delta}). \tag{13.125}$$

where $\Delta = \beta + \gamma$. That is

$$m = \varepsilon^{\beta}\phi(H\varepsilon^{-\Delta}), \tag{13.126}$$

where $\phi(x) = -\mathscr{G}(+1, x)$ is a function of one variable. Similarly we can obtain the relation valid in the regime $\varepsilon < 0$.

The result above tell us that the isotherms m versus H must coincide if we perform a change of scale such that the axis-m is multiplied by $|\varepsilon|^{-\beta}$ and the axis-H by $|\varepsilon|^{-\Delta}$. The experimental data of magnetization m as a function of the field H and of the temperature T can then be analyzed in the following way. In a diagram x-y we place $y = m|\varepsilon|^{-\beta}$ versus $x = H|\varepsilon|^{-\Delta}$. The data will collapse in a curve $y = \phi(x)$ which for that reason is called universal function. Figure 13.9 shows such collapse obtained from the experimental data presented in Fig. 13.2 for the isotherms of nickel.

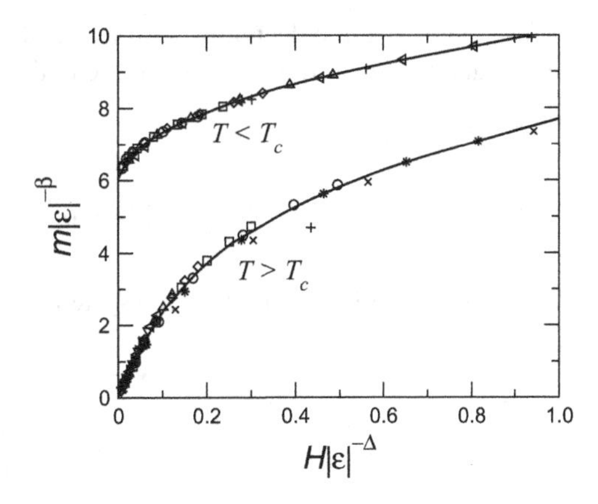

Fig. 13.9 Graph of $m|\varepsilon|^{-\beta}$ versus $H|\varepsilon|^{-\Delta}$ for nickel obtained from the data of Fig. 13.2. The *upper* and *lower branches* correspond, respectively, to the points below and above the critical temperature

Similar results for the susceptibility χ and for the molar heat capacity c_H can be obtained from (13.107) and (13.108). For $\varepsilon > 0$ we find

$$\chi = \varepsilon^{-\gamma}\psi(H\varepsilon^{-\Delta}) \tag{13.127}$$

and

$$c_H = \varepsilon^{-\alpha} f(H\varepsilon^{-\Delta}), \tag{13.128}$$

or

$$c_c - c_H = \varepsilon^{-\alpha} f(H\varepsilon^{-\Delta}), \tag{13.129}$$

in case c_H is finite, where $\psi(x) = -\mathcal{G}_{22}(1,x)$ and $f(x) = -\mathcal{G}_{11}(1,x)$ are universal functions. Analogous results can be derived for $\varepsilon < 0$.

Problems

13.1 Obtain the Maxwell relations for a magnetic system:

$$\left(\frac{\partial T}{\partial M}\right)_S = \left(\frac{\partial H}{\partial S}\right)_M, \qquad \left(\frac{\partial T}{\partial H}\right)_S = -\left(\frac{\partial M}{\partial S}\right)_H,$$

$$\left(\frac{\partial S}{\partial M}\right)_T = -\left(\frac{\partial H}{\partial T}\right)_M, \qquad \left(\frac{\partial S}{\partial H}\right)_T = \left(\frac{\partial M}{\partial T}\right)_H.$$

13.2 Magnetocaloric effect. Show that $(\partial T/\partial H)_S = -TA/C_H$, where A is the magnetocaloric effect, defined by $A = (\partial M/\partial T)_H$. Since $C_H \geq 0$, materials whose magnetization decreases with temperature at constant field ($A < 0$) has the temperature decreased with the decrease in the field in a adiabatic process.

13.3 Show that

$$C_H = C_M + \frac{TA^2}{X_T}.$$

Verify this result for a paramagnetic system.

13.4 Shown that

$$C_H = C_M + T\left(\frac{\partial H}{\partial T}\right)_M^2 X_T.$$

13.5 Equation (13.40), for the magnetization of an ideal paramagnetic, can be written as

$$m = \mu \mathsf{B}(\frac{\mu H}{RT}),$$

where $\mathsf{B}(x)$ is the Brillouin function, defined by

$$\mathsf{B}(x) = \frac{2J+1}{2J}\coth(\frac{2J+1}{2J}x) - \frac{1}{2J}\coth(\frac{x}{2J}).$$

Show that (a) it is a monotonic increasing function and such that $\mathsf{B}(x) \to 1$ when $x \to \infty$; (b) for small values of the argument, it behaves according to $\mathsf{B}(x) = (J+1)x/3J$; (c) for $J = 1/2$ it reduces to $\tanh(x)$; and (d) for $J \to \infty$ it reduces to the Langevin equation $\mathsf{L}(x)$, defined by

$$\mathsf{L}(x) = \coth x - \frac{1}{x}.$$

13.6 Show that the critical temperature of a ferromagnetic system described by the equation of state

$$m = \mu \mathsf{B}(\frac{\mu}{RT}(H + \lambda m))$$

is given by

$$T_c = \frac{(J+1)\lambda\mu^2}{3JR}.$$

Determine the behavior of the spontaneous magnetization, susceptibility and heat capacity around the critical point.

13.7 Near the critical point, the molar Helmholtz free energy of a ferromagnetic system is described by

$$f = f_0 + \frac{1}{2}a(T - T_c)m^2 + \frac{1}{4}b\,m^4,$$

where a and b are constant and f_0 depends only on temperature. Determine the spontaneous magnetization and susceptibility at zero field as functions of temperature. Determine the critical exponents β, γ and δ.

13.8 Show that the spontaneous magnetization, given by (13.58), has the following expansion

$$\left(\frac{m^*}{\mu}\right)^2 = 3\left(1 - \frac{T}{T_c}\right) - \frac{12}{5}\left(1 - \frac{T}{T_c}\right)^2.$$

Use this expansion to determine the behavior of the molar heat capacity just below T_c.

13.9 Show that

$$m = H^{1/\delta}\Phi(\varepsilon H^{-1/\Delta}).$$

13.10 Show that (13.68) yields the following form for the universal function

$$y = \phi(x) = \pm x + \frac{1}{3}x^3.$$

where the signs $+$ and $-$ are valid for temperatures above and below the critical temperature, respectively. The variables x and y are defined by $x = |\varepsilon|^{-1/2}m/\mu$ and $y = |\varepsilon|^{-3/2}/(\lambda\mu)$ where $\varepsilon = (T - T_c)/T_c$.

Chapter 14
Magnetic Ordering

14.1 Antiferromagnetism

Antiferromagnetic Materials

The antiferromagnetic crystals undergo a transition to the paramagnetic state at a temperature called the Néel temperature (Table 14.1). Above it, in the paramagnetic state, the dipoles point in random directions due to the predominance of thermal agitation, producing a zero total magnetization. Below the Néel temperature in antiferromagnetic state, the total magnetization is also zero, but for another reason. In the antiferromagnetic state, the lattice formed by magnetic atoms is divided into two or more equivalent sublattices, each presenting a spontaneous magnetization in a certain direction. The interaction between the dipoles causes a magnetic ordering such that the vector sum of the spontaneous magnetization of the sublattice vanishes, which defines the antiferromagnetic state.

The simplest antiferromagnetic structure consists of a magnetic lattice that is divided into only two equivalent and interpenetrating sublattices. In this case the interaction between neighboring magnetic dipoles, which belong to different sublattices, favors an antiparallel alignment of them, resulting in the ordering of the magnetizations of the sublattices in opposite directions. The simple antiferromagnetism is represented by the fluorides FeF_2 and MnF_2, whose magnetic ions form a body-centered tetragonal lattice, and the compounds $KMnF_3$, $KNiF_3$ and $RbMnF_3$, which have the structure of perovskite and whose magnetic ions form a simple cubic lattice. Other compounds have more complex antiferromagnetic structure, consisting of various magnetic sublattices as is the case of the compounds MnO, FeO, CoO, NiO, whose magnetic lattices are face-centered cubic. In these compounds, the magnetization of the sublattices are all collinear.

A magnetic structure in which the magnetization of sublattices are not collinear is represented by the compound NiF_2. The magnetic ions of this compound form a body-centered tetragonal lattice that is divided into two equivalent sublattices.

M.J. de Oliveira, *Equilibrium Thermodynamics*, Graduate Texts in Physics,
DOI 10.1007/978-3-662-53207-2_14

Table 14.1 Néel temperature T_N, in K, and type of ordering of substances that become antiferromagnetic just below the the Néel temperature. Some have weak ferromagnetism (wf). *Source*: AIP

Material	T_N	Type
FeF_2	90	Simple
MnF_2	75	Simple
$KMnF_3$	95	Simple
$KNiF_3$	275	Simple
$RbMnF_3$	82	Simple
MnO	120	Complex
FeO	198	Complex
CoO	291	Complex
NiO	530	Complex
NiF_2	83	Simple, wf
Cr_2O_3	318	Alternating
α-Fe_2O_3	948	Alternating, wf
$MnCO_3$	32	Alternating, wf
$CoCO_3$	20	Alternating, wf
CrSb	723	Metamagnet
MnTe	323	Metamagnet
$CrCl_2$	20	Metamagnet
$FeCl_2$	24	Metamagnet
$CoCl_2$	25	Metamagnet
MnO_2	84	Helical
$FeCl_3$	15	Helical
$MnAu_2$	363	Helical
Cr	312	
Mn	95	

The magnetization of the sublattices are not exactly opposite, but form an angle giving rise to a nonzero total magnetization. As the deviation from the ideal antiferromagnet state is small this compound exhibits a weak ferromagnetism.

The oxide Cr_2O_3, the hematite α-Fe_2O_3 and the compounds $MnCO_3$ and $CoCO_3$ have a structure in which the magnetic dipoles alternate along a particular crystal direction. The last three are also examples of antiferromagnetic substances that have weak ferromagnetism. The metamagnets CrSb, MnTe, $CrCl_2$, $FeCl_2$ and $CoCl_2$ are constituted by a succession of ferromagnetic layers of alternating signs.

More complex magnetic structures include the periodic in which the magnetization varies, in magnitude or direction, along a crystal axis. The compounds MnO_2, $FeCl_3$ and MnAu are examples of substances that exhibit antiferromagnetic periodic helical structures.

Among the metals, chromium (Cr) is found in the antiferromagnetic state at room temperature, becoming a paramagnetic 39 °C. Manganese (Mn), which is paramagnetic at room temperature, becomes antiferromagnetic below 95 K.

Order Parameter and Free Energy

For a suitable description of the antiferromagnetic state, as well as other types of ordering, it is necessary to use an expanded thermodynamic space relatively to that of a ferromagnetic. The most natural way is to use a space composed by the magnetization of each magnetic sublattice and the temperature. To this end, let us examine the simplest case of a system that is ordered according to two equivalent and interpenetrating magnetic sublattices. Suppose further that the magnetizations of the sublattices occur along a certain crystal direction. Denoting by m_1 and m_2 the magnetizations corresponding to the sublattices 1 and 2, respectively, then the molar Helmholtz free energy $f(T, m_1, m_2)$ is a function of m_1 and m_2 in addition to temperature. Denoting by H_1 and H_2 the fields conjugated to m_1 and m_2, then

$$df = -sdT + H_1 dm_1 + H_2 dm_2. \tag{14.1}$$

The antiferromagnetic state means that the magnetization of the sublattices are different. They need not necessarily have the same magnitude. This happens only in restricted antiferromagnetic state, which occurs in the absence of external field, for which $m_1 = -m_2$ and the total magnetization

$$m = \frac{1}{2}(m_1 + m_2) \tag{14.2}$$

vanishes. In the presence of an external field it is possible the occurrence of a generic antiferromagnetic state for which $m \neq 0$.

Suppose that in an antiferromagnetic phase the values of the magnetization of the sublattices are $m_1 = a$ and $m_2 = b$ with $a \neq b$. Another antiferromagnetic phase equally possible is that for which $m_1 = b$ and $m_2 = a$, as the sublattices are equivalent. Therefore, an antiferromagnetic system displays two different thermodynamic states that should be understood as two thermodynamic phases in coexistence. We note that these two thermodynamic phases have the same total magnetization $m = (a + b)/2$ and therefore can not be distinguished by m. They are distinguished by the order parameter η defined by

$$\eta = \frac{1}{2}(m_1 - m_2), \tag{14.3}$$

that takes a positive value in one phase and a negative value in the other. The order parameter η makes also the distinction of the antiferromagnetic phase, ordered, from the paramagnetic phase, disordered. In the paramagnetic phase the sublattices cannot have different magnetizations, so that $m_1 = m_2$, that is, $\eta = 0$.

An alternative and more interesting description is the one in which the thermodynamic space is formed by the total magnetization m and the order parameter η in addition to the temperature. From (14.2) and (14.3) we get

$$m_1 = m + \eta \qquad \text{and} \qquad m_2 = m - \eta, \tag{14.4}$$

which replaced in (14.1) gives

$$df = -sdT + Hdm + H_* d\eta, \tag{14.5}$$

where

$$H = H_1 + H_2 \qquad \text{and} \qquad H_* = H_1 - H_2. \tag{14.6}$$

the quantity H is the variable conjugated to m and thus is identified with the external field. The quantity H_* is the field conjugated to η, which we call staggered field. Experimentally, this field is identically zero.

In certain applications, it is convenient to use the molar Gibbs free energy $g(T, H, \eta)$, obtained from $f(T, m, \eta)$ by the Legendre transformation

$$g(T, H, \eta) = \min_m \{f(T, m, \eta) - Hm\} \tag{14.7}$$

for which

$$dg = -sdT - mdH + H_* d\eta. \tag{14.8}$$

In the study of phase coexistence is particularly useful the potential $\phi(T, H, H_*)$ obtained by the Legendre transformation

$$\phi(T, H, H_*) = \min_{m,\eta} \{f(T, m, \eta) - Hm - H_*\eta\}. \tag{14.9}$$

As ϕ is a function only of thermodynamic fields then, at the coexistence of phases it must have the same value in each phase.

Néel Theory

The Néel theory corresponds to the application of molecular field theory of Weiss to antiferromagnetic systems. According to the Néel theory the magnetization of each of the sublattices of a simple antiferromagnetic system is given by

$$m_1 = \mu \tanh(\frac{\mu}{RT}(H - \lambda m_2)) \tag{14.10}$$

and

$$m_2 = \mu \tanh(\frac{\mu}{RT}(H - \lambda m_1)), \tag{14.11}$$

where $-\lambda m_2$ and $-\lambda m_1$ are the molecular fields of Weiss acting on the dipole moments corresponding to the sublattices 1 and 2, respectively, because the magnetic dipoles in a sublattice interact only with those belonging to the other sublattice. That is, the molecular field acting on the dipoles of the sublattice 1 is due to the dipoles of sublattice 2 and vice versa. The negative sign induces an antiparallel dipole alignment between different sublattices.

Equations (14.10) and (14.11) introduced by Néel can be derived from the following molar Helmholtz free energy

$$f = \frac{\lambda}{2} m_1 m_2 + \frac{RT}{2}\left(\frac{\mu + m_1}{2\mu} \ln \frac{\mu + m_1}{2\mu} + \frac{\mu - m_1}{2\mu} \ln \frac{\mu - m_1}{2\mu}\right) +$$

$$+ \frac{RT}{2}\left(\frac{\mu + m_2}{2\mu} \ln \frac{\mu + m_2}{2\mu} + \frac{\mu - m_2}{2\mu} \ln \frac{\mu - m_2}{2\mu}\right). \tag{14.12}$$

Indeed, from $H_1 = \partial f / \partial m_1$ and $H_2 = \partial f / \partial m_2$, we obtain

$$H_1 = \frac{1}{2}(\lambda m_2 + \frac{RT}{2\mu} \ln \frac{\mu + m_1}{\mu - m_1}) \tag{14.13}$$

and

$$H_2 = \frac{1}{2}(\lambda m_1 + \frac{RT}{2\mu} \ln \frac{\mu + m_2}{\mu - m_2}). \tag{14.14}$$

Taking into account that $H_1 = H_2 = H/2$, because $H_* = 0$, the results (14.10) and (14.11) follows directly from these equations.

To determine the magnetizations of the sublattices, we should solve (14.10) and (14.11) for m_1 and m_2. When $H = 0$, it is easy to see that the solution is such that $m_1 = -m_2$ and that the order parameter $\eta = m_1 = -m_2$ obeys the equation

$$\eta = \mu \tanh(\frac{\mu\lambda}{RT}\eta), \tag{14.15}$$

which is equivalent to the equation

$$\lambda\eta = \frac{RT}{2\mu} \ln \frac{\mu + \eta}{\mu - \eta}, \tag{14.16}$$

obtained from (14.13) and (14.14).

The solution of these equations is such that for temperatures above the critical temperature T_N, given by

$$T_N = \frac{\lambda \mu^2}{R}, \tag{14.17}$$

we have $\eta = 0$ and $m_1 = m_2 = 0$. Below this temperature, $\eta \neq 0$ and therefore $m_1 = -m_2 \neq 0$, that define the antiferromagnetic state.

Susceptibility

To determine the susceptibility $\chi = \partial m/\partial H$ it is convenient to calculate first $\chi_1 = \partial m_1/\partial H$ and $\chi_2 = \partial m_2/\partial H$ and then determine $\chi = (\chi_1 + \chi_2)/2$. Deriving both members of (14.10) and (14.11) with respect to H, we obtain two equations which can be solved for χ_1 and χ_2. From the expressions for χ_1 and χ_2 we reach the following result for χ at zero field

$$\frac{1}{\chi} = \lambda + \frac{RT}{\mu^2 - \eta^2}, \tag{14.18}$$

valid both in the paramagnetic and antiferromagnetic phases.

Above the Néel temperature, $\eta = 0$, so that

$$\frac{1}{\chi} = \lambda + \frac{RT}{\mu^2}. \tag{14.19}$$

Using (14.17), we get the result

$$\chi = \frac{T_N}{\lambda(T + T_N)}, \tag{14.20}$$

which is the Curie-Weiss law for antiferromagnetic systems.

Below the Néel temperature, we should replace the order parameter η, nonzero solution of (14.15), into (14.18). The result is shown in Fig. 14.1. Just below the Néel temperature,

$$\eta = \mu \sqrt{3 \frac{T_N - T}{T_N}}, \tag{14.21}$$

yielding the result

$$\chi = \frac{T}{2\lambda T_N}. \tag{14.22}$$

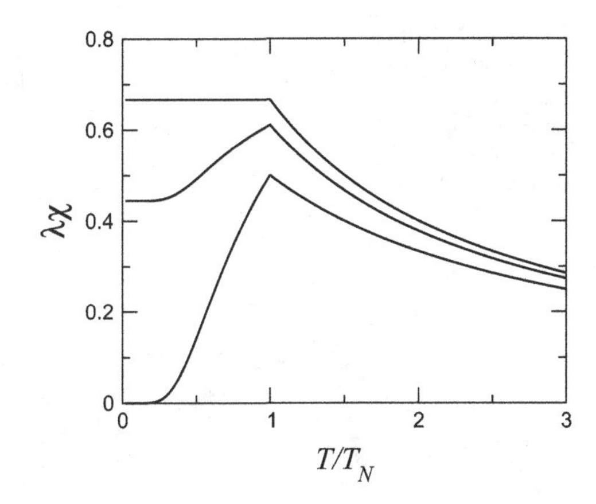

Fig. 14.1 Susceptibility of an antiferromagnetic system according to the Néel theory. The *lower and upper curves* are respectively the parallel and perpendicular susceptibilities. The *middle curve* corresponds to the weighted average (14.54)

valid for temperatures near T_N. According to these results, the susceptibility is finite at the critical temperature and has the value $\chi = 1/2\lambda$, as seen in Fig. 14.1.

Criticality

Near the Néel temperature, the order parameter behaves according to

$$\eta \sim (T_N - T)^\beta. \tag{14.23}$$

As we see from (14.21), the value predicted by the Néel theory for the exponent β is 1/2. Experimentally, however, the observed values are different and are shown in Table 14.2 for various antiferromagnetic compounds.

Other exponents can be defined for the phase transition between paramagnetic and antiferromagnetic phase. The exponent α is related to the molar heat capacity c_H, which behaves according to

$$c_H \sim |T_N - T|^{-\alpha}, \tag{14.24}$$

with $\alpha > 0$ for the case where c_H diverges. If c_H is finite at the critical temperature, with the value c_0, then the behavior is given by

$$c_0 - c_H \sim |T_N - T|^{-\alpha}, \tag{14.25}$$

where, in this expression, $\alpha < 0$. Table 14.2 shows the value of this exponent for various antiferromagnetic compounds. Note that for some compounds α is positive and for others α is negative.

The critical exponent γ, also shown in Table 14.2, governs the critical behavior of the susceptibility χ_* associated to the order parameter η and defined by

Table 14.2 Critical exponents and Néel temperature, in K, of antiferromagnetic compounds, obtained experimentally

Compound	α	β	γ	T_N
$RbMnF_3$	−0.14	0.32	1.37	83.1
MnF_2		0.34	1.27	67.3
FeF_2	+0.14	0.33	1.25	78.3
$DyPO_4$		0.31		3.39
$DyAlO_3$		0.31		3.52

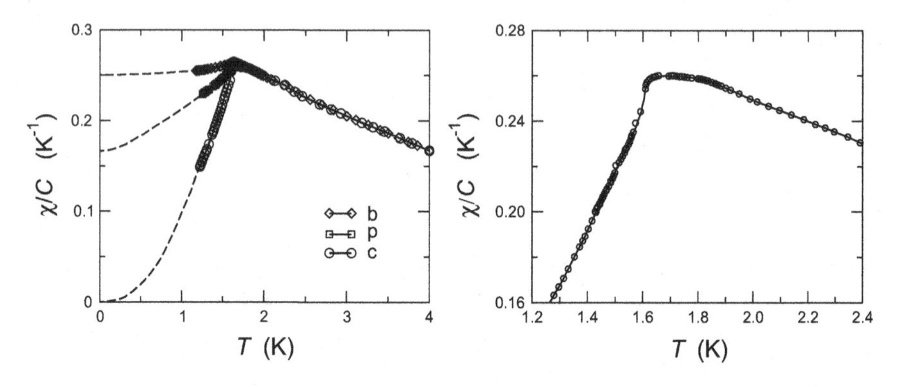

Fig. 14.2 Susceptibility of the antiferromagnetic compound $MnCl_2{\cdot}4H_2O$ obtained experimentally by Lasher et al. [153]. The curves c and b correspond to the susceptibility of a single crystal determined along the directions parallel and perpendicular to the preferential axis, respectively. The curve p represents the susceptibility of the powder. According to Fisher the critical temperature occurs at the inflection point immediately below the maximum of the susceptibility, as shown in detail for the parallel susceptibility. C is the Curie constant

$\chi_* = \partial\eta/\partial H_*$. It is possible to show from the Néel theory that the exponent γ is equal to 1. The experimental values are distinct from those predicted by the Néel theory, as seen in Table 14.2.

The susceptibility χ_* of antiferromagnetic systems must be distinguished from the susceptibility $\chi = \partial m/\partial H$, related to the magnetization m. The susceptibility χ is finite, as predicted by the Néel theory. However, the Néel temperature does not correspond to a maximum of χ. According to Fisher, the Néel temperature corresponds to the inflection point in the susceptibility χ, which occurs just below the maximum of χ, as seen in Fig. 14.2. Yet according to Fisher, near the critical point, χ is related to the molar heat capacity as follows

$$\frac{\partial}{\partial T}(T\chi) \sim c_H, \tag{14.26}$$

from which we conclude that the singular part of χ is governed by the exponent $1-\alpha$.

14.2 Metamagnetic Systems

Free Energy

The behavior of antiferromagnetic systems in the presence of an external field is very different from that observed in ferromagnetic systems. In the latter, an external field as small as it is destroys the phase transition. In antiferromagnetic systems, by contrast, the transition is still present, unless the field is high enough. The different behaviors are due to the role of H with respect to the order parameter. In ferromagnetic systems, H is the field conjugate to the order parameter. In antiferromagnetic systems that role is reserved to the staggered field H_* and not the external field H.

An antiferromagnetic system subject to an external field can display a phase diagram containing various types of magnetic ordering. In order to determine the phase diagram in the space H versus T we start from the following molar Helmholtz free energy

$$f = \frac{\lambda_{12}}{2} m_1 m_2 - \frac{\lambda_{11}}{4}(m_1^2 + m_2^2) +$$

$$+ \frac{RT}{2}\left(\frac{\mu + m_1}{2\mu} \ln \frac{\mu + m_1}{2\mu} + \frac{\mu - m_1}{2\mu} \ln \frac{\mu - m_1}{2\mu}\right) +$$

$$+ \frac{RT}{2}\left(\frac{\mu + m_2}{2\mu} \ln \frac{\mu + m_2}{2\mu} + \frac{\mu - m_2}{2\mu} \ln \frac{\mu - m_2}{2\mu}\right). \qquad (14.27)$$

where $\lambda_{12} > 0$ and $\lambda_{11} \geq 0$ are parameters and m_1 and m_2 are the magnetizations of the sublattices. The first term is the interaction between magnetic dipoles located in different sublattices and favors the antiparallel alignment of them. The second is the interaction between magnetic dipoles located on the same sublattice and favors the parallel alignment of them. Considering that the sublattices are equivalent, the free energy is invariant by permutation of m_1 and m_2.

Several systems are described by the antiferromagnetic free energy (14.27). When $\lambda_{11} = 0$, we recover the expression (14.12), which describes a simple antiferromagnetic system, consisting of two interpenetrating sublattices. When $\lambda_{11} > 0$, the free energy (14.27) describes metamagnetic systems that are antiferromagnetic systems consisting of magnetic layers. The odd layers constitute one sublattice and the even layers the other sublattice. The interaction between layers is antiferromagnetic. Within the layer the interaction is ferromagnetic. Examples of metamagnetic systems are shown in Table 14.1.

It is convenient here to use the variables m and η which are, respectively, the total magnetization of the sample and the order parameter, related to m_1 and m_2 by (14.2) and (14.3). Using (14.4) we may write

$$f = \frac{\lambda_1}{2}m^2 - \frac{\lambda_0}{2}\eta^2 +$$

$$+\frac{RT}{2}\left(\frac{\mu+m+\eta}{2\mu}\ln\frac{\mu+m+\eta}{2\mu} + \frac{\mu-m-\eta}{2\mu}\ln\frac{\mu-m-\eta}{2\mu}\right) +$$

$$+\frac{RT}{2}\left(\frac{\mu+m-\eta}{2\mu}\ln\frac{\mu+m-\eta}{2\mu} + \frac{\mu-m+\eta}{2\mu}\ln\frac{\mu-m+\eta}{2\mu}\right). \tag{14.28}$$

where $\lambda_0 = \lambda_{12} + \lambda_{11}$ and $\lambda_1 = \lambda_{12} - \lambda_{11}$.

To determine the thermodynamic phases we consider the thermodynamic potential $\phi(T, H, H_*)$ defined by (14.9). For $H_* = 0$,

$$\phi(T, H, 0) = \min_{m,\eta}\{f(T, m, \eta) - Hm\}, \tag{14.29}$$

that is, the quantities m and η correspond to the minima of the expression between curls. Deriving this expression with respect to m and η and equating both results to zero, we get the following conditions for a minimum

$$H = \lambda_1 m + \frac{RT}{4\mu}\ln\frac{(\mu+m)^2 - \eta^2}{(\mu-m)^2 - \eta^2} \tag{14.30}$$

and

$$\lambda_0\eta = \frac{RT}{4\mu}\ln\frac{(\mu+\eta)^2 - m^2}{(\mu-\eta)^2 - m^2}. \tag{14.31}$$

These two equations of state implicitly provide m and η as functions of T and H. The solutions are of two types: (a) $\eta = 0$, corresponding to the paramagnetic state, (b) $\eta \neq 0$, corresponding to the antiferromagnetic state. This type of solution appears in pairs, since the equations of state are invariant under the change of sign of η. Thus the antiferromagnetic region is a region of coexistence of two antiferromagnetic phases, defined by η and $-\eta$, nonzero.

Tricritical Point

The two kinds of solution divide the phase diagram into two regions, as shown in Fig. 14.3a for the case $\lambda_{11} = 2\lambda_{12}$. The line separating the two regions corresponds to a continuous phase transition or critical line for low field. For high fields, it corresponds to a discontinuous phase transition in which three phases coexist: the paramagnetic and two antiferromagnetic. The point separating the continuous

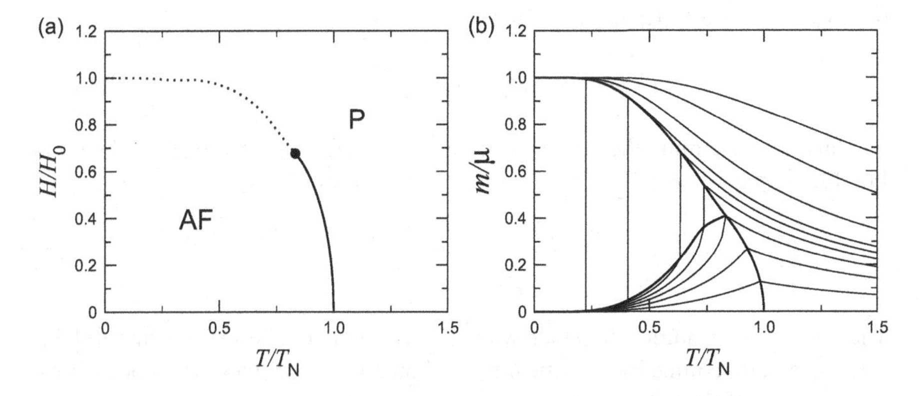

Fig. 14.3 Phase diagram of a metamagnetic system described by the free energy (14.27) for $\lambda_{11} = 2\lambda_{12}$. (**a**) Diagram field H versus temperature T, where $H_0 = \mu\lambda_{12}$. The transition line separates the paramagnetic phase (P) and the two antiferromagnetic phases (AF) in coexistence. The *solid curve* represents a critical line and the *dotted line* the coexistence of three phases: the paramagnetic and two antiferromagnetic. The *full circle* represents the tricritical point. (**b**) Magnetization m versus temperature T. Along the continuous curves the field is constant. The *thick lines* represent the coexistence curve and the critical line

and discontinuous transitions is the tricritical point where the three phases become identical.

The critical line occurs when the nonzero solution $\eta \to 0$. Dividing both sides of (14.31) by η and taking the limit $\eta \to 0$, we get

$$\lambda_0 = \frac{RT}{\mu^2 - m^2} \tag{14.32}$$

This equation together with the equation

$$H = \lambda_1 m + \frac{RT}{2\mu} \ln \frac{\mu + m}{\mu - m} \tag{14.33}$$

obtained from (14.30) by setting $\eta = 0$, give the critical line in the diagram H versus T in parametric form. In particular, for $H = 0$ and therefore $m = 0$, we get the Néel temperature

$$\lambda_0 = \frac{RT_N}{\mu^2}. \tag{14.34}$$

To determine the line of coexistence of three phases we must look to the thermodynamic potential ϕ given by (14.29), which is a function of thermodynamic fields only and therefore, at the coexistence of phases, has the same value in each phase. Denoting by m_p and m_a the magnetizations corresponding to the solutions paramagnetic ($\eta = 0$) and antiferromagnetic ($\eta \neq 0$) of (14.30) and (14.31), then the condition for coexistence is given by

$$f(T, m_a, \eta) - Hm_a = f(T, m_p, 0) - Hm_p. \tag{14.35}$$

Thus, we determine the coexistence line in the diagram H versus T, shown in Fig. 14.3a. In particular, one can show that at the tricritical point

$$\frac{m^2}{\mu^2} = \frac{\lambda_{12}}{3\lambda_{11}}. \tag{14.36}$$

Therefore, this condition together with the equations of the critical line (14.32) and (14.33), determine the tricritical point. Since $|m| \leq \mu$, then the condition for the existence of tricritical point is $\lambda_{12} \leq 3\lambda_{11}$. If this condition is not satisfied, the transition line is reduced to a critical line only.

The magnetization m as a function of temperature for constant H is shown in Fig. 14.3b. Defining the variable $\xi = m_p - m_a$, then it is possible to show that along the coexistence of three phases, ξ behaves near the tricritical point according to

$$\xi \sim |T - T_t|, \tag{14.37}$$

where T_t the temperature of the tricritical point. The variable ξ gives the length of the tie lines which, according to (14.37), varies linearly with distance from the tricritical point, as shown in Fig. 14.3b. Experimental results for a metamagnetic system are shown in Fig. 14.4.

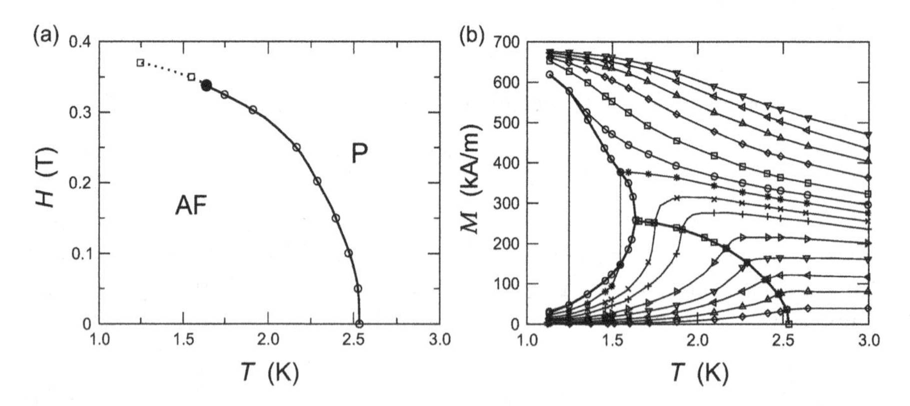

Fig. 14.4 Phase diagram of the metamagnetic compound $Dy_3Al_5O_{12}$ obtained experimentally by Landau et al. [152]. (**a**) Diagram field H versus temperature T. The transition line separates the paramagnetic phase (P) and the antiferromagnetic phase (AF). The *continuous curve* represents a critical line and a *dotted line* represents a coexistence line. The *full circle* represents the tricritical point. (**b**) Volumetric magnetization $\mathscr{M}$ versus temperature T. Along the continuous curves the field is constant. The *thick lines* represent the coexistence curve and the critical line

Landau Theory

Landau theory allows the determination of the topology of the phase diagrams. An important ingredient of the Landau theory is related to the symmetries of the system which are reflected in the symmetries of the free energy. In antiferromagnetic systems, the equivalence of sublattices implies that the free energy is invariant by permutation of the magnetization of sublattices or, equivalently, invariant by the change of sign of the order parameter η.

According to Landau theory, we postulate for the molar free energy $g(T, H, \eta)$, defined by (14.7), the following form

$$g = a_0 + a_2\eta^2 + a_4\eta^4 + a_6\eta^6, \tag{14.38}$$

which is an even function of η due to the symmetry $g(T, H, \eta) = g(T, H, -\eta)$. The parameters a_i are linear functions of T and H and a_6 is strictly positive. As the staggered field $H_* = \partial g/\partial\eta$ vanishes then according to (14.9), the order parameter η corresponds to an absolute minimum of the polynomial (14.38).

The number of coexisting phases equals the number of absolute minima of g. To determine the conditions for existence of a certain number of minima, it is convenient to write the free energy in the form

$$g = g_0 + a_6\psi(\eta), \tag{14.39}$$

where g_0 is the value of the absolute minimum and $\psi(\eta)$ is a polynomial of sixth degree in η. Thus, the minima of g become the double roots of $\psi(\eta)$.

Since ψ is a pair polynomial of sixth-degree, then we can have at most three phases in coexistence. Initially, we determine the conditions for the coexistence of three phases. For this to occur it is necessary that

$$\psi = \eta^2(\eta - \eta_0)^2(\eta + \eta_0)^2 = \eta^2(\eta^2 - \eta_0^2)^2, \tag{14.40}$$

from which we conclude that

$$\frac{a_4}{a_6} = -2\eta_0^2 \quad \text{and} \quad \frac{a_2}{a_6} = \eta_0^4, \tag{14.41}$$

or

$$a_2 = \frac{a_4^2}{4a_6} \quad \text{and} \quad a_4 < 0. \tag{14.42}$$

The three phases in coexistence are: $\eta = 0$, the paramagnetic phase, and $\eta = \pm\eta_0$, the two antiferromagnetic phases.

A critical line (two identical phases) is described by a polynomial ψ having a quadruple root, that is, $\psi = \eta^4(\eta^2 + a^2)$, from which we conclude that

$$a_2 = 0 \qquad \text{and} \qquad a_4 > 0. \tag{14.43}$$

Finally, the tricritical point (three identical phases) corresponds to a sixfold root so that $\psi = \eta^6$, from which follows $a_4 = a_2 = 0$. The phase diagram in space a_2 versus a_4 is shown in Fig. 11.11a.

To determine the magnetization $m = -\partial g/\partial H$, we assume that the coefficients a_i depend linearly on T and H. Thus up to quadratic order in η we obtain $m = c_0 - c_2\eta^2$, where c_0 and c_2 are constant. On the three-phase coexistence line, the magnetization of the paramagnetic phase $m_p = c_0$ and that of the antiferromagnetic phase $m_a = c_0 - c_2\eta_0^2$. Therefore, the variable $\xi = m_p - m_a$ is related with η_0 by $\xi = c_2\eta_0^2$. But according to (14.41), η_0 behaves as

$$\eta_0 \sim |a_4|^{1/2}, \tag{14.44}$$

which yields the result

$$\xi \sim |a_4|, \tag{14.45}$$

that is, ξ approaches linearly the tricritical point, as shown in Fig. 11.11b, which justify result (14.37).

14.3 Anisotropy

Perpendicular Susceptibility

The calculations developed in the previous sections are appropriate for a system with uniaxial anisotropy, that is, such that the magnetizations of the sublattices are parallel to the preferred direction determined by the crystalline anisotropy and applied field along the preferred direction. For a more general treatment that takes into account both applied fields in other directions and the possible emergence of magnetization in other directions, we start with the following molar Helmholtz free energy

$$f = \frac{\lambda\gamma}{2}(m_{1x}m_{2x} + m_{1y}m_{2y}) + \frac{\lambda}{2}m_{1z}m_{2z} +$$

$$+\frac{RT}{2}\left(\frac{\mu + m_1}{2\mu}\ln\frac{\mu + m_1}{2\mu} + \frac{\mu - m_1}{2\mu}\ln\frac{\mu - m_1}{2\mu}\right) +$$

$$+\frac{RT}{2}\left(\frac{\mu + m_2}{2\mu}\ln\frac{\mu + m_2}{2\mu} + \frac{\mu - m_2}{2\mu}\ln\frac{\mu - m_2}{2\mu}\right). \tag{14.46}$$

where $m_1 = (m_{1x}^2 + m_{1y}^2 + m_{1z}^2)^{1/2}$ and $m_2 = (m_{2x}^2 + m_{2y}^2 + m_{2z}^2)^{1/2}$. The components of the total magnetization are given by $m_x = (m_{1x} + m_{2x})/2$, $m_y = (m_{1y} + m_{2y})/2$ and $m_z = (m_{1z} + m_{2z})/2$. The parameter γ measures the anisotropy with respect to the preferred axis z. The value $\gamma = 1$ corresponds to the isotropic case and $\gamma = 0$ corresponds to the maximum anisotropy. We consider here the anisotropic case, $0 \leq \gamma < 1$, so that the spontaneous magnetization of the sublattices will emerge along the preferential axis z.

If the external field is applied along the z direction, then $m_{1x} = m_{1y} = m_{2x} = m_{2y} = 0$ and the expression (14.46) is reduced to (14.12). The results obtained previously for the susceptibility correspond in the present case to the parallel susceptibility $\chi_\parallel = \partial m_z/\partial H_z$. To determine the perpendicular susceptibility $\chi_\perp = \partial m_x/\partial H_x$ we consider a field applied in the x direction. In this case $m_{1x} = m_{2x} = m_x$ and, in addition, $m_{1y} = m_{2y} = 0$ and $m_{1z}^2 = m_{2z}^2 = \eta^2$. The free energy (14.46) becomes then

$$f = \frac{\lambda\gamma}{2} m_x^2 - \frac{\lambda}{2}\eta^2 +$$

$$+ \frac{RT}{2}\left(\frac{\mu + m_2}{2\mu} \ln \frac{\mu + m_2}{2\mu} + \frac{\mu - m_2}{2\mu} \ln \frac{\mu - m_2}{2\mu}\right). \tag{14.47}$$

where $m = (\eta^2 + m_x^2)^{1/2}$, so that the field $H_x = \partial f/\partial m_x$ is given by

$$H_x = \lambda\gamma m_x + \frac{m_x}{m}\frac{RT}{2\mu} \ln \frac{\mu + m}{\mu - m}. \tag{14.48}$$

To determine $\chi_\perp$ at zero field, it suffices to use the above expression for small values of H_x.The magnetization m_x will also be small and will vanish when $H_x \to 0$. For small values of h_x, we get

$$H_x = \lambda\gamma m_x + \frac{m_x}{\eta}\frac{RT}{2\mu} \ln \frac{\mu + \eta}{\mu - \eta}. \tag{14.49}$$

Above the Néel temperature, $\eta = 0$ and we obtain

$$H_x = \lambda\gamma m_x + \frac{RT}{\mu^2} m_x = \lambda\gamma m_x + \lambda \frac{T}{T_N} m_x, \tag{14.50}$$

where we used the result (14.17). Below the Néel temperature, we use the result (14.16) to obtain

$$H_x = \lambda\gamma m_x + \lambda m_x, \tag{14.51}$$

From these two equation, we reach the results

$$\chi_\perp = \frac{T_N}{\lambda(T + \gamma T_N)}, \tag{14.52}$$

valid above the Néel temperature and

$$\chi_\perp = \frac{1}{\lambda(1 + \gamma)}, \tag{14.53}$$

valid below the Néel temperature.

The results for $\chi_\parallel$ and $\chi_\perp$ at zero field are shown in Fig. 14.1 for $\gamma = 1/2$. Below the Néel temperature, $\chi_\perp$ is constant whereas $\chi_\parallel$ increases monotonically from the zero value at $T = 0$. Above T_N both susceptibilities decrease monotonically. For samples of antiferromagnetic substances in the form of powder the measured susceptibility χ corresponds to a weighted mean of the two susceptibilities. For uniaxial anisotropies with a preferred axis the following relation is valid

$$\chi = \frac{1}{3}\chi_\parallel + \frac{2}{3}\chi_\perp, \tag{14.54}$$

which is shown in Fig. 14.1. In Fig. 14.2 we shown experimental data of the susceptibility of the compound $MnCl_2{\cdot}4H_2O$ both in the single crystal and powder forms

Phase Diagram

Let us determine next the possible thermodynamic phases predicted by the free energy (14.46) for the field H applied along the preferential axis. In addition to the paramagnetic and antiferromagnetic phases, a system described by (14.46) exhibits a new magnetic phase, as seen in the phase diagram of Fig. 14.5. This phase, called spin flop, is such that the magnetization of the sublattices point to directions distinct that of the preferential axis although the applied field is parallel to the preferential direction. The phase transitions from the antiferromagnetic state to the paramagnetic state and from the spin flop state to the paramagnetic state are continuous. However, the transition from the antiferromagnetic state to the spin flop state is discontinuous. The phase diagram has two critical lines and one coexistence line, as seen in Fig. 14.5 and Fig. 14.6.

To determined the phase diagram, it is more convenient to consider the thermodynamic potential ϕ, which is function only of the thermodynamic fields, given by

$$\phi = \min\{f - Hm_z\}, \tag{14.55}$$

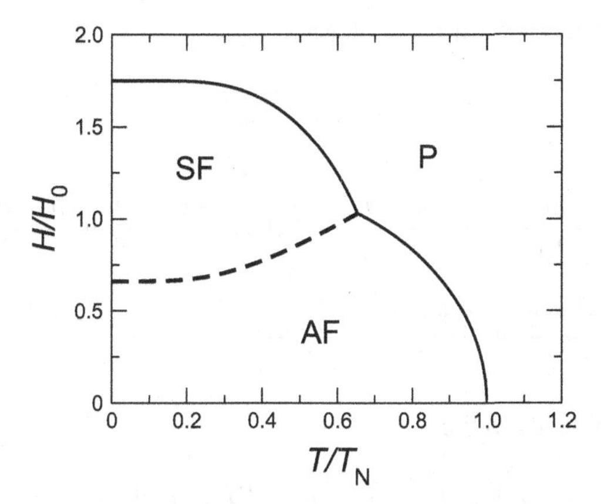

Fig. 14.5 Phase diagram of a magnetic system described by the free energy (14.46) for $\gamma = 0.75$. The system displays three phases: paramagnetic (P), antiferromagnetic (AF) and spin flop (SF). The *thick line* between the phases AF and P and between SF and P are critical lines. The *dashed line* between the phases AF and SF is a coexistence line. The three lines meet at the bicritical point

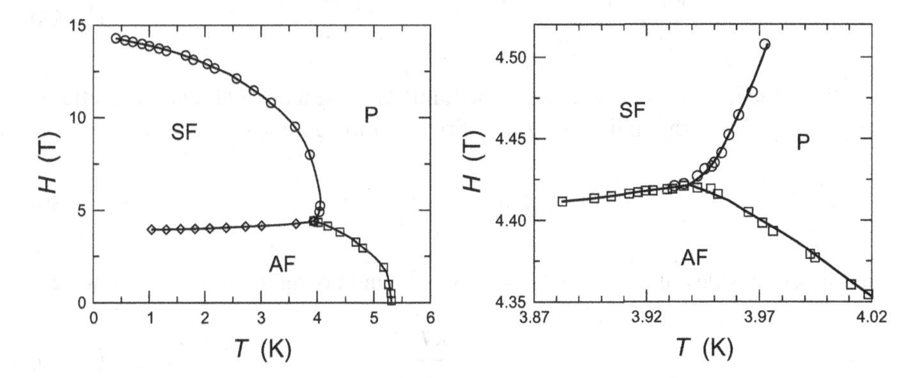

Fig. 14.6 Phase diagram of the antiferromagnetic salt $NiCl_2{\cdot}6H_2O$ obtained experimentally by Oliveira et al. [155]. The system presents three phases: paramagnetic (P), antiferromagnetic (AF) and spin flop (SF). The *lines* between the phases AF and P and between SF and P are critical lines. The *line* between the phases AF and SF is a coexistence line. The three lines meet at the bicritical point

where the minimization is done on the Cartesian components of the sublattice magnetizations. With the help of the potential ϕ we can calculate the coexistence line since ϕ must have the same value on the two phases in coexistence.

In the antiferromagnetic and paramagnetic phases, the magnetizations are parallel to the preferential direction, so that $m_{ix} = m_{iy} = 0$ and $m_{1z} = m_z + \eta$ and

$m_{2z} = m_z - \eta$. Therefore, for these phases,

$$f = \frac{\lambda}{2}(m_z^2 - \eta^2)+$$

$$+\frac{RT}{2}\left(\frac{\mu + m_z + \eta}{2\mu}\ln\frac{\mu + m_z + \eta}{2\mu} + \frac{\mu - m_z - \eta}{2\mu}\ln\frac{\mu - m_z - \eta}{2\mu}\right)+$$

$$+\frac{RT}{2}\left(\frac{\mu + m_z - \eta}{2\mu}\ln\frac{\mu + m_z - \eta}{2\mu} + \frac{\mu - m_z + \eta}{2\mu}\ln\frac{\mu - m_z + \eta}{2\mu}\right). \tag{14.56}$$

The minimization leads to the equations

$$H = \lambda m_z + \frac{RT}{4\mu}\ln\frac{\mu + m_z + \eta}{\mu - m_z - \eta} + \frac{RT}{4\mu}\ln\frac{\mu + m_z - \eta}{\mu - m_z + \eta} \tag{14.57}$$

and

$$\lambda\eta = \frac{RT}{4\mu}\ln\frac{\mu + m_z + \eta}{\mu - m_z - \eta} - \frac{RT}{4\mu}\ln\frac{\mu + m_z - \eta}{\mu - m_z + \eta}. \tag{14.58}$$

The critical line that separates the antiferromagnetic and paramagnetic is obtained by taking the limit $\eta \to 0$. The first equation gives

$$H = \lambda m_z + \frac{RT}{2\mu}\frac{\mu + m_z}{\mu - m_z}. \tag{14.59}$$

Dividing both sides of the second equation by η and taking the limit $\eta \to 0$, we get

$$\lambda = \frac{RT}{\mu^2 - m_z^2} \tag{14.60}$$

These two equations give parametrically the critical line shown in Fig. 14.5.

The spin flop phase is such that the magnetizations have the same magnitude but have distinct direction from each other and from the preferential direction. The components are given parametrically by $m_{1z} = m_{2z} = m\cos\theta$, $m_{1x} = -m_{2x} = m\sin\theta$ and $m_{1y} = m_{2y} = 0$. Using this parametrization we get

$$f = \frac{\lambda}{2}m^2(\cos^2\theta - \gamma\sin^2\theta)+$$

$$+RT\left(\frac{\mu + m}{2\mu}\ln\frac{\mu + m}{2\mu} + \frac{\mu - m}{2\mu}\ln\frac{\mu - m}{2\mu}\right) \tag{14.61}$$

and

$$\phi = \min_{m,\theta}\{f - Hm\cos\theta\}. \tag{14.62}$$

The minimization in m and θ lead us to the equations

$$\lambda m\gamma = \frac{RT}{2\mu}\ln\frac{\mu+m}{\mu-m} \tag{14.63}$$

and

$$H = \lambda m(1+\gamma)\cos\theta. \tag{14.64}$$

The critical line of transition from the spin flop and the paramagnetic phases is obtained when $\theta = 0$ which gives

$$H = \lambda m(1+\gamma). \tag{14.65}$$

This equation together with (14.63) gives parametrically the critical line shown in Fig. 14.5.

To obtain the coexistence line of the antiferromagnetic and the spin flop we compare the potential ϕ obtained on the two phases. The result is shown in Fig. 14.5 for $\gamma = 0.75$. The meeting point of the three lines determine the bicritical point.

Bicritical Point

The bicritical point is located at the end of the coexistence line of the antiferromagnetic and spin flop phases. The antiferromagnetic phase is described by the order parameter $\eta = (m_{1z} - m_{2z})/2$, which vanishes both in the paramagnetic and the spin flop phases. The spin flop phase can also be defined by an order parameter ω, given by $\omega = (m_{1x} - m_{2x})/2$, which vanishes both in the paramagnetic and antiferromagnetic phases. Thus the adequate description of the bicritical point requires the concurrence of two order parameters.

According to the Landau theory, we postulate for the molar Gibbs free energy $g(T, H, \eta, \omega)$ the following form

$$g = a_{00} + a_{20}\eta^2 + a_{02}\omega^2 + a_{40}\eta^4 + a_{22}\eta^2\omega^2 + a_{04}\omega^4, \tag{14.66}$$

which is an even function both in η and ω. The parameters a_{ij} are linear functions of T and H and a_{40} and a_{04} are strictly positive. The conjugate fields to η and ω must vanish so that, according to (14.9), the order parameters η and ω correspond to the minima of (14.66).

The possible phase diagrams are shown in Fig. 14.7. For $a_{22}^2 < 4a_{40}a_{04}$ the diagram presents three phases: paramagnetic and other two which we identify as the antiferromagnetic and the spin flop. The transition between the last two are discontinuous. The other two lines are continuous transition and they meet at the bicritical point. For $a_{22}^2 > 4a_{40}a_{04}$, on the other hand, there is a fourth phase,

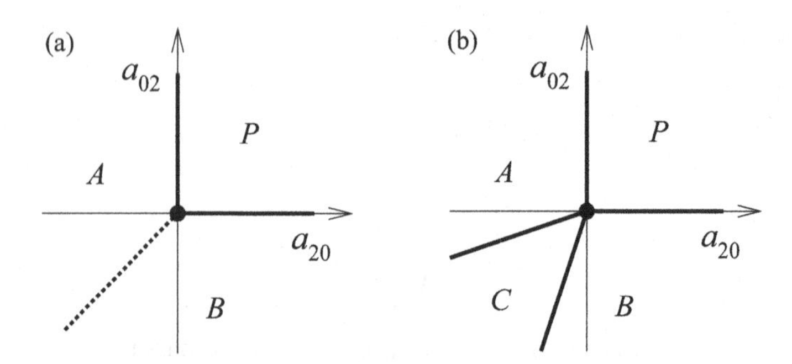

Fig. 14.7 Phase diagram around the bicritical and tetracritical points in the space a_{20} versus a_{02}. The *dashed line* is a coexistence line. The *thick lines* are critical lines. The phases are such that $\eta = \omega = 0$ in P phase, $\eta \neq 0$ and $\omega = 0$ in A phase, $\eta = 0$ and $\omega \neq 0$ in B phase, $\eta \neq 0$ and $\omega \neq 0$ in C phase

characterized by $\eta \neq 0$ and $\omega \neq 0$. In this case the transitions are continuous. The four lines meet at a point called tetracritical point.

14.4 Ferrimagnetism

Ferrimagnetic Materials

The ferrimagnetic materials also order in sublattices that present each of them spontaneous magnetization in distinct directions. However, unlike the antiferromagnetic materials, the sublattices are not equivalent resulting in a non vanishing sum of the magnetizations. The ferrimagnetic materials present therefore, a spontaneous magnetization. This is the case of the magnetite, Fe_3O_4, mineral known since ancient times by its natural magnetism, which disappears at the temperature of 585 °C.

Magnetite is the most notorious example of a class of ferrimagnetic materials known as ferrites which have chemical formula MFe_2O_4 where M is a divalent metal and has the same crystalline structure of spinel. The properties of some ferrites are presented in Table 14.3. In these compounds, the metallic ions, are located in two types of sites, denoted by A and B, that constitute the two nonequivalent sublattices. The site of type A is surrounded by four oxygen atoms located at the vertices of a tetrahedron whereas site B is surrounded by six oxygen atoms located at the vertices of an octahedron. The number of atoms of type B is equal to twice the number of atoms of type A.

Another class of ferrimagnetic materials is that constituted by the compounds with the same crystalline structure of the garnet described by the chemical formula $M_3Fe_5O_{12}$, where M is a trivalent metal and Fe is the trivalent ferric ion. The properties of some ferrimagnetic garnets are presented in Table 14.3 and in

Table 14.3 Critical temperature T_c, of ferrimagnetic materials. The table shows also the magnetization $\mathcal{M}$ (magnetic dipole moment per unit volume) at 20 °C and $b = \mu/\mu_B$ of ferrites; and compensating temperature T_{comp} of the iron garnet. *Source*: AIP

Ferrite	T_c (T)	$\mathcal{M}$ (kA/m)	b
Fe_3O_4	858	477	4.2
$CoFe_2O_4$	793	422	3.7
$NiFe_2O_4$	858	257	2.4
$CuFe_2O_4$	728	135	1.3
$MnFe_2O_4$	568	390	4.5
$MgFe_2O_4$	713	115	1.8
$Li_{1/2}Fe_{5/2}O_4$	943	310	2.6

Garnet	T_{comp} (K)	T_c (K)
$Y_3Fe_5O_{12}$		560
$Yb_3Fe_5O_{12}$	0	548
$Er_3Fe_5O_{12}$	83	556
$Ho_3Fe_5O_{12}$	137	567
$Dy_3Fe_5O_{12}$	226	563
$Tb_3Fe_5O_{12}$	246	568
$Gd_3Fe_5O_{12}$	286	564

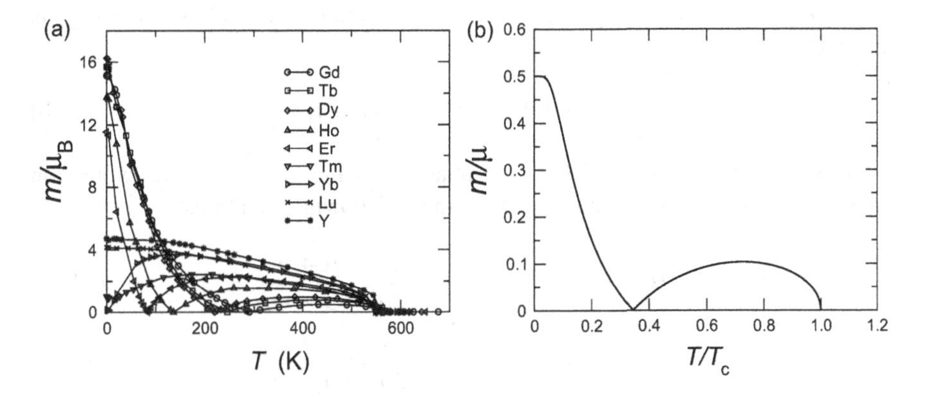

Fig. 14.8 (**a**) Molar spontaneous magnetization m versus temperature of iron garnets $M_3Fe_5O_{12}$, where M is a rare-earth metal or yttrium, obtained experimentally by Pauthenet [156]. (**b**) Molar spontaneous magnetization m versus temperature according to the Néel theory for $p = 0.25$, $\lambda_{22} = 0$ and $\lambda_{11} = 2\lambda_{12}$

Fig. 14.8a. The garnet of iron and yttrium, $Y_3Fe_5O_{12}$, is particularly interesting because the only magnetic ions are the ferric ions. The total magnetization is the result of the sum of opposite magnetizations corresponding to the two sublattices with distinct number of magnetic atoms. This compound loses the spontaneous magnetization at the temperature of 287 °C. At room temperature, the magnetization is equal to 130 kA/m and the saturation magnetization is equal to 200 kA/m. In addition $b = \mu/\mu_B = 5.0$.

Néel Theory

Let us consider the simplest case of a ferrimagnetic system in which the magnetic lattice is divided into two sublattice, in general not equivalent. The total molar magnetization is the weighted mean of the magnetization of sublattices 1 and 2,

$$m = pm_1 + qm_2, \tag{14.67}$$

where p and $q = 1 - p$ are the mole fractions of the magnetic atoms that are in sublattices 1 and 2, respectively. We start from the following molar Helmholtz free energy

$$f = pq\lambda_{12}m_1m_2 - p\frac{\lambda_{11}}{2}m_1^2 - q\frac{\lambda_{22}}{2}m_2^2 +$$

$$+pRT\left(\frac{\mu + m_1}{2\mu}\ln\frac{\mu + m_1}{2\mu} + \frac{\mu - m_1}{2\mu}\ln\frac{\mu - m_1}{2\mu}\right) + \tag{14.68}$$

$$+qRT\left(\frac{\mu + m_2}{2\mu}\ln\frac{\mu + m_2}{2\mu} + \frac{\mu - m_2}{2\mu}\ln\frac{\mu - m_2}{2\mu}\right), \tag{14.69}$$

where λ_{12}, λ_{11} and λ_{22} are positive. The corresponding molar Gibbs free energy is given by

$$g = \min\{f - H(pm_1 + qm_2)\}. \tag{14.70}$$

The magnetizations m_1 and m_2 are obtained as the minima of the expression between curls. The minimization leads us to the following equations

$$H = q\lambda_{12}m_2 - \lambda_{11}m_1 + \frac{RT}{2\mu}\ln\frac{\mu + m_1}{\mu - m_1}, \tag{14.71}$$

$$H = p\lambda_{12}m_1 - \lambda_{22}m_2 + \frac{RT}{2\mu}\ln\frac{\mu + m_2}{\mu - m_2}, \tag{14.72}$$

which can be written in the form

$$m_1 = \mu\tanh(\frac{\mu}{RT}(H - q\lambda_{12}m_2 + \lambda_{11}m_1)) \tag{14.73}$$

and

$$m_2 = \mu\tanh(\frac{\mu}{RT}(H - p\lambda_{12}m_1 + \lambda_{22}m_2)). \tag{14.74}$$

At zero field $H = 0$, in addition to the trivial solution $m_1 = m_2 = 0$ these equations give nontrivial solutions such that $m_1 \neq 0$ and $m_2 \neq 0$. Figure 14.8b presents curves of spontaneous magnetization m versus temperature for the case $p = 0.25$, $\lambda_{22} = 0$ and $\lambda_{11} = 2\lambda_{12}$. It is worth mentioning that, in this case, the total magnetization m vanishes at a temperature, called compensating temperature T_{comp}, smaller than the critical temperature T_c, as seen in Fig. 14.8b. However, this point does not represent any phase transition. Experimentally, this point is indeed observed, as shown in Fig. 14.8a. Table 14.3 shows the compensating temperature as well as the critical temperature of several ferrimagnetic garnets.

Chapter 15
Dielectrics

15.1 Dielectric Materials

Ordinary Dielectrics

Dielectric materials are insulators characterized by the response to an electric field. An ordinary dielectric, such as water, placed in the presence of an electric field acquires an electric polarization that disappears when the field is removed. For small values of the applied electric field E, the polarization (electric dipole moment per unit volume) of an ordinary dielectric is proportional to the field, that is,

$$\mathscr{P} = \bar{\chi} E, \tag{15.1}$$

where $\bar{\chi}$ is the constant of proportionality. If an ordinary dielectric is placed between the plates of a capacitor, the capacitance increases by a factor κ. This factor, which is the dielectric constant is related to $\bar{\chi}$ by means of

$$\bar{\chi} = \epsilon_0(\kappa - 1), \tag{15.2}$$

where ϵ_0 is the permittivity of vacuum. The dielectric constant κ is a characteristics of the dielectric substance and in general depends on the temperature. The dielectric constant of water at room temperature is equal to 80.

The electric polarization of a dielectric substance is due to the molecular microscopic dipoles of the molecules. Such molecular dipoles can be permanent, as happens with polar molecules, or can be induced by the applied field. In a dielectric composed of non-polar molecules, the polarization is only due to the effect of electrical induction, and therefore should disappear when the field vanish. In a dielectric composed by molecules with permanent dipole moment, the polarization also vanishes in the absence of field, but this is due to thermal agitation, which causes a disordering of the electric dipoles. The dipoles point to arbitrary directions

M.J. de Oliveira, *Equilibrium Thermodynamics*, Graduate Texts in Physics,
DOI 10.1007/978-3-662-53207-2_15

in space resulting in a null vector sum of dipole moments. An applied field causes the partial orientation of the dipoles, giving rise to a polarization.

Ferroelectrics

Ferroelectrics are materials that have a polarization even in the absence of an applied electric field. This spontaneous polarization generally decreases with increasing temperature and disappears above a temperature, called transition temperature T_0, as shown in Fig. 15.1. Above the transition temperature the material becomes paraelectric. In this state the polarization behaves according to (15.1), that is, the material becomes an ordinary dielectric: the polarization arises if we apply an electric field and disappears if we turn off the field. Rochelle salt or potassium sodium tartrate tetrahydrate ($NaKC_4H_4O_6{\cdot}4H_2O$), the substance in which ferroelectricity has been observed for the first time, is ferroelectric at room temperature, but loses polarization spontaneously if heated above 24 °C. Above this temperature, it becomes paraelectric. Interestingly, the Rochelle salt also ceases to be ferroelectric below −18 °C. Other ferromagnetic materials are listed on Table 15.1.

The ferroelectric crystals can basically be classified into two groups. In the first group are those consisting of permanent dipoles. The transition to the ferroelectric state occurs through the ordering of the permanent dipoles and so the transition is of the order-disorder type. In the second group are those consisting of induced dipoles. The transition to the ferroelectric is due to the relative displacement of two sublattices, one composed by anions and the other by cations.

The crystals of the first group are represented by potassium dihydrogen phosphate (KH_2PO_4) or KDP. In this substance phosphate groups are connected by hydrogen bonds. Each hydrogen ion can be located near one or the other of two

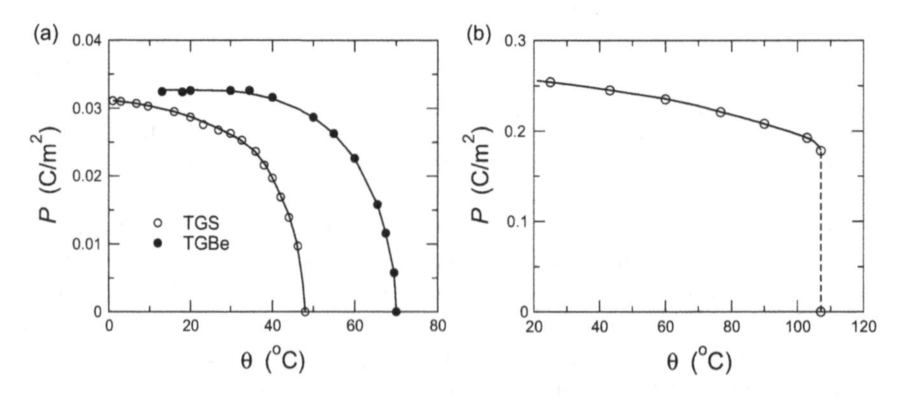

Fig. 15.1 (**a**) Spontaneous polarization of triglycine sulfate (TGS) and triglycine fluoberyllate (TGFB) obtained experimentally by Hoshino et al. [161]. (**b**) Spontaneous polarization of barium titanate obtained experimentally by Merz [163]

Table 15.1 Transition temperature T_0, in K, to the paraelectric phase of ferroelectric and antiferroelectric crystalline substances. Some substances (d) decompose before losing the ferroelectricity. *Source*: LB

Ferroelectrics	T_0
$BaTiO_3$	380
$KNbO_3$	691
$PbTiO_3$	763
$LiNbO_3$	1483
$LiTaO_3$	938
SbSI	295
HCl	98
HBr	90
$NaNO_2$	437
KNO_3	398
KH_2PO_4	123
RbH_2PO_4	147
CsH_2PO_4	152
KH_2AsO_4	97
RbH_2AsO_4	110
CsH_2AsO_4	143
KD_2PO_4	213
$(NH_4)_2SO_4$	224
$(NH_4)_2BeF_4$	176
$CH_3NH_3Al(SO_4)_2{\cdot}12H_2O$	177
$C(NH_2)_3Al(SO_4)_2{\cdot}6H_2O$	d
$Ca_2B_6O_{11}{\cdot}5H_2O$	266
$SC(NH_2)_2$	169
$N(CH_3)_4{\cdot}HgCl_3$	d
$(NH_2CH_2COOH)_3{\cdot}H_2SO_4$	321
$(NH_2CH_2COOH)_3{\cdot}H_2SeO_4$	295
$(NH_2CH_2COOH)_3{\cdot}H_2BeF_4$	343
$NaKC_4H_4O_6{\cdot}4H_2O$	297
Antiferroelectrics	T_0
$PbZrO_3$	503
$NaNbO_3$	638
$NH_4H_2PO_4$	148
$NH_4H_2AsO_4$	216
$ND_4D_2PO_4$	242

phosphate groups, which determines the orientation of the dipole attached to it. The ordering of the dipoles, which characterizes the ferroelectric state, is a consequence of the ordering of hydrogen ions. At high temperatures, thermal agitation causes the disordering of hydrogen ions and therefore of the dipoles, which characterizes the paraelectric state. To observe the ferroelectric state in KDP it is necessary to reduce its temperature below −150 °C. Other representatives of this group are the triglycine sulphate ($(NH_2CH_2COOH)_3{\cdot}H_2SO_4$) or TGS and triglycine fluoberyllate

$((NH_2CH_2COOH)_3{\cdot}H_2BeF_4)$ or TGFB whose spontaneous polarizations are shown in Fig. 15.1a as functions of temperature.

The ferroelectrics of the second group are represented by the barium titanate ($BaTiO_3$). In this substance there are two sublattices, one of them consisting of negative ions and the other of positive ions. The ferroelectric state arises by displacement of an entire sublattice in relation to the other. At room temperature $BaTiO_3$ is ferroelectric and has a tetragonal structure. If heated, it undergoes a transition to the paraelectric phase, at a temperature of 107 °C, as shown in Fig. 15.1b, passing to have a cubic structure. If cooled, however, it passes through two ferroelectric transitions. One at the temperature of 0 °C, acquiring the orthorhombic structure, and another at −90 °C acquiring the rhombohedral structure. All three transitions are discontinuous.

The ferroelectric transitions represent a special subclass of the class of structural transitions which are those in which the crystalline structure undergoes a change. As an example of a structural transition, we mention the compound $SrTiO_3$ which undergoes such a transition at a temperature of 110 K. Above this temperature, the compound has the structure of perovskite. Below that, the compound displays a structure that is distorted in relation to the original structure. This distortion, however, does not entail the emergence of a spontaneous polarization in $SrTiO_3$. If the distortion is accompanied by the emergence of the spontaneous polarization, then it is also a ferroelectric transition, as happens with the compound $BaTiO_3$. Is also worth mentioning that some crystals have other types of electrical ordering, with zero total polarization. Such states are called antiferroelectrics as is the case, for example, of $PbZrO_3$ crystal.

15.2 Dielectric Properties

Electric Work

To study the effects of the electric field on a dielectric we use a capacitor within which the dielectric sample is placed. The capacitor is maintained at a potential difference which generates an electric field within the capacitor. The electric dipole moment per unit volume or volumetric polarization $\mathscr{P}$ and the electric field $\mathbf{E}$ are related to the electric displacement $\mathbf{D}$ by

$$\mathbf{D} = \epsilon_0 \mathbf{E} + \mathscr{P}, \tag{15.3}$$

where ϵ_0 is the permittivity of vacuum. For a uniform electric field, we can show that the infinitesimal work performed on the dielectric dW_{ele} is given by

$$dW_{\text{ele}} = \mathbf{E} \cdot d\mathbf{P} \tag{15.4}$$

where $\mathbf{P}$ is the total electric dipole of the sample.

The demonstration of this result is as follows. Consider a capacitor of parallel plates of area A and distance between plates equal to ℓ within which is the dielectric. The capacitor is maintained at a potential difference equal to $\mathscr{V}$. We assume that the electric field E and the electric displacement D inside the capacitor as well as the volumetric polarization $\mathscr{P}$ are perpendicular to the plates of the capacitor and neglect border effects. The work dW required to carry a free charge dq_{liv} against the potential is given by

$$dW = \mathscr{V} dq_{\text{liv}}. \tag{15.5}$$

But $\mathscr{V} = E\ell$ and the free charges are related to the electric displacement D by $q_{\text{liv}} = DA$ so that

$$dW = E\ell A dD = EV dD, \tag{15.6}$$

where $V = A\ell$ is the volume of the capacitor. On the other hand, if the capacitor was empty the work would be

$$dW_0 = \epsilon_0 EV dE. \tag{15.7}$$

Therefore, the electric energy transferred to the electric material inside the capacitor will be $dW_{\text{ele}} = dW - dW_0$ or

$$dW_{\text{ele}} = EV d\mathscr{P}, \tag{15.8}$$

or yet

$$dW_{\text{ele}} = E dP, \tag{15.9}$$

where $P = V\mathscr{P}$ is the total electric dipole moment of the sample.

In the International System of Units, the electric dipole moment P is measured in coulomb meter (C m) and the electric field E in volts per meter (V/m). The electric displacement D and the volumetric polarization $\mathscr{P}$ are measured in C/m^2. The electric dipole moment per mole, or molar polarization, is given in C m/mol.

Thermodynamic Potentials and Coefficients

The infinitesimal change in internal energy of a dielectric equals the sum of the heat introduced $dQ = TdS$ and the work performed on the electrical system $dW_{\text{ele}} = EdP$, that is,

$$dU = TdS + EdP. \tag{15.10}$$

We remark that the total dipole moment P is an extensive quantity. Thus, we can understand that the internal energy U of a sample of dielectric is a function of the extensive variables S and P. Taking into account the number of moles N of the sample, we consider $U(S, P, N)$ to be also a function of N, what represents the fundamental relation of a dielectric in the energy representation. The entropy $S(U, P, N)$ is also a fundamental relation for which

$$dS = \frac{1}{T}dU - \frac{E}{T}dP. \tag{15.11}$$

From $U(S, P, N)$ we can get other representations by means of Legendre transformations. The enthalpy $H(S, E, N)$, the Helmholtz free energy $F(T, P, N)$ and the Gibbs free energy $G(T, E, N)$ are defined by the Legendre transformations

$$H = \min_P(U - EP), \tag{15.12}$$

$$F = \min_P(U - TS), \tag{15.13}$$

$$G = \min_P(F - EP). \tag{15.14}$$

From them, we obtain the results

$$dF = -SdT + EdP, \tag{15.15}$$

$$dH = TdS - PdE, \tag{15.16}$$

$$dG = -SdT - PdE. \tag{15.17}$$

From (15.17), we see that

$$S = -\left(\frac{\partial G}{\partial T}\right)_E, \qquad P = -\left(\frac{\partial G}{\partial E}\right)_T. \tag{15.18}$$

From these quantities we can determine certain thermodynamic coefficients including the isothermal electric susceptibility and the pyroelectric coefficient A, defined by

$$X_T = \left(\frac{\partial P}{\partial E}\right)_T, \qquad A = \left(\frac{\partial P}{\partial T}\right)_E, \tag{15.19}$$

and the heat capacities at constant field and at constant polarization, defined by

$$C_E = T\left(\frac{\partial S}{\partial T}\right)_E, \qquad C_P = T\left(\frac{\partial S}{\partial T}\right)_P, \tag{15.20}$$

The property of convexity of the thermodynamic potentials implies that the susceptibility and the heat capacities are nonnegative quantities, $X_T \geq 0$, $C_E \geq 0$ and $C_P \geq 0$.

The variation of the electrical polarization of a ferroelectric material caused by the change in temperature is called pyroelectric effect. A measure of this effect is given by the pyroelectric coefficient defined above.

Molar Quantities

The thermodynamic properties of a dielectric can be derived from the fundamental relation $u(s, p)$, where $u = U/N$ is the molar internal energy, $s = S/N$ is the molar entropy and $p = P/N$ is the molar dipole moment or simply polarization. For a dielectric

$$du = Tds + Edp. \tag{15.21}$$

Inverting the relation $u(s, p)$ we get the fundamental relation in the entropy representation $s(u, p)$. We can also define other molar quantities. Among them we find the molar thermodynamic potentials: $h = H/N, f = F/N$ and $g = G/N$ for which

$$dh = Tds - pdE, \tag{15.22}$$

$$df = -sdT + Edp, \tag{15.23}$$

$$dg = -sdT - pdE. \tag{15.24}$$

The electric susceptibility and the molar heat capacities are given by

$$\chi = \left(\frac{\partial p}{\partial E}\right)_T, \qquad c_E = T\left(\frac{\partial s}{\partial T}\right)_E, \qquad c_P = T\left(\frac{\partial s}{\partial T}\right)_P. \tag{15.25}$$

Debye Law

Let us first examine the properties of an ideal dielectric, understood as a system consisting of noninteracting microscopic dipoles or such that the interactions are negligible. Examples of such systems are gases at low pressures. In this state, the molecules are sufficiently far from each other so that the electric field acting on a molecule, due to the other, can be neglected when compared with the applied field.

The molar polarization p of an ideal dielectric is composed of two parts: an induced, p_1, and another, $p2$, permanent. The induced part is independent of temperature and simply proportional to the applied electric field

$$p_1 = \alpha E, \tag{15.26}$$

where α is the molar polarizability. The permanent part depends on the temperature. By analogy with ideal paramagnetism we postulate the following form due to Langevin

$$p_2 = p_0 \left(\coth \frac{p_0 E}{RT} - \frac{RT}{p_0 E} \right), \tag{15.27}$$

where p_0 is the permanent molar polarization. Therefore, the molar polarization $p = p_1 + p_2$ of an ideal dielectric is given by

$$p = \alpha E + p_0 \left(\coth \frac{p_0 E}{RT} - \frac{RT}{p_0 E} \right). \tag{15.28}$$

For small values of the electric field, the expression in parentheses behaves as $p_0 E/3RT$, so that the polarization varies with the field in accordance with

$$p = \chi_0 E, \tag{15.29}$$

where

$$\chi_0 = \alpha + \frac{p_0^2}{3RT}, \tag{15.30}$$

which is the Debye law. The second term, which is analogous to the Curie law, corresponds to the contribution of permanent dipoles. The first term is due to induced dipoles that may also be present in the polar molecules.

In an ordinary dielectric, as we have seen, the relation between the volumetric polarization (dipole moment per unit volume) $\mathscr{P}$ and the field is $\mathscr{P} = \bar{\chi} E$. Given that the dipole moment per unit volume $\mathscr{P}$ and dipole moment per mole p are connected by $p = v\mathscr{P}$, where v is the molar volume, then $\chi_0 = v\bar{\chi}$, from which follows the relation between χ_0 and the dielectric constant κ

$$\chi_0 = v\epsilon_0(\kappa - 1). \tag{15.31}$$

Thus we see that χ_0 can be experimentally determined from the measurement of dielectric constant and the molar volume.

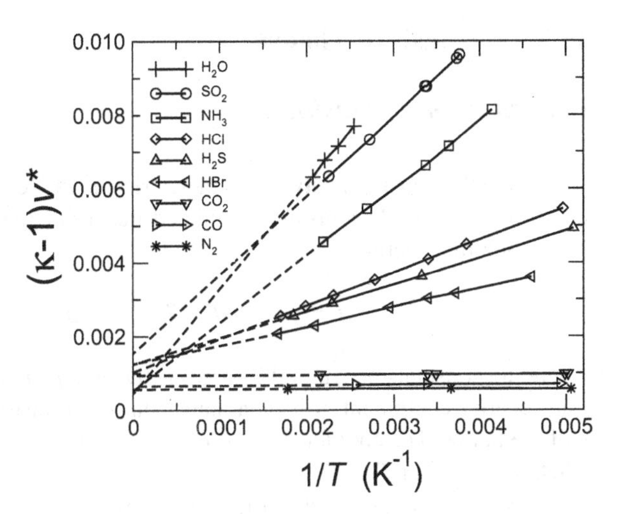

Fig. 15.2 Dielectric constant κ of several gases obtained experimentally by Zahn [167], and Zahn and Miles [167] and of the water vapor obtained experimentally by Sanger et al. [164]. The graph shows the variation of $(\kappa - 1)v^*$ with the inverse of temperature T, where $v^* = v/v_0$ and v_0 is the molar volume of an ideal gas at the pressure of 1 atm and at 0 °C

Replacing (15.31) into (15.30), Debye law can be written in the form

$$(\kappa - 1)v = \frac{1}{\epsilon_0}(\alpha + \frac{p_0^2}{3RT}). \tag{15.32}$$

Figure 15.2 shows $(\kappa - 1)v$ versus $1/T$ for various gases. By Debye law (15.32), the slope of the straight line fitted to the experimental data gives the permanent polarization p_0.

The dense gases as well as the liquids and solids deviate from ideal behavior and Debye law ceases to be valid. In these states, due to the proximity between molecules, one can not neglect the mutual action between the dipoles. A dipole does not feel only the external field but also the field generated by the other molecule dipoles. An approximate form for the dielectric properties of these systems is that given by the Clausius-Mossotti relation, which is obtained by replacing the field E in (15.29) by $E + \mathscr{P}/3\epsilon_0$, where the second term is the Lorentz field. With the Lorentz correction, Debye law becomes

$$\frac{\kappa - 1}{\kappa + 2}v = \frac{1}{3\epsilon_0}(\alpha + \frac{p_0^2}{3RT}). \tag{15.33}$$

For nonpolar molecules, $p_0 = 0$, and this equation is reduced to the Clausius-Mossotti relation

$$\frac{\kappa - 1}{\kappa + 2}v = \frac{\alpha}{3\epsilon_0}. \tag{15.34}$$

15.3 Devonshire Theory

Discontinuous Transition

To explain the transition between paraelectric and the ferroelectric phases occurring in barium titanate, Devonshire postulated the following expression for the molar Helmholtz free energy

$$f = f_0 + b(T - T_1)p^2 - cp^4 + dp^6, \tag{15.35}$$

where the constants b, c, d and T_1 are positive and f_0 depends only on temperature. The negative coefficient of p^4 guarantees that the transition is discontinuous as we will see below. The coefficient of p^2 can be either positive or negative according to whether $T > T_1$ or $T < T_1$.

The electric field $E = (\partial f/\partial p)_T$ is given by

$$E = 2b(T - T_1)p - 4cp^3 + 6dp^5, \tag{15.36}$$

and the susceptibility $\chi = (\partial p/\partial E)_T$ is given by

$$\frac{1}{\chi} = 2b(T - T_1) - 12cp^2 + 30dp^4. \tag{15.37}$$

Therefore, in the paraelectric phase, for which $p = 0$, the susceptibility is given by

$$\chi = \frac{1}{2b(T - T_1)}. \tag{15.38}$$

Expression (15.35) is not always convex in p. In the cases where it is not convex, the free energy of the system should be understood as the convex hull of expression (15.35) obtained by the minimization of $f - Ep$.

At zero field, $E = 0$, the thermodynamic properties are therefore obtained by the minimization of expression (15.35). That is, the polarization corresponds to the absolute minimum of (15.35).

In the paraelectric phase $p = 0$ is a absolute minimum of f. In the ferroelectric phase, the absolute minimum occurs at $p = p_*$, where p_* is the spontaneous polarization, which is the solution of (15.36) for $E = 0$, that is,

$$b(T - T_1) - 2cp_*^2 + 3dp_*^4 = 0, \tag{15.39}$$

whose solution is

$$p_* = \sqrt{\frac{c}{3d} + \frac{c}{3d}\sqrt{1 - \frac{3bd}{c^2}(T - T_1)}}. \tag{15.40}$$

which gives the variation of p_* with temperature.

The phase transition occurs when the Helmholtz free energy of the ferromagnetic phase $f(p_*)$ becomes equal to that of the paraelectric phase $f(0)$, that is, when $f(p_*) = f(0)$, which leads us to equation

$$b(T_0 - T_1) - cp_*^2 + dp_*^4 = 0. \tag{15.41}$$

This equation together with equation (15.39) determine the transition temperature, which we denote by T_0, and the value of p_* at the transition, which we denote by p_0, that is,

$$T_0 = T_1 + \frac{c^2}{4bd}, \tag{15.42}$$

$$p_0 = \sqrt{\frac{c}{2d}}. \tag{15.43}$$

Eliminating T_1 from expression (15.40), we can write the dependence of p_* with temperature in the following form

$$p_* = p_0\sqrt{\frac{2}{3} + \frac{1}{3}\sqrt{1 - \frac{12bd}{c^2}(T - T_0)}}, \tag{15.44}$$

valid for $T \leq T_0$. Above T_0, the polarization vanishes so that the polarization has a jump at $T = T_0$ equal to p_0, characterizing a discontinuous transition.

If the system described by (15.35) is subject to an electric field, the phase transition does not disappear but is displaced to another value of temperature as long as the field does not exceed a certain value E_c. For fields above E_c the transition disappears. Thus the phase diagram E versus T has a discontinuous transition line that starts at the point $T = T_0$ and $E = 0$ and ends at the critical point $T = T_c$ and $E = E_c$, as seen in Fig. 15.3

The properties described above are obtained by the minimization of $\psi(p) = f(T,p) - Ep$, given by

$$\psi(p) = f_0 + b(T - T_1)p^2 - cp^4 + dp^6 - Ep. \tag{15.45}$$

Along the transition line two phase coexist: a paraelectric phase characterized by $p = p_1$ and another characterized by $p = p_2$. Therefore, these two values of p must be the absolute minima of $\psi(p)$ what results in the following form

$$\psi(p) = \psi_0 + d(p - p_1)^2(p - p_2)^2[(p + r_1)^2 + r_2^2], \tag{15.46}$$

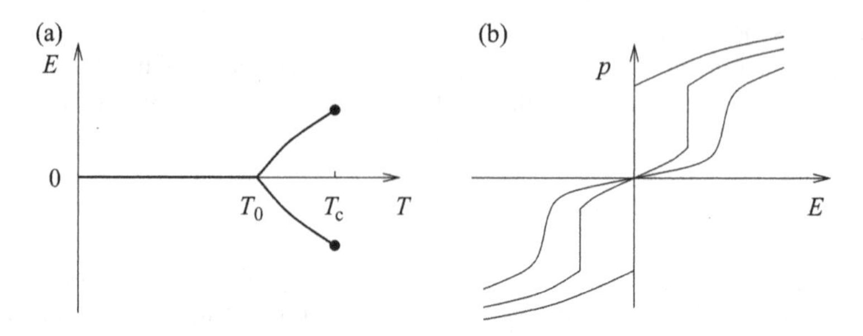

Fig. 15.3 (**a**) Phase diagram electric field E versus temperature T of a system described by the free energy (15.35). The *thick lines* are discontinuous transitions that meet at the triple point $T = T_0$ and $E = 0$. The *full circles* are critical points. (**b**) Isotherms of polarization p versus E. Three isotherms are shown: one below the temperature of the triple point T_0, another between T_0 and the critical temperature T_c and another one above T_c

where r_1 and r_2 are nonzero. Developing the product and comparing with (15.45), we get the relations

$$r_1 = p_1 + p_2, \qquad r_2^2 = p_1 p_2, \tag{15.47}$$

$$c = d(2r_1^2 - 3r_2^2), \qquad b(T - T_1) = d(r_1^4 + 3r_2^4 - r_1^2 r_2^2) \tag{15.48}$$

and

$$E = 2dr_1^3 r_2^2. \tag{15.49}$$

These equations give implicitly the coexistence line between the paraelectric and ferroelectric phases. Near the transition point $T = T_0$ at zero field $E = 0$, we get the following behavior

$$b(T - T_0) = \frac{1}{p_0}|E|. \tag{15.50}$$

The terminal point of the line define the critical point, which is obtained form equations above when $p_2 = p_1$, that is, when $r_2 = r_1^2/4$, which gives the results

$$T_c = T_1 + \frac{3c^2}{5bd} \qquad \text{and} \qquad E_c = 16d\left(\frac{c}{5d}\right)^{5/2}, \tag{15.51}$$

or yet

$$T_c = T_0 + \frac{7c^2}{20bd}. \tag{15.52}$$

Continuous Transition

The majority of ferroelectric substances pass to the paraelectric phase through a first order transition as happens, for example, with the barium titanate. Some ferroelectrics, however, undergoes a continuous transition. According the Devonshire theory, the free energy of such systems is give by

$$f = f_0 + b(T - T_c)p^2 + cp^4, \tag{15.53}$$

where the constants b and c are positive and f_0 depends only on temperature.

The electric field E is given by

$$E = 2b(T - T_c)p + 4cp^3, \tag{15.54}$$

from which we get the inverse of susceptibility

$$\frac{1}{\chi} = 2b(T - T_c) + 12cp^2. \tag{15.55}$$

At zero field, $p = 0$ for $T \geq 0$ and in this case the susceptibility is

$$\chi = \frac{1}{2b(T - T_c)}. \tag{15.56}$$

For $T < T_c$, a spontaneous polarization emerges, given by

$$p = \sqrt{\frac{b}{2c}(T_c - T)}, \tag{15.57}$$

which vanishes continuously when $T \to T_c$, characterizing a continuous phase transition. In this case the susceptibility is

$$\chi = \frac{1}{4b(T_c - T)}. \tag{15.58}$$

The molar pyroelectric coefficient $\alpha = A/N = (\partial p/\partial T)_E$ at zero field has therefore the following behavior

$$\alpha = -\sqrt{\frac{b}{8c(T_c - T)}}, \tag{15.59}$$

diverging when $T \to T_c$.

Ferroelectric Transitions

In addition to the transition between the paraelectric and ferroelectric phases which occurs at 107 °C, the barium titanate undergoes two more transitions that occur at 0 °C and at −90 °C. These transitions are discontinuous and occur among ferroelectric phases. In the paraelectric phase the barium titanate has cubic crystal structure. In the ferroelectric phases the structures are tetragonal, orthorhombic and rhombic.

The three discontinuous phase transitions occurring in $BaTiO_3$ can be obtained from the following free energy

$$f = f_0 + b(T - T_1)(p_x^2 + p_y^2 + p_z^2) - c(p_x^4 + p_y^4 + p_z^4)$$

$$+ a(p_x^2 p_y^2 + p_x^2 p_z^2 + p_y^2 p_z^2) + d(p_x^6 + p_y^6 + p_z^6), \qquad (15.60)$$

where f_0 depends only on temperature and the coefficients a, b, c and d are positive. The paraelectric phase is characterized by the absence of spontaneous polarization, that is, $p_x = p_y = p_z = 0$. The three ferroelectric phases are characterized by

$$p_x = p_y = 0 \quad \text{and} \quad p_z \neq 0, \qquad (15.61)$$

$$p_x = 0 \quad \text{and} \quad p_y = p_z \neq 0, \qquad (15.62)$$

$$p_x = p_y = p_z \neq 0. \qquad (15.63)$$

If we set $p_x = p_y = 0$ and $p_z = p$, the free energy is reduced to the expression

$$f = f_0 + b(T - T_1)p^2 - cp^4 + dp^6, \qquad (15.64)$$

which we have used previously to describe the ferroelectric to paraelectric transition. Setting $p_x = 0$ and $p_y = p_z = p$ we get

$$f = f_0 + 2b(T - T_1)p^2 - (2c - a)p^4 + 2dp^6. \qquad (15.65)$$

Finally, setting $p_x = p_y = p_z = p$ we obtain

$$f = f_0 + 3b(T - T_1)p^2 - 3(c - a)p^4 + 3dp^6. \qquad (15.66)$$

At zero field, the spontaneous polarization are obtained by the minimization of the above expressions and the transition temperatures by comparison of the free energies corresponding to the several thermodynamic phases.

15.4 Order-Disorder Transition

Hydrogen Bonds

In ferroelectric materials such that the hydrogen bonds have a relevant role, the ferroelectric state is described by the ordering of the hydrogen ions and consequently by the ordering of the permanent dipoles which result in a spontaneous polarization. In the paraelectric phase the dipoles are disordered so that the polarization vanishes. Among ferroelectrics whose mechanism of transition is of the type order-disorder we find the potassium dihydrogen phosphate dihydrate (KH_2PO_4) or KDP and its isomorphs and the triglycine sulfate, $(NH_2CH_2COOH)_3{\cdot}H_2SO_4$, and its isomorphs.

In analogy with ferromagnetic systems we postulate the following free energy for such systems

$$f = -\alpha\frac{p^2}{2} + RT\left(\frac{p_0+p}{2p_0}\ln\frac{p_0+p}{2p_0} + \frac{p_0-p}{2p_0}\ln\frac{p_0-p}{2p_0}\right), \tag{15.67}$$

where α is a positive constant. The electric field $E = \partial f/\partial p$ is given by

$$E = -\alpha p + \frac{RT}{2p_0}\ln\frac{p_0+p}{p_0-p}. \tag{15.68}$$

At zero field, we get

$$\alpha p = \frac{RT}{2p_0}\ln\frac{p_0+p}{p_0-p}. \tag{15.69}$$

or equivalently

$$p = p_0\tanh(\frac{p_0\alpha}{RT}p). \tag{15.70}$$

From this equation, we conclude that above the critical temperature T_c, given by

$$T_c = \frac{p_0^2\alpha}{R}, \tag{15.71}$$

the polarization vanishes and the system finds itself in the paraelectric phase. Below this temperature, $p \neq 0$ and the system exhibits a spontaneous polarization. Near the critical temperature, p behaves as

$$p = p_0\sqrt{3(1-\frac{T}{T_c})}. \tag{15.72}$$

Isotopic Effect

An important effect observed in phase transitions of compounds of the KDP type is that the critical temperature becomes approximately twice greater when hydrogen is replaced by deuterium, called isotopic effect. This means that the quantum tunneling of the hydrogen between two equilibrium positions located at the hydrogen bonds has a relevant role. When deuterium, which has a larger mass, replaces hydrogen, the tunneling becomes less frequent causing a greater stability of the ferroelectric state.

According to de Gennes the tunneling corresponding to the model introduced by Blind can be written by means of a transverse field, that is, by means of a field that acts perpendicularly to the direction of the electric dipoles. This field give rise to a polarization in the transversal direction, which we denote by q. Thus in analogy with a ferromagnetic system in the presence of perpendicular field we postulate the following free energy

$$f = -\frac{\alpha}{2}p^2 + RT\left(\frac{p_0 + r}{2p_0}\ln\frac{p_0 + r}{2p_0} + \frac{p_0 - r}{2p_0}\ln\frac{p_0 - r}{2p_0}\right), \tag{15.73}$$

where r is given by

$$r = \sqrt{q^2 + p^2}. \tag{15.74}$$

In fact, the free energy must be understood as the convex hull of the right hand side of (15.73) which is equivalent to minimize the expression $f - Ep - \Gamma q$ where Γ is the transverse field.

The transverse field $\Gamma = \partial f/\partial q$ is given by

$$\Gamma = \frac{RTq}{2p_0 r}\ln\frac{p_0 + r}{p_0 - r}, \tag{15.75}$$

and the electric field $E = \partial f/\partial p$, by

$$E = -\alpha p + \frac{RTp}{2p_0 r}\ln\frac{p_0 + r}{p_0 - r}. \tag{15.76}$$

At zero field, the solution of this last equation is $p = 0$ and corresponds to the paraelectric phase. In this case

$$\Gamma = \frac{RT}{2p_0}\ln\frac{p_0 + q}{p_0 - q} \qquad \text{or} \qquad q = p_0 \tanh\frac{p_0\Gamma}{RT}. \tag{15.77}$$

The spontaneous polarization, which characterizes the ferroelectric phase is the nonzero solution of (15.76) for $E = 0$, that is, solution of equation

$$\alpha = \frac{RT}{2p_0 r} \ln \frac{p_0 + r}{p_0 - r}. \qquad (15.78)$$

Comparing with (15.75), we see that the transverse field Γ is related with q through

$$q = \frac{\Gamma}{\alpha}, \qquad (15.79)$$

which replaced in the previous equation gives

$$\sqrt{\Gamma^2 + \alpha^2 p^2} = \alpha p_0 \tanh\left(\frac{p_0}{RT}\sqrt{\Gamma^2 + \alpha^2 p^2}\right). \qquad (15.80)$$

The spontaneous polarization vanishes continuously characterizing a continuous transition between the ferroelectric and paraelectric phases. Taking the limit $p \to 0$ in (15.80) we get the critical temperature T_c, given by

$$\Gamma = \alpha p_0 \tanh(\frac{p_0 \Gamma}{RT_c}). \qquad (15.81)$$

Near the critical point the spontaneous polarization behaves as

$$p \sim (T_c - T)^{1/2}. \qquad (15.82)$$

It is worth noticing that in the limit $T \to 0$, we get from (15.80) the following expression for the polarization

$$p = \sqrt{p_0^2 - \frac{\Gamma^2}{\alpha^2}}, \qquad (15.83)$$

that is, the saturation polarization depends on the ratio Γ/α between the transverse field Γ and the interaction α between dipoles.

Slater-Takagi Theory

As we have seen, the ferroelectric state of KDP is related with the ordering of the protons located at the hydrogen bonds that connect the phosphate groups. Each phosphate group is represented by a tetrahedron which is characterized according to the configurations of the protons in its four vertices. To each one of the 16 possible configurations of protons in the vertices of the tetrahedron, the Slater-Takagi theory associates a certain dipole moment and a certain energy as follows. Among the six configurations that have two protons, four have moment equal to zero and energy ϵ_0, one has moment equal to $+2\mu$ and energy zero and one has moment equal to -2μ and energy zero. Among the eight configurations that have one or three protons, four have moment equal to $+\mu$ and energy ϵ_1 and four have moment equal to $-\mu$ and energy ϵ_1. The other configurations are forbidden.

The molar Gibbs free energy $g(T, E)$ is determined by

$$g = \min_p \{f - Ep\}, \tag{15.84}$$

where the molar Helmholtz free energy f and the molar polarization p depend on the mole fractions of each type of configurations of the tetrahedrons. According to Ishibashi, the molar Helmholtz free energy f that is obtained from the Slater-Takagi theory is given by

$$f = u - Ts, \tag{15.85}$$

where u is the molar energy, given by

$$u = 4\epsilon_0 c_0 + 4\epsilon_1 (d_+ + d_-), \tag{15.86}$$

and s is the molar entropy, given by

$$\begin{aligned} s = R\{&2(c_+ + 2c_0 + 3d_+ + d_-) \ln(c_+ + 2c_0 + 3d_+ + d_-) \\ &+2(c_- + 2c_0 + d_+ + 3d_-) \ln(c_- + 2c_0 + d_+ + 3d_-) \\ &- c_+ \ln c_+ - c_- \ln c_- - 4c_0 \ln c_0 - 4d_+ \ln d_+ - 4d_- \ln d_-\}, \end{aligned} \tag{15.87}$$

where c_0, c_+, c_-, d_+ and d_- are the mole fractions of each type of configuration of the tetrahedrons and are connected by the relation

$$4c_0 + c_+ + c_- + 4d_+ + 4d_- = 1. \tag{15.88}$$

The polarization p is given by

$$p = [(c_+ - c_-) + 2(d_+ - d_-)]p_0, \tag{15.89}$$

where p_0 is the saturation polarization.

At zero field, which interest us here, g is obtained by the minimization of the expression (15.85) for f, subjected to the restriction (15.88). The minimization leads to the following results

$$c_0 = c_+^{1/2} c_-^{1/2} \eta_0, \tag{15.90}$$

$$d_+ = c_+^{3/4} c_-^{1/4} \eta_1, \tag{15.91}$$

$$d_- = c_-^{3/4} c_+^{1/4} \eta_1, \tag{15.92}$$

where $\eta_0 = e^{-\epsilon_0/RT}$ and $\eta_1 = e^{-\epsilon_1/RT}$ and

$$(c_+ + 2c_0 + d_-)d_- = (c_- + 2c_0 + d_+)d_+. \tag{15.93}$$

These relations together with (15.88) determine the five mole fractions as functions of temperature.

One solution is such that $c_+ = c_-$ and corresponds to the paraelectric phase since in this case $d_- = d_+$ and $p = 0$. The corresponding free energy is

$$g = -RT \ln \frac{1}{2}(1 + 2\eta_0 + 4\eta_1). \tag{15.94}$$

The solution such that $c_+ \neq c_-$ corresponds to the ferroelectric phase. It leads us to the following value for the polarization

$$p = \frac{\sqrt{(1 - 2\eta_0)^2 - (2\eta_1)^2}}{1 - 2\eta_0} p_0, \tag{15.95}$$

and to the following free energy

$$g = -RT \ln \left(1 + \eta_1 \sqrt{1 - (p/p_0)^2}\right). \tag{15.96}$$

The transition from the ferroelectric phase to the paraelectric phase occurs when the expressions (15.94) and (15.96) for the Gibbs free energy become equal. The equality between them leads to the following expression for the transition temperature T_0

$$2e^{-\epsilon_0/RT_0} + 2e^{-\epsilon_1/RT_0} = 1. \tag{15.97}$$

The original Slater theory corresponds to the case in which only the six configurations with two protons are allowed. The other are forbidden what is equivalent to saying that $\epsilon_1 \to \infty$ ($\eta_1 = 0$). In this case the ferroelectric phase has constant polarization $p = p_0$ for any temperature $T < T_0$ and the transition temperature is given by $RT_0/\epsilon_0 = 1/\ln 2$. When ϵ_1 is finite, which corresponds to the Slater-Takagi theory, the polarization decreases with temperature and vanishes continuously at the transition temperature T_0.

It is worth mentioning that the Slater theory ($\eta_1 = 0$), for the case in which $\epsilon_0 = 0$ ($\eta_0 = 1$), is reduced to the theory introduced by Pauling to explain the residual entropy of ordinary ice. In this case, the six allowed configurations of the tetrahedron are associated to the same energy and the system exhibits only the paraelectric phase. From (15.94) we see that $g = -RT\ln(3/2)$, from which we conclude that $s = -\partial g/\partial T = R\ln(3/2)$ which is the result due to Pauling for the residual entropy of the ordinary ice.

Problems

15.1 Electrocaloric effect. Show that $(\partial T/\partial E)_S = -TA/C_E$, where A is the pyroelectric effect defined by $A = (\partial P/\partial T)_E$. Since $C_E \geq 0$, materials whose polarization decreases with temperatures at constant field ($A < 0$) have the temperature decreased when the field is decreased adiabatically.

15.2 Show that the molar free energy $g(T, E)$ of a dielectric that obeys the equation of state (15.28) is given by

$$g = -\frac{1}{2}\alpha E^2 - RT\left(\ln\sinh\frac{p_0 E}{RT} - \ln\frac{p_0 E}{RT}\right).$$

Determine the molar entropy s and show that the molar energy $u = g + Ts + pE$ is given by

$$u = \frac{1}{2}\alpha E^2 + u_0,$$

where u_0 depends only on temperature.

15.3 At zero temperature the free energy (15.73) is reduced to

$$f = -\frac{\alpha}{2}p^2, \qquad p^2 + q^2 \leq p_0^2.$$

Show that the minimization of $f - Ep - \Gamma q$ results in the polarization given by (15.83).

15.4 Find the transition temperatures among the three ferromagnetic phases and the paramagnetic phase of a system described by the free energy given by the expression (15.60). Determine also the spontaneous polarization of the three ferroelectric phases.

Chapter 16
Solids

16.1 Stress and Strain

Anisotropy

A solid substance is characterized by having a spatial structure consisting of a three-dimensional lattice at whose vertices the atoms are located. Because the atoms are in continuous motion due to thermal agitation, the vertices of the lattice are defined, more properly, as the equilibrium positions around which the atoms vibrate. This structure gives to the solid the rigidity that opposes not only the compression and stretch but also shear. In other words, offering resistance to the volume and shape changes. The spatial structures can be ordered as those of the crystalline solid or disordered such as those of the amorphous solids and glasses.

Crystal structures are classified according to the symmetries they exhibit. The set of symmetry operations of a crystalline structure comprises a symmetry group. The symmetry groups are brought together in classes and the classes in crystal systems. There are seven crystal systems: triclinic, monoclinic, orthorhombic, tetragonal, trigonal, hexagonal and cubic. The structures of the cubic system can be simple cubic (sc), body-centered cubic (bcc) and face-centered cubic (fcc). In a simple solid, composed of one kind of atom, that orders according to the sc structure, the atoms are located only at the vertices of the cubes forming the ordered spatial structure. In a simple solid with ccc structure the atoms are at the vertices and at the centers of the cubes and in a simple solid with fcc structure, they lie at the vertices and centers of the sides of the cubes.

Due to the spatially ordered structure, crystalline solids, in contrast to gases and liquids, are anisotropic. Elasticity may depend on the direction along which the solid is pulled or compressed. The thermal expansion on a direction may differ from that on a different direction. It should be noted that variations in properties arising from the anisotropy are observed in single crystals. Polycrystals, and the amorphous and glassy solids, are from the macroscopic point of view, examples of isotropic solids.

M.J. de Oliveira, *Equilibrium Thermodynamics*, Graduate Texts in Physics,
DOI 10.1007/978-3-662-53207-2_16

This does not mean that they behave, as to the elastic properties, like liquids and gases, which are also isotropic. Solids, isotropic or not, are distinguished from fluids by resisting shear.

Stress Tensor

External forces acting on the surface of a solid can induce the appearance of stress and shear inside the solid. In a fluid external forces only induce hydrostatic compressions and possibly internal motions. The hydrostatic compressions are forces perpendicular to the surface of any internal region and have the same magnitude per unit area at any point on the surface. In a solid, in general, the forces on an internal region are not perpendicular to the surface and do not have the same magnitude per unit area. This complexity of the forces requires a more detailed description of the mechanical equilibrium than that given simply by hydrostatic compression. The internal forces in a solid are described by means of a quantity similar to the pressure, called stress. Unlike pressure, which is a scalar, stress is a tensor quantity and is defined as follows.

Consider a sample of a solid in the form of a rectangular parallelepiped of edges ℓ_1, ℓ_2 and ℓ_3, with edges parallel to a Cartesian coordinate system, as shown in Fig. 16.1a. We denote by $\vec{f}_1, \vec{f}_2$ and $\vec{f}_3$ the forces acting on the faces A, B and C, respectively. The forces acting on opposite sides are, respectively, $-\vec{f}_1$, $-\vec{f}_2$ and $-\vec{f}_3$ so that the resultant of the forces on the sample is zero. The decomposition of these Cartesian vectors

$$\vec{f}_1 = f_{11}\hat{x} + f_{12}\hat{y} + f_{13}\hat{z}, \tag{16.1}$$

$$\vec{f}_2 = f_{21}\hat{x} + f_{22}\hat{y} + f_{23}\hat{z}, \tag{16.2}$$

$$\vec{f}_3 = f_{31}\hat{x} + f_{32}\hat{y} + f_{33}\hat{z}, \tag{16.3}$$

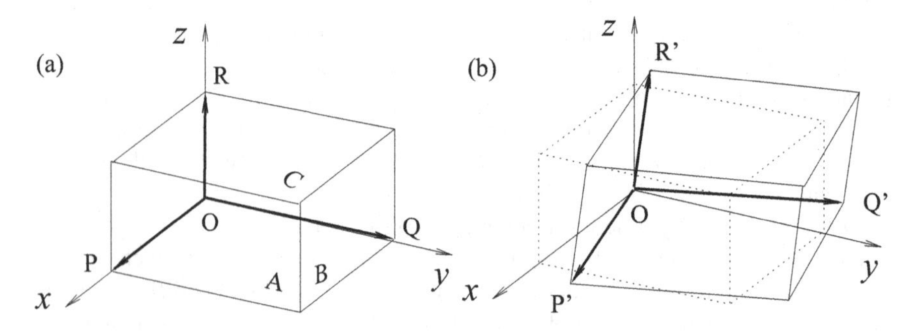

Fig. 16.1 Sample of a homogeneous solid in the form of a parallelepiped. (**a**) Before the deformation the length and direction of the edges are described by the vectors OP= $\ell_1\hat{x}$, OQ= $\ell_2\hat{y}$ and OR= $\ell_3\hat{z}$. (**b**) After the deformation they are described by OP'= $\vec{\ell}_1$, OQ'= $\vec{\ell}_2$ and OR'= $\vec{\ell}_3$

defines the Cartesian components f_{ij} of these vectors. Since the sample is in mechanical equilibrium torques on it should vanish. This requirement applied to each of the three Cartesian axes results in the following condition

$$\ell_i f_{ij} = \ell_j f_{ji}. \tag{16.4}$$

In order to describe efforts which are independent of sample size, what can be done by considering homogeneous samples, we introduce forces per unit area

$$\sigma_{ij} = \frac{f_{ij}}{A_i}, \tag{16.5}$$

where A_i are the area of the three faces of the parallelepiped, given by $A_1 = \ell_2\ell_3$, $A_2 = \ell_3\ell_1$ and $A_3 = \ell_1\ell_2$. They have the unit of pressure and comprise the elements of a tensor σ called *stress tensor*. The diagonal elements describe the stretches and are positive in a proper stretch and negative in a compression. The off-diagonal elements describe the shear stresses. Due to property (16.4) we see that

$$\sigma_{ij} = \sigma_{ji}, \tag{16.6}$$

which means that the stress tensor is symmetrical.

The stress tensor can be decomposed into a tensor of zero trace and a tensor that represents a hydrostatic compression. Indeed, from the identity

$$\sigma_{ij} = (\sigma_{ij} + p\delta_{ij}) - p\delta_{ij}, \tag{16.7}$$

where $p = -(\sigma_{11} + \sigma_{22} + \sigma_{33})/3$, we see that the tensor represented by the first term has zero trace. The second term constitutes the components of a tensor that has only diagonal elements and are all equal. This second tensor, defined by $-p\delta_{ij}$, represents a hydrostatic compression at a pressure p since the forces $\vec{f}_i$ that describe a hydrostatic compression are perpendicular to the faces of the parallelepiped of Fig. 16.1a and are proportional to the areas of the faces, that is, $f_{ij} = -pA_i\delta_{ij}$.

Strain Tensor

The description of the mechanical state of a sample of a solid can not be made only by its volume as happens to a fluid in thermodynamic equilibrium. Due to anisotropy, the deformations resulting from forces of the same intensity may have different magnitudes and may depend on the direction of the forces. In other words, the deformations are not isotropic. To properly describe the mechanical state of a solid we introduce a quantity called strain tensor, defined as follows.

Consider again the sample of a homogeneous solid in the form of a rectangular parallelepiped with edges ℓ_1, ℓ_2, and ℓ_3, as shown in Fig. 16.1a. Then consider a deformation of this body as shown in Fig. 16.1b. The lengths and directions of edges of the parallelepiped are described by the vectors $\vec{\ell}_1$, $\vec{\ell}_2$ and $\vec{\ell}_3$, whose Cartesian decomposition

$$\vec{\ell}_1 = \ell_{11}\hat{x} + \ell_{12}\hat{y} + \ell_{13}\hat{z}, \tag{16.8}$$

$$\vec{\ell}_2 = \ell_{21}\hat{x} + \ell_{22}\hat{y} + \ell_{23}\hat{z}, \tag{16.9}$$

$$\vec{\ell}_3 = \ell_{31}\hat{x} + \ell_{32}\hat{y} + \ell_{33}\hat{z}, \tag{16.10}$$

define the Cartesian components ℓ_{ij} of these vectors.

In order to describe the deformation of the sample regardless of size we introduce the relative displacements ξ_{ij} defined by

$$\ell_{ij} = \ell_i(\delta_{ij} + \xi_{ij}). \tag{16.11}$$

They are dimensionless and comprise the elements of a tensor ξ called *strain tensor*. The diagonal elements describe stretch deformations and the off-diagonal elements describe shear deformations. Given that the sample is homogeneous, the strain tensor does not depend on the size of the sample used to determine it. The volume V_0 of the region before the deformation is

$$V_0 = \ell_1\ell_2\ell_3. \tag{16.12}$$

After deformation the volume V is

$$V = |\vec{\ell}_1 \times \vec{\ell}_2 \cdot \vec{\ell}_3|. \tag{16.13}$$

The mechanical state of a solid body in equilibrium is defined not only by the strain tensor ξ but also by a reference state from which we define this tensor. We adopt as the reference state a rectangular parallelepiped of edges ℓ_1, ℓ_2 and ℓ_3 like that shown in Fig. 16.1a. Any other state, like the one shown in Fig. 16.1b, is defined by the vectors $\vec{\ell}_1$, $\vec{\ell}_2$ and $\vec{\ell}_3$, which are related to the strain tensor and the reference state by means of (16.11).

The components of the strain tensor describe deformations of any magnitude. However, elastic deformations of a solid body are small in general which allows us to restrict ourselves to the study of the case in which the components of the strain tensor is small compared with the unit, that is, $\xi_{ij} << 1$. In this regime of small strain the sample volume, given by (16.13) reduces to

$$V = V_0(1 + \xi_{11} + \xi_{22} + \xi_{33}), \tag{16.14}$$

expression valid up to linear terms in ξ_{ij} and depends only on the diagonal components of the strain tensor.

An equivalent description of the mechanical state of a solid in equilibrium can be made through the variable V given by (16.14) together with the variables η_{ij} defined by

$$\eta_{ij} = \xi_{ij} - \frac{1}{3}\delta_{ij}(\xi_{11} + \xi_{22} + \xi_{33}), \tag{16.15}$$

which constitute the components of the tensor η. This tensor has zero trace and for this reason only two of its diagonal components can be chosen as independent. In a process in which the volume remains constant, it describes the change in shape.

Work of Elastic Forces

To determine the work of elastic forces we look at the forces $\vec{f}_1, \vec{f}_2$ and $\vec{f}_3$ acting on the faces A, B and C, whose Cartesian components are defined by (16.1)–(16.3). The forces acting on opposite faces are considered to be equal in magnitude but in opposite directions. These forces are related to the stress tensor in accordance with

$$f_{ij} = A_i \sigma_{ij}, \tag{16.16}$$

where A_i are the areas of the projections of the faces of the oblique parallelepiped of Fig. 16.1b on the three planes of the Cartesian system. As we are in the regime of small deformations the areas of the sides A, B and C can be approximated by $A_1 = \ell_2\ell_3$, $A_2 = \ell_3\ell_1$ and $A_3 = \ell_1\ell_2$.

To determine the work of forces we consider the displacements of the faces. Since the forces acting on opposite faces are antiparallel, we assume that the displacements of the faces are antiparallel. Denoting by $d\vec{\ell}_i/2$ the infinitesimal displacement of a face, on which acts the force $\vec{f}_i$, then the infinitesimal displacement of the opposite face will be $-d\vec{\ell}_i/2$ and will be subject to a force $-\vec{f}_i$. The infinitesimal work of the forces acting on the six faces of the sample is then

$$dW = \sum_i \vec{f}_i \cdot d\vec{\ell}_i = \sum_{ij} f_{ij} d\ell_{ij}. \tag{16.17}$$

According to the definition of the strain tensor, $d\ell_{ij} = \ell_i d\xi_{ij}$. Using this result and the relation (16.16) between the forces and the stress tensor, we obtain the following result for the infinitesimal work

$$dW = V_0 \sum_{ij} \sigma_{ij} d\xi_{ij}. \tag{16.18}$$

where $V_0 = \ell_1\ell_2\ell_3$ is the volume of the sample in the reference state. The volume V in state ξ_{ij}, in the regime of small deformations, is given by (16.14) from which we conclude that

$$dV = V_0(d\xi_{11} + d\xi_{22} + d\xi_{33}). \tag{16.19}$$

When the stress tensor σ corresponds to a hydrostatic pressure, that is, $\sigma_{ij} = -p\delta_{ij}$, we see that the work is reduced to $dW = -pdV$, which is the form used in gases and liquids.

Given that the strain tensor is symmetric, there are only six independent components of the tensor. The expression for the work will have as a consequence only six independent terms. Thus it is appropriate to introduce the *stress components* σ_i, defined by $\sigma_1 = \sigma_{11}$, $\sigma_2 = \sigma_{22}$, $\sigma_3 = \sigma_{33}$, $\sigma_4 = \sigma_{23}$, $\sigma_5 = \sigma_{13}$ and $\sigma_6 = \sigma_{12}$, and the *strain components* ξ_i, defined by $\xi_1 = \xi_{11}$, $\xi_2 = \xi_{22}$, $\xi_3 = \xi_{33}$, $\xi_4 = \xi_{23} + \xi_{32}$, $\xi_5 = \xi_{13} + \xi_{31}$ and $\xi_6 = \xi_{12} + \xi_{21}$. The infinitesimal work reads then

$$dW = V_0 \sum_{k=1}^{6} \sigma_k d\xi_k, \tag{16.20}$$

and the infinitesimal volume reads

$$dV = V_0 \sum_{k=1}^{3} d\xi_k. \tag{16.21}$$

Thermodynamic Potentials

The variation of the internal energy U of a sample is the sum of the heat introduced and the work of the forces acting on the sample, given above. In differential form

$$dU = TdS + V_0 \sum_{k=1}^{6} \sigma_k d\xi_k. \tag{16.22}$$

The internal energy $U(S, \xi)$ is thus a function of entropy S and the strain components ξ_k. From U we get other thermodynamic potentials such as the Helmholtz free energy $F = U - TS$, whose differential is

$$dF = -SdT + V_0 \sum_{k=1}^{6} \sigma_k d\xi_k, \tag{16.23}$$

from which we get the relations

$$\sigma_i = \frac{1}{V_0}\left(\frac{\partial F}{\partial \xi_i}\right)_T, \tag{16.24}$$

where the derivative is carried out by keeping constant the other strain components and the temperature. The Helmholtz free energy $F(T, \xi)$ is a function of temperature T and the strain components ξ_k.

The Gibbs free energy $G = F + pV$ is obtained by a Legendre transformation from F. Using the description of the mechanical state defined by the tensor η and the volume V, we get the following differential form

$$dG = -SdT + Vdp + V_0 \sum_i \sigma_i d\eta_i. \tag{16.25}$$

We must remember however that η_1, η_2 and η_3 are not independent because $\eta_1 + \eta_2 + \eta_3 = 0$.

In addition to F and G other thermodynamic potentials, obtained by Legendre transformations, can be used to describe the elastic properties a solid. A particularly useful potential is the one defined by $\Phi = F - V_0(\sigma_1\xi_1 + \ldots + \sigma_6\xi_6)$ and whose differential form is

$$d\Phi = -SdT - V_0 \sum_{k=1}^{6} \xi_k d\sigma_k, \tag{16.26}$$

from which follows

$$\xi_i = -\frac{1}{V_0}\left(\frac{\partial \Phi}{\partial \sigma_i}\right)_T, \tag{16.27}$$

where the derivative is performed keeping constant the other stress components and the temperature.

16.2 Thermodynamic Coefficients

Elastic Coefficients

Several thermodynamic coefficients are used to characterize the thermodynamic properties of solids. These coefficients are similar to those used in the description of gases and liquids such as the compressibility, the coefficient of thermal expansion and heat capacity. The coefficients that describe the elasticity relate the variations of stress and variations in strain in the form of quotients. The isothermal *stiffness coefficient* is defined by

$$c_{ij} = \left(\frac{\partial \sigma_i}{\partial \xi_j}\right)_T, \tag{16.28}$$

where the derivative is performed by keeping constant the temperature and the other strain components. The property of convexity of the potential F with respect the set of variables ξ_i implies that matrix with coefficients c_{ij} has nonnegative eigenvalues. particularly the convexity of F implies $c_{ii} \geq 0$.

The isothermal *compliance coefficient* is defined by

$$\kappa_{ij} = \left(\frac{\partial \xi_i}{\partial \sigma_j} \right)_T, \tag{16.29}$$

where the derivative is performed keeping constant the temperature and the other stress components. The property of convexity of the potential Φ with respect to the set of variables σ_i implies that the matrix with the coefficients κ_{ij} has nonnegative eigenvalues. Particularly the convexity of Φ implies $\kappa_{ii} \geq 0$.

The coefficient κ_{ij} is not generally the inverse of c_{ij}. However, they are closely linked. Considering in one hand that the stress components σ_i are functions of strain components ξ_j and on the other hand that latter are functions of the former then the following identities can be achieved

$$\sum_{j=1}^{6} \left(\frac{\partial \sigma_i}{\partial \xi_j} \right)_T \left(\frac{\partial \xi_j}{\partial \sigma_k} \right)_T = \delta_{ik}. \tag{16.30}$$

From this identity is follows immediately the result

$$\sum_{j=1}^{6} c_{ij} \kappa_{jk} = \delta_{ik}, \tag{16.31}$$

which says the matrix with elements κ_{ij} is the inverse of the matrix with elements c_{jk}.

The coefficients c_{ij} are not all independent. Using relations analogous to Maxwell relations

$$\left(\frac{\partial \sigma_i}{\partial \xi_j} \right)_T = \left(\frac{\partial \sigma_j}{\partial \xi_i} \right)_T, \tag{16.32}$$

it follows $c_{ij} = c_{ji}$. Similarly we may conclude that $\kappa_{ij} = \kappa_{ji}$. Therefore, from the 36 coefficients c_{ij}, or κ_{ij}, only 21 are independent. The number of independent coefficients can be further reduced if we take into account the symmetries of the crystalline structure.

The triclinic system does not impose any restriction and has therefore 21 independent coefficients. However, in this case it is possible to orient the sample such that three coefficients vanish reducing the number of independent coefficients to 18. In the monoclinic system, the number of independent coefficients is 13 but the orientation of the sample can reduce them to 12. In the orthorhombic system the number is nine. Some classes of the trigonal and tetragonal systems have 7 independent coefficients but they may be reduced to 6 by the orientation of the sample. Other classes of these two systems have six independent coefficients. The hexagonal system has 5 and cubic system has only 3.

In the cubic system we take as independent c_{11}, c_{12} and c_{44}. The other are given by

$$c_{33} = c_{22} = c_{11}, \qquad c_{23} = c_{13} = c_{12}, \qquad c_{66} = c_{55} = c_{44}. \tag{16.33}$$

The remaining coefficients vanish. These results are summarized in the matrix

$$\begin{pmatrix} c_{11} & c_{12} & c_{12} & 0 & 0 & 0 \\ c_{12} & c_{11} & c_{12} & 0 & 0 & 0 \\ c_{12} & c_{12} & c_{11} & 0 & 0 & 0 \\ 0 & 0 & 0 & c_{44} & 0 & 0 \\ 0 & 0 & 0 & 0 & c_{44} & 0 \\ 0 & 0 & 0 & 0 & 0 & c_{44} \end{pmatrix}. \tag{16.34}$$

The convexity of F implies that the matrix (16.34) has nonnegative eigenvalues which result in the following restrictions

$$c_{11} - c_{12} \geq 0, \qquad c_{11} + 2c_{12} \geq 0, \qquad c_{44} \geq 0. \tag{16.35}$$

From the firs two we get $c_{11} \geq 0$.

Relations similar to those of the stiffness coefficients c_{ij} related to the cubic system are valid for the compliance coefficients

$$\kappa_{33} = \kappa_{22} = \kappa_{11}, \qquad \kappa_{23} = \kappa_{13} = \kappa_{12}, \qquad \kappa_{66} = \kappa_{55} = \kappa_{44}. \tag{16.36}$$

The other coefficients κ_{ij} vanish. The convexity of the potential Φ implies the conditions

$$\kappa_{11} - \kappa_{12} \geq 0, \qquad \kappa_{11} + 2\kappa_{12} \geq 0, \qquad \kappa_{44} \geq 0, \tag{16.37}$$

and $\kappa_{11} \geq 0$, which follows from the first two.

Compressibility

A solid body immersed in a liquid undergoes compression which are perpendicular to its surface and of the same intensity anywhere in the surface. Under these conditions the solid body undergoes a hydrostatic compression represented by the pressure p. The compressibility κ is defined as the ratio of the relative decrease of the volume and the increase in hydrostatic pressure

$$\kappa = -\frac{1}{V_0}\left(\frac{\partial V}{\partial p}\right)_T. \tag{16.38}$$

This expression defined the isothermal compressibility. The adiabatic compressibility is defined similarly.

The compressibility of solids is related with the elasticity coefficients. Indeed, using the result $(\partial V/\partial \xi_k) = V_0$, $k = 1, 2, 3$, which follows from (16.21), we see that

$$\kappa = -\sum_{i=1}^{3}\left(\frac{\partial \xi_i}{\partial p}\right)_T = -\sum_{i,j=1}^{3}\left(\frac{\partial \xi_i}{\partial \sigma_j}\right)_T\left(\frac{\partial \sigma_j}{\partial p}\right). \tag{16.39}$$

Since $\sigma_1 = \sigma_2 = \sigma_3 = -p$ for a hydrostatic compression and using the definition of the coefficient κ_{ij}, we get the relation

$$\kappa = \sum_{i,j=1}^{3}\kappa_{ij}. \tag{16.40}$$

The bulk modulus B is defined as the inverse of the compressibility, $B = 1/\kappa$.

For a solid with cubic symmetry

$$\kappa = 3(\kappa_{11} + 2\kappa_{12}). \tag{16.41}$$

Using the relation between κ_{ij} and c_{ij} we get $B = 1/\kappa$ in terms of c_{ij} for the cubic system

$$B = \frac{1}{3}(c_{11} + 2c_{12}). \tag{16.42}$$

Thermal Expansion and Heat Capacity

The coefficients of thermal expansion are defined by

$$\alpha_i = \left(\frac{\partial \xi_i}{\partial T}\right), \tag{16.43}$$

where the derivative is carried out by keeping all the six stress components σ_j fixed. Recalling that the stress tensor is symmetric, we can make a rotation of the Cartesian axes so as to transform the stress tensor in diagonal form. From this result we conclude that only three coefficients of expansion are independent. This happens with crystalline solids of the systems triclinic, monoclinic and orthorhombic. Solids of the tetragonal, trigonal and hexagonal crystalline systems have greater symmetry and are described by two independent coefficients of expansion. The crystalline solids of the cubic system, the highest symmetry, are described by only one coefficient of expansion and accordingly behave like isotropic solids. Therefore, for a solid cubic symmetry, $\alpha_1 = \alpha_2 = \alpha_3$. An immediate consequence of this result is that thermal expansion of polycrystalline samples of cubic structure do not produce tensions between monocrystalline grains because each expands isotropically. In contrast, the thermal expansion of polycrystalline samples of noncubic structure causes internal tensions because the monocrystalline grains expand in distinct forms along a same direction.

In analogy with fluid, for which one defines heat capacity at constant volume and at constant pressure, here we define the heat capacity for which all the strain components ξ are kept constant, given by

$$C_\xi = T\left(\frac{\partial S}{\partial T}\right)_\xi, \tag{16.44}$$

and the heat capacity for which all stress components σ_i are kept constant, given by

$$C_\sigma = T\left(\frac{\partial S}{\partial T}\right)_\sigma. \tag{16.45}$$

Hooke Law

Consider a sample of a crystalline solid in a nondeformed state at a temperature T_0, that is, such that the stresses and compressions are absent. This nondeformed state is described by the vanishing of the strain components, which we consider to be the reference state. Next we consider a deformation of the sample which passes to a be in a state defined by the strain components ξ_i and temperature T.

Assuming the strain to be small, that is, such that $\xi_i << 1$, we expand the free energy $f = F/V_0$ in power of the strain components up to quadratic terms to obtain the result

$$f = f_0 - (T - T_0)\sum_i a_i\xi_i + \frac{1}{2}\sum_i c_{ii}\xi_i^2 + \sum_{i<j} c_{ij}\xi_i\xi_j, \tag{16.46}$$

where f_0 depends only on T. The coefficients a_i and c_{ij} are determined at the temperature T_0.

From this expansion for f and using (16.24) we get the stress components,

$$\sigma_i = -a_i(T - T_0) + \sum_{j=1}^{6} c_{ij}\xi_j. \tag{16.47}$$

If the process is isothermal, $T = T_0$, this equation is the expression of the Hooke law for crystalline solids.

Taking into account that the matrix κ_{ij} is the inverse of the matrix c_{ij}, we reach from (16.47) the result

$$\xi_i = \alpha_i(T - T_0) + \sum_{j=1}^{6} \kappa_{ij}\sigma_j, \tag{16.48}$$

where

$$\alpha_i = \sum_{j=1}^{6} a_i\kappa_{ij}, \tag{16.49}$$

is the coefficient of thermal expansion. If the process is isothermal equation (16.48) is again the expression of the Hooke law.

Suppose next a thermal expansion carried out in the absence of tensions and compressions. This means a process such that $\sigma_i = 0$. In this case,

$$\xi_i = \alpha_i(T - T_0). \tag{16.50}$$

For solids with cubic symmetry, the deformation is isotropic, $\alpha_3 = \alpha_2 = \alpha_1$ and $\xi_3 = \xi_2 = \xi_1$.

Isotropic Solids

We have seen that the compliance coefficients that describe a crystalline solid are not all independent. Each crystal system imposes constraints arising from the symmetries that reduce the number of the independent coefficients. The lowest number is three and corresponds to the crystal system with highest symmetry which is the cubic system. The isotropic solids impose an additional restriction reducing the number to only two. Many solid behave as isotropic with respect to the elastic properties. Examples of isotropic solids include the glassy and amorphous solids and somehow also the polycrystalline solids.

The relation among the stiffness coefficients are the same as that valid for solids of the cubic system, given by (16.33), with the additional restriction $c_{11} = c_{12}+2c_{44}$ which is equivalent to the parametrization $c_{11} = \lambda + 2\mu$, $c_{12} = \lambda$ and $c_{44} = \mu$, where λ and μ are the Lamé coefficients. The matrix of the coefficients is

$$\begin{pmatrix} \lambda + 2\mu & \lambda & \lambda & 0 & 0 & 0 \\ \lambda & \lambda + 2\mu & \lambda & 0 & 0 & 0 \\ \lambda & \lambda & \lambda + 2\mu & 0 & 0 & 0 \\ 0 & 0 & 0 & \mu & 0 & 0 \\ 0 & 0 & 0 & 0 & \mu & 0 \\ 0 & 0 & 0 & 0 & 0 & \mu \end{pmatrix}. \tag{16.51}$$

The coefficient $\mu = c_{44}$ is identified with the ratio between the increase in the shear stress and the shear strain, determined by keeping constant the other strain components. The coefficient $\lambda = c_{12}$ is the ratio between the increase in the stretch stress and the deformation perpendicular to it, keeping constant the other strain components. Since the eigenvalues of the matrix (16.51) should be nonnegative, the Lamé coefficients obey the conditions

$$\mu \geq 0, \qquad 3\lambda + 2\mu \geq 0. \tag{16.52}$$

Similarly the compliance coefficients of an isotropic solid have the same relations valid for the cubic system, given by (16.36), with the additional restriction $\kappa_{11} = \kappa_{12} + \kappa_{44}/2$, which is equivalent to the parametrization

$$\kappa_{11} = \frac{1}{E}, \qquad \kappa_{12} = -\frac{\nu}{E}, \qquad \kappa_{44} = \frac{2}{E}(1+\nu), \tag{16.53}$$

where E is the Young modulus and ν is the Poisson ratio. The Young modulus is the ratio between the increase in the stretch stress and the deformation along the stretch carried out keeping constant the other stress components. That is,

$$E = \frac{1}{\kappa_{11}} = \left(\frac{\partial \sigma_1}{\partial \xi_1}\right)_T, \tag{16.54}$$

where the derivative is performed at constant σ_i, $i \neq 1$, and should not be confused with c_{11}. The Poisson ratio is the ratio between the transverse and longitudinal deformation when the sample is pulled longitudinally keeping constant the pressure and other stress components, that is,

$$\nu = -\frac{\kappa_{12}}{\kappa_{11}} = -\left(\frac{\partial \xi_2}{\partial \xi_1}\right)_T, \tag{16.55}$$

where the derivative is carried at constant σ_i, $i \neq 1$. The relations between these two quantities and the Lamé coefficients are given by

$$E = \frac{\mu(3\lambda + 2\mu)}{\lambda + \mu}, \qquad \nu = \frac{\lambda}{2(\lambda + \mu)}, \tag{16.56}$$

which are obtained recalling that the matrices c_{ij} and κ_{ij} are the inverses of each other.

The conditions of stability (16.51) lead us to the following restriction on E and ν

$$E \geq 0, \qquad -1 \leq \nu \leq \frac{1}{2}. \tag{16.57}$$

Although the Poisson ratio ν can be negative there is no known substance with such a property.

Using relation (16.41) between the compressibility and the compliance coefficients and the parametrization (16.53) we see that the compressibility is given by

$$\kappa = \frac{3}{E}(1 - 2\nu). \tag{16.58}$$

The bulk modulus $B = 1/\kappa$ is given by

$$B = \lambda + \frac{2}{3}\mu, \tag{16.59}$$

expression obtained by using (16.56).

16.3 Structural Phase Transition

Polymorphism

Solids of the same chemical composition but with different crystal structures are called structural polymorphic forms. Iron, for example, can be found in three polymorphic forms, known as α, γ and δ. Phase α, or ferrite, has a bcc structure is the stable phase up to 910 °C. From this temperature up to 1400 °C, the stable phase is the phase γ, or austenite, which has a fcc structure. Above 1400 °C to the melting point, which occurs at 1540 °C, the stable phase is the phase δ, which is also bcc. Silica (SiO_2) can also be found in various polymorphic forms, among which we cite quartz, tridymite and cristobalite. The diamond and graphite are polymorphic forms of carbon. It should be noted that many polymorphic forms are metastable, some with decay time so great that may be considered stable, as diamond. At high pressures however diamond becomes stable and graphite metastable. Some polymorphic forms on the other hand are always metastable, as happens to the phase of iron-carbon known as martensite, obtained by rapid cooling of austenite.

When subjected to temperature and pressure variations, a solid can undergo a structural transformation, passing from a polymorphic form to another. The change in structural form is identified with a phase transition, but should be distinguished from that occurring between liquid and vapor. These two phases are quantitatively different, because they differ by the density, but are qualitatively similar in structure, because the two are isotropic. This similarity allows for example the vapor to change continuously into liquid, that is, without undergoing a phase transition, if it is conducted by an appropriate trajectory in the phase diagram that bypasses the liquid-vapor critical point. Two phases with different structures on the other hand are qualitatively distinct because they differ by the symmetries they display. The passage from one structure to another occurs always through a phase transition, which is characterized by the occurrence of singularities in thermodynamic properties.

The structural transitions are governed by different microscopic mechanisms. The transition of order-disorder type as that which occurs in the copper-zinc alloy causes a change in crystal structure. The ferroelectric transition of certain substances such as KH_2PO_4 or $NaNO_2$ are also of order-disorder type and are accompanied by a crystallographic change. A structural transition of another kind is that caused by coordinated movement of atoms of a sublattice of the crystal structure and called displacive transition. This type of transition can involve two or more sublattices moving in different directions as happens to $SrTiO_3$. If a sublattice is composed of ions, its displacement can produce an electric polarization giving rise to a ferroelectric phase, as with $BaTiO_3$.

Another type of structural transition is that caused by the elastic distortion, such as with TeO_2. This compound undergoes a continuous transition by pressure increase from a tetragonal structure to an orthorhombic structure. The latter, considered distorted with respect to the former, is called ferroelastic. The undistorted structure is called paraelastic. The class of compounds which undergo this same type of structural transition includes the compound V_3Si, which undergoes a discontinuous transition from a cubic to a tetragonal structure below 17 K, and the compound $LaNbO_4$, that undergoes a continuous transition from a tetragonal to a monoclinic structure below 495 °C.

Ferroelastic Transition

We wish to analyze here the ferroelastic-paraelastic transition occurring in a crystalline solid when the temperature and pressure are varied. To this end we examine a sample at a given temperature T and subject to a hydrostatic compression p. The only mechanical stress on the sample is the hydrostatic compression that is to say the stretch components of the stress are equal and the shear stress components are zero. Here it is convenient to use the Gibbs potential because these conditions are naturally satisfied when the tensor η takes the values corresponding to minimum of $G(\eta)$ subject to the constraint $\eta_1 + \eta_2 + \eta_3 = 0$.

The order parameter of the ferroelastic transition is defined as the tensor η. In the paraelastic phase the tensor η vanishes and in the ferroelastic phase one of its components becomes nonzero. According to Landau theory, which we adopt here, the function $g(\eta) = G/V_0$ has a polynomial form and should be invariant under the symmetry operations corresponding to the paraelastic phase. Since η has trace zero we use the following parameterization for diagonal components:

$$\eta_1 = -\frac{1}{2}x + \frac{\sqrt{3}}{2}y, \qquad \eta_2 = -\frac{1}{2}x - \frac{\sqrt{3}}{2}y, \qquad \eta_3 = x. \tag{16.60}$$

Initially we consider a transition such that the paraelastic phase has cubic structure and the ferroelastic phase has tetragonal structure, described by $\eta_4 = \eta_5 = \eta_6 = 0$. In this case it suffices to consider g as a function of x and y only. A polynomial of fourth degree with cubic invariance is given by

$$g = g_0 + A_1(x^2 + y^2) + B_1 x(x^2 - 3y^2) + C_1(x^2 + y^2)^2, \tag{16.61}$$

where g_0, A_1, B_1 and C_1 depend only on T and p. The free energy g should be understood as the convex hull of the expression on the right hand side of this equation. Using the parametrization $x = r\cos\theta$ and $y = r\sin\theta$, $r > 0$,

$$g = g_0 + A_1 r^2 + B_1 r^3 \cos 3\theta + C_1 r^4. \tag{16.62}$$

The minimum of g related to the ferroelastic phase occurs when $\cos 3\theta = -1$, for $B > 0$, and $\cos 3\theta = 1$, for $B < 0$. In both cases,

$$g = g_0 + A_1 r^2 - |B_1| r^3 + C_1 r^4, \tag{16.63}$$

where r is given by

$$2A_1 - 3|B_1| r + 4C_1 r^2 = 0. \tag{16.64}$$

The three solutions corresponding to the three possible values of θ are equivalent, that is, they have the same value of g and describe a tetragonal structural. The absolute minimum of g occurs at $r = 0$ for $A_1 > A_{10} = B_1^2/4C_1$ and at a value $r \neq 0$ given by (16.64) for $A_1 < A_{10}$. The transition occurs when $A_1 = A_{10}$ and corresponds to a discontinuous transition, that is, the order parameter r has a jump at $A_1 = A_{10}$.

Next we consider the case in which the paraelastic phase has cubic structure and the ferroelastic phase has trigonal structure, described by $\eta_1 = \eta_2 = \eta_3$ that is by $x = y = 0$. A polynomial of the fourth degree with cubic invariance in the variables η_4, η_5 and η_6 is given by

$$g = g_0 + A_2(\eta_4^2 + \eta_5^2 + \eta_6^2) + B_2\eta_4\eta_5\eta_6 + C_2(\eta_4^4 + \eta_5^4 + \eta_6^4). \tag{16.65}$$

The minimum of g corresponding to the ferroelastic phase occurs when $|\eta_4| = |\eta_5| = |\eta_6| = z$ with $\eta_4\eta_5\eta_6 < 0$ if $B_2 > 0$ and vice-versa. In both case g is

$$g = g_0 + 3A_2 z^2 - |B_2| z^3 + 3C_2 z^4, \tag{16.66}$$

and z is given by

$$2A_2z - |B_2|z^2 + 4C_2z^3 = 0. \tag{16.67}$$

There are four ferroelastic solutions and they have the same value of g. The absolute minimum of g occurs at $z = 0$ for $A_1 > A_{10} = B_1^2/36C_1$ and at a value $z \neq 0$ given by (16.67) for $A_1 < A_{10}$. The transition occurs at $A_1 = A_{10}$ and corresponds to a discontinuous transition, that is, the order parameter z has a jump at $A_1 = A_{10}$.

Devonshire Theory

In the structural transitions of displacive type as those occurring in the compounds $SrTiO_3$ and $BaTiO_3$, the order parameter is not identified with the strain tensor η. In displacive transitions, the order parameter is identified with the vector $\vec{q}$ that defines the equilibrium position of a particular atom relative to the equilibrium position it possessed in the structure of higher symmetry. When the displacement $\vec{q}$ becomes nonzero it induces the appearance of distortions in the crystalline solid, characterized by the strain tensor η. We therefore assume the existence of a coupling between the order parameter $\vec{q}$ and the strain tensor η. If the displacement $\vec{q}$ gives rise to a polarization then the phase of lower symmetry is also a ferroelectric phase with a spontaneous polarization proportional to $\vec{q}$.

To describe the structural phase transitions we assume that g is a function not only of T, p and η, but also of $\vec{q}$. The values taken by η and $\vec{q}$ in each phase are those that minimize g. According to Landau g must be invariant under the operations of symmetry of the phase of highest symmetry, which we assume to be the cubic structure. We assume that the potential g has three terms:

$$g = g_q + g_e + g_a. \tag{16.68}$$

The first depends only on $\vec{q}$ and in accordance with Devonshire we suppose that g_q is a polynomial of the sixth degree in q_i. Since g_q should be invariant by the symmetry operation $\vec{q} \leftrightarrow -\vec{q}$, the polynomial contains only monomials of even degree. Other symmetry operations include the permutations among the Cartesian components q_1, q_2 and q_3 of the vector $\vec{q}$. The polynomial g contains only one monomial of the second degree and two monomials of the fourth order. As to monomials of the sixth order, we consider just one of them, although there are others. In accordance with these considerations the thermodynamic potential g_q reads

$$g_q = g_0 + b(T - T_1)(q_1^2 + q_2^2 + q_3^2) - c(q_1^4 + q_2^4 + q_3^4) +$$
$$+ a(q_1^2q_2^2 + q_1^2q_3^2 + q_2^2q_3^2) + d(q_1^6 + q_2^6 + q_3^6), \tag{16.69}$$

where the coefficients a, b, c and d do not depend on temperature. The second term g_e depends only on the strain components η_i of zero trace. We assume that it is given only by the invariants of the second order in η,

$$g_e = A(\eta_1^2 + \eta_2^2 + \eta_3^2) + B(\eta_4^2 + \eta_5^2 + \eta_6^2). \tag{16.70}$$

The third term describes the coupling between the strain and the order parameter, which we assume to be linear in η and quadratic in $\vec{q}$,

$$g_a = -k(\eta_1 q_1^2 + \eta_2 q_2^2 + \eta_3 q_3^2) - h(\eta_4 q_2 q_3 + \eta_5 q_1 q_3 + \eta_6 q_1 q_2). \tag{16.71}$$

Minimizing g with respect to η_i, under the restriction $\eta_1 + \eta_2 + \eta_3 = 0$, we get

$$\eta_1 = \frac{k}{6A}(2q_1^2 - q_2^2 - q_3^2), \qquad \eta_4 = \frac{h}{2B} q_2 q_3, \tag{16.72}$$

and similar equations for η_2, η_3, η_5 and η_6. The equations obtained by minimizing g with respect to the variables q_i together with (16.72) lead us to the following solutions:

(a) $q_i = 0$ and $\eta_i = 0$, that describes the cubic phase;
(b) $q_1 = q_2 = 0$, $q_3 \neq 0$, $\eta_1 = \eta_2 \neq \eta_3$, $\eta_4 = \eta_5 = \eta_6 = 0$, that describes the tetragonal phase;
(c) $q_1 = 0$ and $q_2 = q_3 \neq 0$, $\eta_2 = \eta_3 \neq \eta_1$ $\eta_4 \neq 0$, $\eta_5 = \eta_6 = 0$, that describe the orthorhombic phase; and
(d) $q_1 = q_2 = q_3 \neq 0$, $\eta_1 = \eta_2 = \eta_3$, $\eta_3 = \eta_4 = \eta_6$, that describes the trigonal phase.

Let us examine in more detail the transition from cubic to tetragonal phase. To this end it suffices to take into account the description of the tetragonal phase presented in item (b) and consider g as as function of q_3 and η_3. From the expression for g_q, g_e and g_a, we get

$$g = g_0 + b(T - T_1)q_3^2 - cq_3^4 + dq_3^6 + \frac{3}{2}A\eta_3^2 - k\eta_3 q_3^2. \tag{16.73}$$

The minimization leads us to $\eta_3 = kq_3^2/3A$ and

$$b(T - T_1)q_3 - 2cq_3^3 + 3dq_3^5 - k\eta_3 q_3 = 0, \tag{16.74}$$

or

$$b(T - T_1)q_3 - 2c'q_3^3 + 3dq_3^5 = 0, \tag{16.75}$$

where

$$c' = c + \frac{k^2}{6A}. \tag{16.76}$$

Equation (16.75) can be understood as coming from the minimization of an effective potential given by

$$g_{ef} = g_0 + b(T - T_1)q_3^2 - c'q_3^4 + dq_3^6. \qquad (16.77)$$

As seen in the previous chapter, a free energy of this type predicts a discontinuous transition that occurs at a temperature T_0 given by

$$T_0 = T_1 + \frac{(c')^2}{4bd}, \qquad (16.78)$$

with a jump in the order parameter equal to $q_0 = \sqrt{c'/2d}$.

Problems

16.1 Show that a crystalline solid belonging to a system of cubic symmetry has only three independent stiffness coefficients and that they are given by (16.34).

16.2 Obtain the compliance coefficients κ_{ij} of a solid with cubic structure from the stiffness coefficients c_{ij}.

16.3 Show that an isotropic solid has only two independent stiffness coefficients and that they are given by (16.51).

16.4 Show that the Young modulus and the Poisson ration are related with the Lamé coefficients by the expression (16.56).

16.5 Use the Hooke law to determined the deformations of a cylindric isotropic solid subjected only to stretch stress applies at its ends. Suppose next that the lateral surface of the cylinder is fixed, that is, that there are no lateral deformations. Determine in this case the deformation along the axis of the cylinder and the lateral compressions. In both cases determine also the variations in the volume of the cylinder. Present the results in terms of the Young modulus E and the Poisson ratio ν.

16.6 The Gibbs free energy density

$$g = g_0 + ay^2 + bx^2 + cxy^2 + dy^4,$$

where x and y are related to the components of the traceless tensor η by (16.60), describes the transition from the paraelastic phase with tetragonal structure to the ferroelastic phase with orthorhombic structure. Show that the transition is continuous an determine x and y as functions of the parameter a. Assume that the coefficients b, c and d are positive and that the coefficient a is a linear function of T and p, taking positive or negative values.

16.7 The Gibbs free energy density

$$g = g_0 + a\eta_6^2 + b\eta_6 y + cy^2 + d\eta_6^4,$$

where y is related to the components of the traceless tensor η by (16.60), describes the transition from the paraelastic phase with tetragonal structure to the ferroelastic phase with monoclinic structure. Show that the transition is continuous an determine x and η_6 as functions of the parameter a. Assume that the coefficients b, c and d are positive and that the coefficient a is a linear function of T and p, taking positive or negative values.

Chapter 17
Liquid Crystals

17.1 Mesophases

Thermotropics

Liquid crystals are substances which exhibit thermodynamic phases intermediate between crystalline solid and ordinary liquid, called mesophases, which are distinct from both ordinary liquid and crystalline solid. They have the property of fluidity, that characterizes the liquids, and at the same time exhibit the phenomenon of birefringence, as happens to crystalline solids. Microscopically, the liquid crystals are distinguished by displaying only one of two types of ordering that characterize the molecular crystalline solids. In a molecular crystal, the positions of the centers of the molecules form a spatial ordered structure and in addition the molecules are ordered as to the orientation. In an ordinary liquid, both positional and orientational orders are absent. In a liquid crystal molecules have orientational order but positional order is absent. In the opposite situation to liquid crystals, the molecules of a plastic crystal exhibit positional order but orientational order is absent.

The shape of the molecules has a fundamental role in the formation the orientational order. The packing of rod-shaped molecules, for example, induces the formation of a structure with orientational order, called nematic, in which the axes of the molecules become parallel. In fact, the axes of the molecules are not exactly parallel but fluctuate around a given direction due to thermal agitation.

Several types of mesophases can be found between the crystalline phase and the ordinary liquid phase, which we call isotropic phase. The simplest mesophase is called nematic and occurs in the compound *p*-azoxyanisole (PAA) which is a solid at room temperature. The transition from solid to the nematic phase occurs at 118 °C and from that to the isotropic phase at 135 °C. A nematic liquid crystal stable around the room temperature is the compound *p*-methoxybenzylidene-*p*-*n*-butylaniline (MBBA). This compound passes from the solid to the nematic phase at 21 °C and from this phase to the isotropic liquid at 45 °C. One of the first

M.J. de Oliveira, *Equilibrium Thermodynamics*, Graduate Texts in Physics,
DOI 10.1007/978-3-662-53207-2_17

liquid crystal studied, the cholesteryl benzoate passes from the solid phase to the mesophase chiral nematic, known also as cholesteric phase, at 145 °C and from this phase to the isotropic phase at 178 °C.

A liquid crystal can display not one but several distinct mesophases. Increasing the temperature, the compound cholesteryl myristate passes successively through the phases solid, smectic-A, chiral nematic and isotropic. The transitions occur at 71 °C, at 79 °C and at 85 °C, respectively. The compound called 4-n-pentylbenzenethio-4′-n-decyloxybenzoate has three mesophases between the solid and isotropic phases, that occur in the sequence: smectic-C, smectic-A and nematic. The transition temperatures from one phase to another are, respectively, 60, 63, 80 and 86 °C.

The liquid crystals mentioned above are known as thermotropic because the simplest way to induce the transitions is by varying the temperature. It is important to note that thermotropic liquid crystals are composed by pure substances. In another class of liquid crystals, made by mixing two or more pure substances, the mesophases can arise not only by changing the temperature but especially by the variation of the concentration of the components of the mixture. Those liquid crystals are known as lyotropic.

Lyotropics

In general, the lyotropics are formed by long molecules, called amphiphilic, whose skeleton is a carbon chain and whose ends have opposite behavior in the presence of water molecules. One end attracts them while the other repels them. When mixed with water, this behavior favors the development of various types of molecular structures. The most common among them is the spherical structure called micelle whose surface is composed by hydrophilic ends of the molecules and whose central part consists of the hydrophobic ends. The amphiphilic compounds are also known as surfactants. Two major groups of amphiphiles are saponaceous and phospholipids. A typical example of the first group is the sodium laurate and of the second group is dipalmitoylphosphatidylcholine. If the concentration of amphiphilic water is small, a solution is formed. From a certain concentration, there is the formation of spherical structures such as micelles or, depending on the type of amphiphilic, of vesicles. If the concentration increase even more other structures are formed such as lamellar structure, the cubic structure and hexagonal structure.

17.2 Nematics

Order Parameter

The orientational ordering of a sample of a liquid crystal which is in the isotropic phase can be achieved submitting the sample to non-uniform external forces. Denoting by F_i the Cartesian components of an external force $\vec{F}$ that acts at a point $\vec{r}$ with Cartesian components x_i, the non-uniformity means that the derivative $\Lambda_{ij} = \partial F_i/\partial x_j$ is nonzero. These forces are irrotational and have zero divergence which is equivalent to saying that Λ_{ij} have the properties

$$\Lambda_{ji} = \Lambda_{ij}, \qquad \sum_i \Lambda_{ii} = 0, \tag{17.1}$$

and can be understood as the components of a traceless symmetric tensor.

We assume that the sample consists of a set of n constituent units and that the action of external forces occurs through the dipole moments and quadrupole moment of the constituent units. A further assumption is that the external forces act on certain points of each constituent unit and that the density of such points is given by the function $\rho(\vec{r})$. From ρ, we define the Cartesian components p_i of the dipole moment vector,

$$p_i = \int x_i \rho \, d^3 r, \tag{17.2}$$

and the Cartesian components q_{ij} of the quadrupole moment tensor,

$$q_{ij} = \int (x_i x_j - \delta_{ij} r^2) \rho \, d^3 r, \tag{17.3}$$

of a given constituent unit.

The energy of interaction of the external forces with the dipoles is

$$E_p = -\sum_i F_i P_i, \tag{17.4}$$

where $P_i = n p_i$ is the total dipole moment, and the interaction energy with the quadrupoles is

$$E_q = -\sum_{ij} \Lambda_{ij} Q_{ij}, \tag{17.5}$$

where $Q_{ij} = n q_{ij}$ is the total quadrupole.

Next we assume that the dipole moments are absent. The infinitesimal work of the external forces due to an infinitesimal variation of the moments of the quadrupole is

$$dW = -\sum_{ij} \Lambda_{ij} dQ_{ij}, \tag{17.6}$$

so that the infinitesimal variation of energy U of the sample is given by

$$dU = TdS + \sum_{ij} \Lambda_{ij} dQ_{ij}. \tag{17.7}$$

An example of external forces $\vec{F}$ having the properties mentioned above are those generated by an electric field. In this case p_i and q_{ij} are identified as the electric dipoles and electric quadrupoles.

The tensorial quantity Q of components Q_{ij} define the order parameter of the nematic phase. In the absence of external forces, $Q_{ij} = 0$ in the isotropic phase. In the nematic phase one of the components of Q becomes nonzero even in the absence of external forces. From the definition of q_{ij} and taking into account that $Q_{ij} = nq_{ij}$, we see that the tensor Q is symmetric and traceless. It is convenient to use a reference system whose axes are parallel to the singular axes of the nematic liquid crystal. In this reference system the tensor Q is diagonal and is given by

$$Q = \begin{pmatrix} Q_1 & 0 & 0 \\ 0 & Q_2 & 0 \\ 0 & 0 & Q_3 \end{pmatrix}, \tag{17.8}$$

where Q_1, Q_2 and Q_3 are the principal components and also the eigenvalues of the tensor Q. Since the trace is invariant under any rotation of the Cartesian axes then the trace of Q vanishes.

$$Q_1 + Q_2 + Q_3 = 0. \tag{17.9}$$

When the three principal components are distinct from each other the tensor Q describes a biaxial nematic phase. When two of them are equal but distinct from the third, the tensor Q describes a uniaxial nematic phase which can be of rod or disc type. When all three are equal, and therefore zero, we are faced with the isotropic phase.

Landau-de Gennes Theory

The theory of Landau-de Gennes describes the transition between the nematic phase and the isotropic phase. According to this theory, the molar Gibbs free energy must be invariable by the symmetry operation relating to the isotropic phase. This means that $g(Q_1, Q_2, Q_3)$ must be invariant by any permutation of variables Q_i. The theory of Landau-de Gennes assumes that the free energy is a linear combination of the invariants I_n of order n. An invariant of order n contains only powers of Q_i of order n.

The invariant of first order is identified with the trace of Q and therefore vanishes identically. The invariant of second and third order are given by

$$I_2 = \mathrm{Tr}\, Q^2, \qquad I_3 = \mathrm{Tr}\, Q^3. \tag{17.10}$$

The other invariants are $I_4 = I_2^2$, $I_5 = I_2 I_3$, $I_6 = I_3^2$ and $I_6' = I_2^3$. Notice that there are two invariants of sixth order. Up to sixth order, the free energy g reads

$$g = g_0 + a_2 I_2 + a_3 I_3 + a_4 I_2^2 + a_5 I_2 I_3 + a_6 I_3^2 + a_6' I_2^3. \tag{17.11}$$

Notice that the free energy should be understood as the convex hull of the expression in the right hand side of (17.11). Taking into account that the thermodynamic field conjugate to the variables Q_i should vanish, then the thermodynamic states are identified as the minima of this expression.

Initially we examine the transition from the isotropic phase to the uniaxial nematic phase which is described by the tensor Q such that two components are equal. Assuming that the two components are Q_1 and Q_2 we can use the following parameterization $Q_1 = Q_2 = -q/2$ and $Q_3 = q$. In this case

$$I_2 = \frac{3}{2}q^2, \qquad I_3 = \frac{3}{4}q^3, \tag{17.12}$$

and using invariants up to fourth order, the function g becomes

$$g = g_0 + \frac{3}{2}a_2 q^2 + \frac{3}{4}a_3 q^3 + \frac{9}{4}a_4 q^4. \tag{17.13}$$

We assume that a_4 is positive and that a_2 is linear with temperature and pressure and takes positive or negative values. The coefficient a_3 is considered to be negative but the solution for the positive case is obtained from the case $a_3 < 0$ by the transformation of q into $-q$.

For values of the parameter a_2 large enough the absolute minimum of g occurs at $q = 0$ which determines the isotropic phase. Decreasing a_2, we obtain the nematic phase for which $q \neq 0$. The value of q for the nematic phase corresponds to the nonzero root of $\partial g/\partial q = 0$, that is,

$$a_2 q + \frac{3}{4}a_3 q^2 + 3a_4 q^3 = 0. \tag{17.14}$$

From the two nonzero solutions the one corresponding to the lower free energy is given by

$$q = \frac{1}{a_4}\left(\frac{|a_3|}{8} + \sqrt{\frac{a_3^2}{64} - \frac{a_2 a_4}{3}}\right). \tag{17.15}$$

The transition from the nematic phase to the isotropic phase is discontinuous due to the presence of a third order invariant in the free energy g. To determine the transition temperature which corresponds to the coexistence of phases isotropic and nematic we impose the following form for the free energy

$$g = g_0 + \frac{9}{4}a_4 q^2 (q - q_0)^2, \tag{17.16}$$

which describes the coexistence between the isotropic phase, $q = 0$, and the nematic phase, $q = q_0$. Comparing with (17.13) we see that

$$a_2 = \frac{a_3^2}{24a_4}, \qquad q_0 = \sqrt{\frac{2a_2}{3a_4}}. \tag{17.17}$$

Thus we can draw the following conclusions. For $a_2 > a_0 = a_3^2/24a_4$, the order parameter vanishes, and the phase is isotropic. For $a_2 < a_0$ the phase is nematic with q given by (17.15) with a jump equal to q_0 at the transition point $a_2 = a_0$.

Landau Point

Next we will analyze the possibility of the occurrence of a biaxial nematic phase characterized by distinct values of Q_1, Q_2 and Q_3. We assume a free energy of the form (17.11) with $a_5 = 0$ and $a_6' = 0$,

$$g = g_0 + a_2 I_2 + a_3 I_3 + a_4 I_2^2 + a_6 I_3^2, \tag{17.18}$$

and such that the coefficients a_2 and a_3 have linear dependence with temperature and pressure taking positive or negative values and that a_4 and a_6 are strictly positive.

Taking into account that Q is a traceless tensor, it is convenient to use the following parametrization

$$Q_1 = -\frac{1}{2}(q - \eta), \qquad Q_2 = -\frac{1}{2}(q + \eta), \qquad Q_3 = q, \tag{17.19}$$

from which we obtain the following results

$$I_2 = \frac{1}{2}(3q^2 + \eta^2), \qquad I_3 = \frac{3}{4}q(q^2 - \eta^2). \tag{17.20}$$

The minima of g are obtained from $\partial g/\partial q = 0$ and $\partial g/\partial \eta = 0$.

To determined the transition from the isotropic to a uniaxial we impose the condition $\eta = 0$. In this case $I_2 = 3q^2/2$ and $I_3 = 3q^3/4$ which replaced in (17.18) results in the following form for the free energy

$$g = g_0 + \frac{3}{2}a_2 q^2 + \frac{3}{4}a_3 q^3 + \frac{9}{4}a_4 q^4 + \frac{9}{16}a_6 q^6. \tag{17.21}$$

The nematic-isotropic phase transition, which is first order, is determined by comparing (17.21) with a sixth degree polynomial that has two minima with the same value of g, one at $a = 0$ (isotropic phase) and the other at $q = q_0$ (uniaxial nematic phase), which describes the coexistence of the two phases. This comparison gives the equation for the nematic-isotropic transition line and the jump in the order parameter along the coexistence line,

$$a_2 = \frac{2a_3^2}{3a_6}, \qquad q_0 = \sqrt{\frac{2a_2}{3a_4}}, \tag{17.22}$$

valid for small values of a_2. The coexistence line is shown in Fig. 17.1a. Notice that q_0 vanishes at the point $a_3 = 0$ and $a_2 = 0$.

As long as a_3 is nonzero, the isotropic phase is contiguous to a uniaxial nematic phase as seen in Fig. 17.1a and the transition is discontinuous. When $a_3 = 0$, the transition ceases to be discontinuous, and becomes a continuous transition that occurs at $a_2 = 0$. The point $a_2 = 0$, $a_3 = 0$ is a special point of the phase diagram and is called Landau point.

When η becomes nonzero there is the emergence of a biaxial phase. The minima of g such that $\eta \neq 0$ and $q \neq 0$ are determined by

$$I_2 = -\frac{a_2}{2a_4}, \qquad I_3 = -\frac{a_3}{2a_6}, \tag{17.23}$$

which is obtained by the minimization of the free energy (17.18). Recall that I_2 and I_3 are related to q and η by (17.20) so that (17.23) gives implicitly q and η as functions of a_2 and a_3. The transition line from the biaxial nematic to the uniaxial nematic phases is obtained by taking the limit $\eta \to 0$ because this transition is

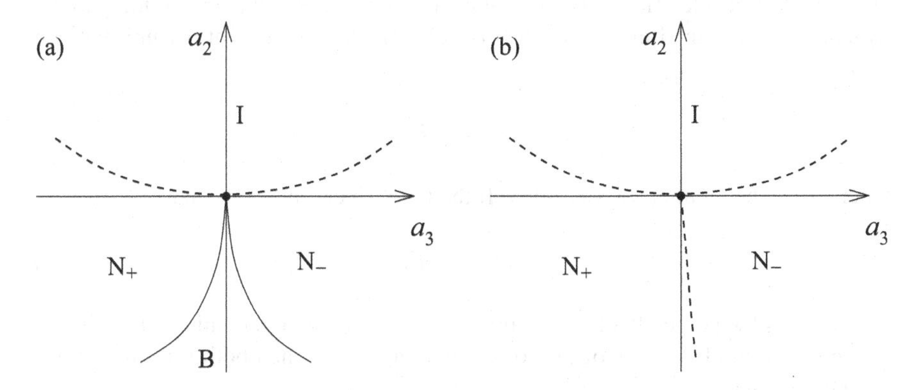

Fig. 17.1 Phase diagram for the transition between the isotropic (I), uniaxial nematic (N_+ and N_-) and biaxial (B) phases according to the Landau-de Gennes theory. The *dotted and continuous lines* describe the discontinuous and continuous transition, respectively. (**a**) Diagram corresponding to $a_5 = 0$. (**b**) Diagram corresponding to a value of a_5 such that $a_5^2 > 4a_4a_6$

continuous, what can be understood noticing that g, given by (17.18), contains only even powers of the order parameter η of the biaxial phase and $a_4 > 0$. Taking the limit $\eta \to 0$ in (17.23) and eliminating q, we get

$$a_2 = -c|a_3|^{2/3}, \tag{17.24}$$

where $c = (12a_4^3/a_6^2)^{1/3}$, that describes the critical line corresponding to the biaxial uniaxial transition as shown in Fig. 17.1a.

If the coefficient a_5 is nonzero and such that $a_5^2 < 4a_4a_6$, the phase diagram has the topology shown in Fig. 17.1a with the critical lines approaching the Landau point with a certain slope. If $a_5^2 > 4a_4a_6$ the biaxial phase no longer exists. In this case the phase diagram exhibits only the two uniaxial phase in addition to the isotropic phase. The two uniaxial phases become contiguous and the transition between them is discontinuous. The line of coexistence between the two nematic phase meet the nematic-isotropic coexistence line at the point $a_3 = a_2 = 0$ as seen in Fig. 17.1b. Notice that in both cases the order parameters q and η vanish continuously at the Landau point.

Coupling to the Electric and Magnetic Fields

Let us examine here the response of a uniaxial nematic liquid crystal to a static electric field. We will restrict ourselves to nematics which behave as an ordinary dielectric, that is, nematics such that the dipole moment of the sample is proportional to the applied field. In an isotropic liquid, the constant of proportionality, which is the electric susceptibility at zero field, is independent of direction. In a nematic it depends on the direction of the applied field. In other terms, the susceptibility has an anisotropy. If the electric field $\mathbf{E}$ is applied along the direction corresponding to the axis of the rod-shaped molecules then the electric dipole moment per unit volume $\mathscr{P}$ is

$$\mathscr{P} = \chi_\parallel E, \tag{17.25}$$

where $\chi_\parallel$ is the parallel susceptibility. If the field is perpendicular then

$$\mathscr{P} = \chi_\perp E, \tag{17.26}$$

where $\chi_\perp$ is the perpendicular susceptibility. Denoting by $\mathbf{n}$ the unit vector parallel to the direction of the axis of the rod-shaped molecules then both formulas above can be written as

$$\mathscr{P} = \chi_\perp \mathbf{E} + \Delta\chi(\mathbf{n} \cdot \mathbf{E})\mathbf{n}, \tag{17.27}$$

where the difference

$$\Delta\chi = \chi_{\parallel} - \chi_{\perp} \tag{17.28}$$

is a measure of the anisotropy, which can be positive or negative.

The electric contribution g_{ele} to the free energy per unit volume will be

$$g_{\text{ele}} = -\frac{\chi_{\perp}}{2}E^2 - \frac{\Delta\chi}{2}(\mathbf{n}\cdot\mathbf{E})^2, \tag{17.29}$$

which is obtained by integration of the relation $\mathscr{P} = -\partial g_{\text{ele}}/\partial\mathbf{E}$. The first term is independent of orientation. The second term instead depends on the orientation of nematic liquid crystal sample. If $\Delta\chi > 0$ then it favors a parallel alignment with the applied field. If $\Delta\chi < 0$, it favors an alignment perpendicular to the applied field. In any case, the electric field becomes a method for the alignment of a nematic sample in a given direction.

We may also consider the response to a magnetic field $\mathbf{B}$. Similarly the magnetization of a sample subject to a magnetic field is given by

$$\mathscr{M} = \chi_{\perp}^{\text{m}}\mathbf{B} + \Delta\chi^{\text{m}}(\mathbf{n}\cdot\mathbf{B})\mathbf{n}, \tag{17.30}$$

where

$$\Delta\chi^{\text{m}} = \chi_{\parallel}^{\text{m}} - \chi_{\perp}^{\text{m}} \tag{17.31}$$

is the difference between $\chi_{\parallel}^{\text{m}}$ and $\chi_{\perp}^{\text{m}}$, the magnetic susceptibilities parallel and perpendicular to the director $\mathbf{n}$, respectively.

The magnetic contribution g_{mag} to the free energy per unit volume is

$$g_{\text{mag}} = -\frac{\chi_{\perp}^{\text{m}}}{2}B^2 - \frac{\Delta\chi^{\text{m}}}{2}(\mathbf{n}\cdot\mathbf{B})^2, \tag{17.32}$$

which is obtained by integration of the relation $\mathscr{M} = -\partial g_{\text{mag}}/\partial\mathbf{B}$.

The order parameter of the nematic phase can be experimentally obtained by means of experimental measurements of the anisotropic part $\Delta\chi$ of the electric susceptibility or the anisotropic part $\Delta\chi^{\text{m}}$ of the magnetic susceptibility. Among other forms of experimental measurement of the order parameter we mention the measurement of birefringence,

$$\Delta n = n_{\parallel} - n_{\perp}, \tag{17.33}$$

the difference between $n_{\parallel}$ and $n_{\perp}$, the index of refraction parallel and perpendicular to the director, respectively.

17.3 Smectics

Order Parameter

In addition to the orientational order, smectic liquid crystals have positional order along one direction. We imagine the smectic liquid crystals as succession of layers inside which the molecules are in the nematic state described by the director **n**. The direction perpendicular to the layers defines the axis z. If the director **n** is parallel to the axis z, the liquid crystal is called smectic -A. If the director **n** form an angle with the axis z, then the liquid crystal is called smectic -C.

Next let us examine the phase transitions occurring in a smectic -A liquid crystal. To this end we need to define the order parameter that describes this phase. Denoting the local density of the liquid crystal by $\rho(z)$, we see that $\rho(z)$ is a periodic function of z with a period equal to L, the distance between two successive layers. Taking into account only the first terms of the expansion of $\rho(z)$ in Fourier series, we write

$$\rho = \rho_0 + \sigma \cos(\frac{2\pi}{L}z + \varphi), \tag{17.34}$$

where σ gives the measure of change in density when one passes from one to the next layer and, therefore, vanishes when the liquid crystal undergoes a transition to the isotropic phase or to the nematic phase.

It is important to note that two states that differ only by a translation along the z direction are described by same value of σ but different values of phase φ. Thus, a smectic -A state is described by the variables σ and φ which can be combined to define the complex variable

$$\psi = \sigma e^{i\varphi}, \tag{17.35}$$

which determined the order parameter of the smetic -A phase. It vanishes both in the isotropic and nematic phases. Equation (17.34) can therefore be written as

$$\rho = \rho_0 + \text{Re}\{\psi e^{i2\pi z/L}\}. \tag{17.36}$$

Phase Transition

To describe the phase transitions in a liquid crystal that can be found in isotropic, nematic and smectic -A, we adopt the following free energy

$$g = g_0 + a_2 q^2 + a_3 q^3 + a_4 q^4 + b_2 \sigma^2 + b_4 \sigma^4 - c_4 q^2 \sigma^2, \tag{17.37}$$

where the coefficients depend on the temperature and possibly on other thermodynamic fields. We consider that the coefficients a_4, b_4 are strictly positive and that $c_4^2 < 4a_4b_4$ to ensure the convexity of g with respect to the set of variables q and σ. The possible thermodynamic phases correspond to the various types of absolute minima of g, obtained when the coefficients are varied. The isotropic phase occurs when $q = 0$ and $\sigma = 0$, the nematic phase, when $q \neq 0$ and $\sigma = 0$, and the smectic-A phase, when $q \neq 0$ and $\sigma \neq 0$. To exclude minima corresponding to $q = 0$ and $\sigma \neq 0$, we assume $b_2 > 0$.

Let us start by determining the transition between the nematic and isotropic phases. In this case, $\sigma = 0$ in both phases so that the free energy (17.37) is reduced to

$$g = g_0 + a_2 q^2 + a_3 q^3 + a_4 q^4. \tag{17.38}$$

The order parameter of the nematic phase q is given by the equation

$$2a_2 + 3a_3 q + 4a_4 q^2 = 0, \tag{17.39}$$

obtained from $\partial g/\partial q = 0$. The free energy (17.38) is similar to that given by (17.13), which describes the nematic isotropic transition. Therefore, using the same reasoning used in that case, we conclude that the transition is first order and occurs when

$$a_2 = \frac{a_3^2}{4a_4}, \tag{17.40}$$

shown as a horizontal line in the phase diagram of Fig. 17.2.

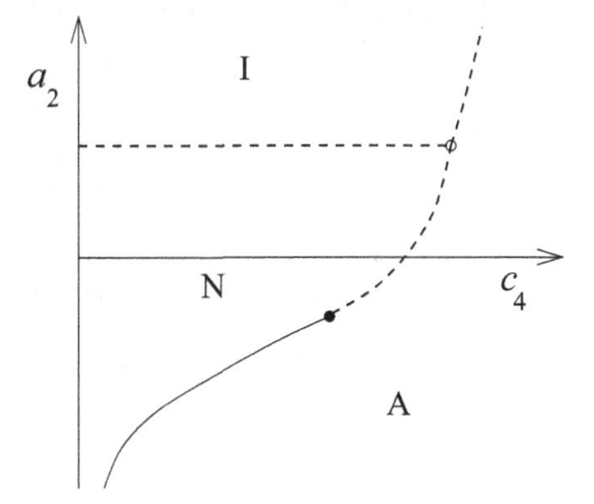

Fig. 17.2 Phase diagram of a liquid crystal which can be found in the isotropic (I), nematic (N) and smectic-A phases, according to the free energy given by (17.37). The *dotted and continuous lines* describe the discontinuous and continuous transitions, respectively. The *filled circle* represents a tricritic point and the *empty circle* represents a triple point

To find the other transitions, we determine the equations that give the minima of g. They are

$$2a_2q + 3a_3q^2 + 4a_4q^3 - 2c_4q\sigma^2 = 0, \tag{17.41}$$

$$2b_2\sigma + 4b_4\sigma^3 - 2c_4q^2\sigma = 0. \tag{17.42}$$

The nematic phase is described by $\sigma = 0$ and q given by the root of (17.39). The smectic-A phase corresponds to the nonzero solutions, $q \neq 0$ and $\sigma \neq 0$, determined by equations

$$(2a_2b_4 + b_2c_4) + 3a_3b_4q + (4a_4b_4 - c_4^2)q^2 = 0, \tag{17.43}$$

$$q^2 = \frac{1}{c_4}(b_2 + 2b_4\sigma^2). \tag{17.44}$$

From the solutions of the equations that determine q and σ for each phase, we determine the respective free energies. When the transition between two phases is discontinuous, it is determined by comparing the free energies. From the solutions of (17.43) and (17.44) and using a numerical procedure to determine the lowest free energy, we construct the phase diagram in the variables a_2 and c_4, shown in Fig. 17.2. The diagram shows lines of coexistence between nematic and smectic-A phases, nematic and isotropic and smectic-A and isotropic. The three coexistence lines meet at a triple point. The diagram also gives a critical line between the smectic-A and the nematic phases that ends at a tricritic point.

To determine the critical line and the tricritical point we replace (17.44) into (17.43) and expand the result in powers of σ. The critical line occurs when $\sigma \to 0$, that is, when the constant term vanishes, which gives us the result

$$2a_2 + 3a_3\sqrt{\frac{b_2}{c_4}} + 4a_4\frac{b_2}{c_4} = 0, \tag{17.45}$$

valid when $c_4 > 0$. The tricritical point occurs when the constant term and the coefficient of σ^2 vanish what occurs when

$$3a_3b_4 + 2(4a_4b_4 - c_4^2)\sqrt{\frac{b_2}{c_4}} = 0. \tag{17.46}$$

Problems

17.1 From the free energy g given by (17.13), which describes the nematic isotropic transition, determined the density $\rho = 1/v$, $v = \partial g/\partial p$, as a function of temperature and shown that ρ has a jump at the transition point. To this end assume that the coefficient a_2 is linear with pressure and temperature and that $g_0 = RT \ln p + bp + a_0$, where b and a_0 are constant.

17.2 A liquid crystalline elastomer is a substance that combines the properties of anisotropy of liquid crystals and the elastic properties of materials such as rubber. Its thermodynamic properties can be described by the free energy

$$g = g_0 + a_2 q^2 + a_3 q^3 + a_4 q^4 - bq\xi + \frac{1}{2} K\xi^2$$

where q is the order parameter of the nematic phase and ξ is the strain along a certain direction, a_2, a_3, a_4 and b are parameters, and K is the Young's modulus. The stress σ is given by $\sigma = \partial g/\partial \xi$. Determine the stress σ as a function of the strain ξ and shown that it has a horizontal segment.

Chapter 18
Thermal Radiation

18.1 Electromagnetic Waves

Radiation

When we are outdoors on a sunny day we feel the effects of heat on our skin even if the environment is at a low temperature. Likewise, when we are near the fire, we feel more heat on the surface of our body that is exposed to fire than on the opposite surface. These effects are caused by radiative heat transfer from the emitting source, the sun or fire, to your body. This radiative transfer is distinguished from other forms of heat transfer because it does not require a material medium to propagate. In material media heat transfer occurs by thermal conduction or convection. The thermal conduction is carried out in the solids by vibration of the crystal lattice and in electrical conductors also by free electrons. In liquids and gases it occurs mainly by the collision between molecules. The heat transfer by convection is that occurring by the macroscopic motion of matter.

The radiative transfer does not require the presence of a material medium and is the main form of heat transfer when the bodies are separated. It occurs by means of electromagnetic waves that carry energy and can be absorbed and emitted by matter. The photons, that are the elementary particles resulting from the quantization of the electromagnetic field, are continually emitted and absorbed by bodies by means of several processes: transitions between electronic, vibrational and rotational levels in atoms and molecules, ionization or combination of free electron with ions, change of the kinetic energy of free electrons, excitations of the crystal lattice in solids. These processes give rise to a continuous spectrum of emission, or absorption, superimposed by spectral lines resulting from transitions between discrete energy levels.

The radiative transfer is the mechanism by which two bodies, physically isolated from each other, come into equilibrium. If they are at different temperatures, the hotter emits more radiation than it absorbs and the colder absorbs more than it

M.J. de Oliveira, *Equilibrium Thermodynamics*, Graduate Texts in Physics,
DOI 10.1007/978-3-662-53207-2_18

emits. In thermal equilibrium, each of the bodies absorbs and emits equal amounts of radiation. A bar of iron heated to redness cools when removed from the heat source and reaches room temperature mainly by radiative energy loss to the environment. The same occurs with a filament of an electric lamp, which is practically materially isolated from the environment due to the vacuum inside the lamp, when one interrupts the electric power supply.

Electromagnetic waves propagate in vacuum at the speed of light, $c = 2.9979 \times 10^8$ m/s. The monochromatic waves are characterized by the frequency ν, or equivalently by the wavelength λ since these two quantities are related by $c = \lambda\nu$. Electromagnetic waves have distinct names according to its wavelength. Visible light corresponds to a very small range of wavelength between 0.38×10^{-6} and 0.76×10^{-6} m. Wavelengths above the visible up to 10^{-3} m correspond to the infrared radiation. Above the infrared are radio waves. Below the visible down to 10^{-9} m, correspond to the ultraviolet radiation. The X-rays have wavelengths between 10^{-8} and 10^{-12} m and the gamma rays, below 10^{-10} m. Thermal radiation comprises the ultraviolet and infrared radiation in addition to the visible light.

Electromagnetic waves carry not only energy but also momentum. There is a close relationship between the momentum $\mathfrak{p}$ of an electromagnetic wave and its energy $\mathscr{E}$,

$$\mathfrak{p} = \frac{\mathscr{E}}{c}, \tag{18.1}$$

a relation due to Maxwell. The same relation is valid for the momentum $\mathfrak{p}$ of a photon and its energy $\mathscr{E}$. Considering that waves have momentum it is natural to imagine that they exert pressure on bodies. Consider a cubic cavity of edge L and the total energy U of the electromagnetic radiation within the cavity. As electromagnetic waves move in three independent spatial directions we can assume that $1/3$ of them move in a particular direction, say along the x direction, parallel to one edge of the cubic cavity. The momentum transferred to one of the walls perpendicular to the x direction during a time interval $2L/c$ is $(1/3)2U/c$. The force on the wall is the ratio between these two quantities and thus equal to $U/3L$. Therefore, the pressure p on the wall, which is the ratio between the force and the area of the face of the cubic cavity, is $p = (U/3L)/L^2 = U/3V$, where $V = L^3$ is the cavity volume, that is,

$$p = \frac{u}{3}, \tag{18.2}$$

which is the relation between the radiation pressure p and the energy per unit volume $u = U/V$ of the radiation, relation due to Maxwell.

Kirchhoff Law

To characterize the spectrum of electromagnetic radiation of a body at a given temperature, we measure the energy emitted in each one of the frequencies of electromagnetic wave. More precisely we measure the energy per unit time and per unit area of the emitter $\phi \Delta \nu$ emitted by the emitter within the frequency range between ν and $\nu + \Delta \nu$. The monochromatic emissive power ϕ, as a function of ν, is different from one substance to another. However, there is a hull that can be considered as the emissive power common to all bodies and therefore independent of the emitter substance, which we call universal spectrum or black body spectrum. The idea of black body was introduced by Kirchhoff who established the universality of the radiation spectrum. According to Kirchhoff, the power spectral density of a black body is independent of the particular characteristics of the emitter, depending only on the frequency of the radiation and the temperature.

Certain bodies can be seen approximately as black bodies as the sun or as the tungsten filament of a lit electric bulb. The black body radiation can be obtained by a cavity whose walls are at a certain temperature T. The emitted radiation is observed through a small hole made in the cavity. Considering that the photons almost do not interact, the thermal equilibrium of radiation contained in a cavity occurs through the mechanism of emission and absorption of photons by the walls.

Stefan-Boltzmann Law

The radiative heat flux emitted by a surface is characterized by emissive power Φ defined as the radiative energy emitted by a surface by unit time and per unit area. It is convenient to define the emissive spectral power $\phi(\nu)$ with respect to a frequency ν. This quantity is defined such that the product $\phi(\nu)\Delta\nu$ is the radiative energy emitted by a surface per unit time and per unit area, by the radiation with frequencies between ν and $\nu + \Delta\nu$. The two quantities are related by

$$\Phi = \int_0^\infty \phi(\nu) d\nu. \tag{18.3}$$

For the analysis of radiation within a cavity, it is convenient to introduce the spectral density of energy $\rho(\nu)$ defined such that $du = \rho \Delta \nu$ is the energy per unit volume of the radiation with frequencies between ν and $\nu + d\nu$. The density of energy, that is, the energy per unit volume u of electromagnetic radiation is related to ρ by means of

$$u = \int_0^\infty \rho(\nu) d\nu. \tag{18.4}$$

Since the radiation that emerges from the cavity through the hole has velocity c, then the emissive powers and the energy densities are related by $\phi = \rho c/4$ and $\Phi = uc/4$. The factor $1/4$ must be introduced by the following reason. Consider a point within the cavity situated near the cavity opening and define the direction z as one that arises at this point and goes out of the cavity through the hole. Since the radiation in the cavity is isotropic, then the fraction of the radiation passing through the vicinity of that point and emerging through the hole should be proportional to the component of the propagation direction $\hat{k}$ of the electromagnetic wave in the direction $z > 0$. Using spherical coordinates and taking into account that the component of $\hat{k}$ in the z direction is $\cos\theta$, this fraction is determined by

$$\frac{\int_0^{2\pi}\int_0^{\pi/2}(\cos\theta)\sin\theta d\theta d\phi}{\int_0^{2\pi}\int_0^{\pi}\sin\theta d\theta d\phi} = \frac{\pi}{4\pi} = \frac{1}{4}. \tag{18.5}$$

The Kirchhoff universal law of radiation states that the spectral emissive power of a blackbody depends only on the temperature T and on the frequency ν. Therefore, the power spectral density $\rho(T, \nu)$ only depends on T and ν and the energy density $u(T)$ depends only on T. As the radiation pressure p is related to u by $p = u/3$ then the pressure radiation $p(T)$ also depends only T. Stefan has shown experimentally that the emissive power Φ is proportional to the fourth power of the absolute temperature,

$$\Phi = \sigma T^4. \tag{18.6}$$

The constant of proportionality constant is called Stefan-Boltzmann constant and is

$$\sigma = 5.6697 \times 10^{-8}\,\mathrm{W\,m^{-2}\,K^{-4}}. \tag{18.7}$$

Relation (18.6), called Stefan-Boltzmann law, was demonstrated by Boltzmann from the expression obtained by Maxwell for the radiation pressure, $p = u/3$. This demonstration will be presented below.

Using relations $u = 4\Phi/c$ and $p = u/3$ we see that

$$u = \frac{4\sigma}{c}T^4, \tag{18.8}$$

$$p = \frac{4\sigma}{3c}T^4. \tag{18.9}$$

Thus, the density of energy and the pressure are also proportional to the fourth power of the temperature.

Free Energy

Consider a cavity of volume V whose walls are at a temperature T. Thermodynamically, radiation within the cavity is characterized by only two independent thermodynamic quantities and not by three as in a system with one chemical component. In this sense, we can say that the radiation contained in a cavity is actually a thermodynamic system with zero component. We can compare the radiation in a cavity with the saturated vapor of a one component system for which the vapor pressure only depends on the temperature. Similarly to what happens with the liquid-vapor coexistence in which a volume variation does not cause change in the pressure if the temperature is kept constant, so does with radiation. By varying the volume of the cavity, a certain amount of radiation will be absorbed or emitted by the walls but the pressure remains unchanged.

In the representation of energy, $U(S, V)$ depends only on entropy S and volume V. The differential form of energy is given by

$$dU = TdS - pdV. \tag{18.10}$$

The extensivity of the energy allows us to reduce the fundamental relation into the form $U(S, V) = Vu(s)$ where $u = U/V$ is the energy density and $s = S/V$ is the entropy density and the system can be described simply by $u(s)$. The relation $T = du/ds$ is also valid. The Euler equation,

$$U = TS - pV, \tag{18.11}$$

implies $u = Ts - p$.

Using the Helmholtz free energy representation $F = U - TS$ we see that $F(T, V)$ depends only on T and V. In addition,

$$dF = -SdT - pdV. \tag{18.12}$$

The extensivity allows us to write $F(T, V) = Vf(T)$ where $f = F/V$ is the free energy density. Notice that $f(T)$ depends only on the temperature. From the Euler equation

$$F = -pV, \tag{18.13}$$

and therefore $p = -f$ from which follows that the pressure depends only on temperature. In addition, $f = u - Ts$ and $s = -df/dT$ and the entropy density depends only on temperature.

We have seen that the radiation pressure is related with energy density by $p = u/3$. Since $f = -p$ then $f = -u/3$. Replacing this last result in $f = u - Ts$, we obtain the result $4f = -Ts$. Deriving both sides of this equation with respect to temperature and taking into account that $s = -df/dT$, we obtain the following differential equation for the entropy

$$3s = T\frac{ds}{dT}. \tag{18.14}$$

Integrating this equation, we get the entropy as a function of temperature

$$s = AT^3, \tag{18.15}$$

result obtained assuming that the entropy vanishes at the absolute zero of temperature.

Since $s = -df/dT$ and $p = -f$ then $s = dp/dT$. Integrating this equation we get the radiation pressure

$$p = \frac{1}{4}AT^4, \tag{18.16}$$

where we assumed that the pressure vanishes at $T = 0$. The energy density is $u = 3p$ so that

$$u = \frac{3}{4}AT^4, \tag{18.17}$$

a result consistent with the Stefan-Boltzmann law. Comparing with (18.9), we see that the constant A is related to the Stefan-Boltzmann constant by

$$A = \frac{16\sigma}{3c}. \tag{18.18}$$

When we vary the volume of a cavity at a constant temperature the pressure remains invariant. However, a certain amount of radiation will be absorbed or emitted by the walls. Consequently, a certain amount of heat will be exchanged with the cavity walls. To determine the amount of heat, we calculate the enthalpy $H = U + pV$ because the variation is enthalpy equals the heat exchanged at constant pressure. Using relation $p = U/3V$, we get $H = 4pV$ and therefore the amount of heat exchanged Q is equal to $Q = \Delta H = 4p\Delta V$. The work done W by the radiation in this process is equal to the decrease in the free energy because the temperature remains invariant. Since $F = -pV$, then $W = -\Delta F = p\Delta V$. The variation of energy is $\Delta U = 3p\Delta V$.

18.2 Planck Law

Wien Displacement Law

According to Kirchhoff the spectral density of energy $\rho(\nu, T)$ of a black body depends only on the frequency ν of the radiation and temperature. Wien showed that the spectral density must satisfy the following scaling law

$$\rho(\nu, T) = \nu^3 \mathscr{F}(\frac{\nu}{T}), \tag{18.19}$$

known as Wien displacement law. The Stefan-Boltzmann law is a direct consequence of this law. Indeed, integrating the spectral density of energy over all frequencies, we get the total density of energy

$$u(T) = \int_0^\infty \rho(\nu, T)d\nu = \int_0^\infty \nu^3 \mathscr{F}(\frac{\nu}{T})d\nu. \tag{18.20}$$

Doing the following change of integration variable, $x = \nu/T$, we get the result

$$u(T) = T^4 \int_0^\infty x^3 \mathscr{F}(x)dx. \tag{18.21}$$

But the integral does not depend on T and we may conclude that $u(T)$ is proportional to the fourth power of temperature.

The spectral density of energy $\rho(\nu, T)$ has a maximum at a certain value of ν that we denote by ν_{max}. To get ν_{max}, we derive $\rho(\nu, T)$ with respect to ν to obtain the following equation

$$3\nu^2 \mathscr{F}(\frac{\nu}{T}) + \frac{\nu^3}{T}\mathscr{F}'(\frac{\nu}{T}) = 0, \tag{18.22}$$

or yet

$$3\mathscr{F}(x) + x\mathscr{F}'(x) = 0, \tag{18.23}$$

where $x = \nu/T$. Suppose that x_0 is the solution of the equation above. We conclude that

$$\nu_{\mathrm{max}} = x_0 T, \tag{18.24}$$

that is, raising the temperature of a black body the frequency value for which the spectral density of energy is maximum moves to the region of high frequencies, known as Wien displacement law.

Planck Radiation Law

The radiation in a cavity can be viewed as a collection of harmonic oscillators vibrating with different frequencies. Each oscillator corresponds to a normal mode of vibration of the electromagnetic waves. Considering a cubic cavity of volume $V = L^3$, the normal modes correspond to waves propagating parallel to the sides of the cube. The wavelength λ must be a divisor of the length L of the edge of the cube, that is, $\lambda = L/n$, where n is an integer. Since λ and the frequency ν are related by $\nu = c/\lambda$ it follows that the normal modes have frequencies $\nu_n = nc/L$. Next we determine how many vibration modes exist or equivalently how many oscillators exist with frequencies between ν and $\nu + \Delta\nu$. Considering just one direction, this number equals $2(L/c)\Delta\nu$, where the factor 2 is due to the two independent modes of propagation of the electromagnetic waves, which are transversal. Considering the three directions this number becomes $8\pi(L/c)^3\nu^2\Delta\nu = 8\pi(V/c^3)\nu^2\Delta\nu$.

The spectral density of energy $\rho(\nu, T)$ is therefore

$$\rho = \frac{8\pi\nu^2}{c^3}E, \tag{18.25}$$

where $E(\nu, T)$ is the mean energy of an oscillator. To determined $E(\nu, T)$ we proceed as Planck.

The formula introduced by Wien for the spectral density is

$$\rho = A\nu^3 e^{-a\nu/T}, \tag{18.26}$$

valid for high frequencies, where A and a are two constants. Comparing with (18.25), we see that it corresponds to the following expression for E,

$$E = b\nu e^{-a\nu/T}, \tag{18.27}$$

where b is another constant, related to A by $A = 8\pi b/c^3$. Writing $1/T$ as a function of E and using the relation $1/T = \partial S/\partial E$ we conclude that

$$\frac{\partial S}{\partial E} = -\frac{1}{a\nu}\ln\frac{E}{b\nu}, \tag{18.28}$$

from which we get

$$\frac{\partial^2 S}{\partial E^2} = -\frac{1}{a\nu E}. \tag{18.29}$$

In the regime of low frequencies (18.27) ceases to be valid and should be replaced by another expression. Following Planck we assume, in this regime, the following expression for the energy of an oscillator

$$E = kT, \tag{18.30}$$

from which we obtain $\partial S/\partial E = 1/T = k/E$ and therefore

$$\frac{\partial^2 S}{\partial E^2} = -\frac{k}{E^2}. \tag{18.31}$$

Proceeding according to Planck, we assume the following interpolation formula between (18.29) and (18.31),

$$\frac{\partial^2 S}{\partial E^2} = -\frac{k}{E(h\nu + E)}, \tag{18.32}$$

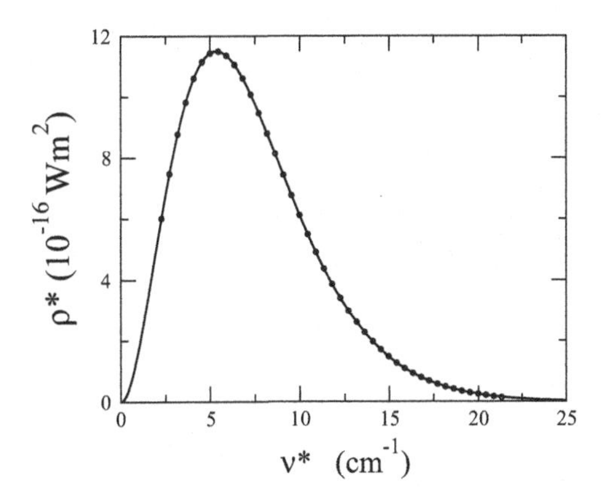

Fig. 18.1 Spectral density of energy for the background cosmic radiation. The graph shows $\rho^* = c^2\rho/4\pi$ as a function of $\nu^* = \nu/c$. The *circles* represent observational data at the microwave band, collected by Fixsen et al. [175]. The *continuous line* represent the function $\rho^* = 2hc^2\nu^{*3}/(e^{hc\nu^*/kT} - 1)$ corresponding to $T = 2.725$ K

which we consider to be valid for any frequency. For large values of E, it reduces to (18.31). For small values of E, it reduces to (18.29), if $a = h/k$. Integrating, we get

$$\frac{1}{T} = \frac{\partial S}{\partial E} = \frac{k}{h\nu} \ln \frac{h\nu + E}{E}, \tag{18.33}$$

or, isolating E,

$$E = \frac{h\nu}{e^{h\nu/kT} - 1}, \tag{18.34}$$

from which we get

$$\rho = \frac{8\pi h\nu^3}{c^3} \frac{1}{e^{h\nu/kT} - 1}, \tag{18.35}$$

which is the spectral density of energy due to Planck for the black body radiation. This formula contains two universal constants. The Boltzmann constant k and the Planck constant h.

We remark that (18.30) should be also considered as the mean energy of and oscillator at high temperature what justifies the identification of the constant k as the Boltzmann constant. As is well known the average energy of a harmonic oscillator is equal to kT only in the classical regime. This result ceases to be valid at low temperatures and must be replaced by the result of Planck, given by (18.34) valid in the quantum context.

The background cosmic radiation, discovered by Penzias and Wilson, occurs at the microwave band and has a black body spectrum corresponding to the temperature of 2.725 K, as can be seen in Fig. 18.1.

Chapter 19
Thermochemistry

19.1 Extent of Reaction and Affinity

Chemical Reactions

A chemical reaction is a process in which one or more substances, the reactants, are transformed into another or other substances, the products. A flame is a striking manifestation of combustion, which consists of the intense combination of a substance with oxygen. The reaction with oxygen can be mild such as rust, spontaneously occurring in iron exposed to atmospheric air. The decomposition of a chemical compound into chemical elements, as in the electrolysis of water, or the formation of a chemical compound from its constituents, as in the synthesis of ammonia from nitrogen and hydrogen, are examples of chemical reactions performed in the laboratory. A vessel with several chemical species inside of which chemical reactions take place, is a thermochemical system. Chemical reactions can be endothermic when heat is absorbed by the thermochemical system, or exothermic when heat is released.

Chemical reactions are represented by chemical equations constituted by the chemical formulas of the reactants and products and the respective stoichiometric coefficients. The formation of water from hydrogen and oxygen is represented by the chemical equation

$$2H_2 + O_2 \rightarrow 2H_2O, \tag{19.1}$$

which means that two moles of hydrogen react with one mole of oxygen to produce two moles of water. A chemical equation shows the quantitative relations between reactants and products but do not necessarily represent the molecular process among the reactants or the reaction mechanism. At the molecular level the process may include intermediate reactions and involve other chemical species. The combustion of hydrogen, represented by (19.1) is actually composed by chain

M.J. de Oliveira, *Equilibrium Thermodynamics*, Graduate Texts in Physics,
DOI 10.1007/978-3-662-53207-2_19

reactions comprising other chemical species in addition to H_2 and O_2. Though relevant to the development of the reaction, the quantities of other chemical species remain invariant and are not represented in the chemical equation, that shows only the species whose quantities change, decreasing for the reactants and increasing for products.

The reversed reaction of that given by (19.1) corresponds to the decomposition of water, represented by

$$2H_2O \rightarrow 2H_2 + O_2, \tag{19.2}$$

which occurs by the process of electrolysis, when an electric current passes through a dilute salt solution. Electrolysis, briefly represented by chemical equation (19.2), consists of several intermediate reactions including those which occur at the cathode, which produces hydrogen, and at the anode, where oxygen is produced.

The occurrence of the equilibrium of a reaction requires that its reverse also occurs as is the case, for example, of the formation of ammonia from nitrogen and hydrogen and its reverse, the decomposition of ammonia, both represented by

$$N_2 + 3H_2 \rightleftharpoons 2NH_3. \tag{19.3}$$

In this equation, conventionally, the chemical species on the left are called reactants and those on the right are known as products. The competition between the reactions leads the system to equilibrium. If the reaction rate to the right is greater than the rate of reaction to the left, there is an increase of the products which causes an increase in the rate of reaction to the left approaching it to the rate of reaction to the right. Eventually, the rates become equal and the chemical equilibrium is established. In equilibrium, the quantities of products and reactants remain invariant. Later we will analyze in detail the conditions of thermochemical equilibrium and will see that in this regime the concentrations are interrelated in a manner consistent with the chemical reactions. We will also examine the behavior of the thermochemical equilibrium with temperature and pressure.

Thermodynamic State

To understand how one defines the thermodynamic state of a thermochemical system in equilibrium, we examine a specific example. A vessel contains a gas mixture of nitrogen, hydrogen and ammonia, within which the chemical reaction represented by (19.3) takes place. The state of thermodynamic equilibrium in the representation of energy U is defined by the entropy S, the volume V and the number of moles N_1 of nitrogen, N_2 of hydrogen and N_3 of ammonia. In differential form the internal energy U is given by

$$dU = TdS - pdV + \mu_1 dN_1 + \mu_2 dN_2 + \mu_3 dN_3, \tag{19.4}$$

where the temperature T, the pressure p and the chemical potentials μ_i of the three species are given by

$$T = \frac{\partial U}{\partial S}, \qquad p = -\frac{\partial U}{\partial V}, \qquad \mu_i = \frac{\partial U}{\partial N_i}. \tag{19.5}$$

This description, however, is only suitable in the absence of reaction, in which case N_1, N_2 and N_3 may in fact be considered independent variables. In a closed system and in the absence of reaction, they remain invariant. If the system is closed but there are chemical reactions, they can vary. The variations, however, can not be arbitrary but must be in accordance to the rules imposed by the chemical reaction. In any event, it is possible to find combinations of N_1, N_2 and N_3 that, in a closed system, can be invariant and therefore be considered independent. Since there is only one reaction and three chemical species, we should find two independent variables, as shown below.

Equation (19.3) means that one mole of nitrogen reacts with three moles of hydrogen to produce two moles of ammonia. If, due to chemical reaction, the amount of nitrogen is decreased by one mole then the amount of hydrogen will decrease by three moles which means that

$$X_2 = N_2 - 3N_1 \tag{19.6}$$

remains invariant under the chemical reaction. The amount of ammonia, in turn, will increase by two moles and

$$X_3 = N_3 + 2N_1 \tag{19.7}$$

will also remain invariant. Next we use the two variables X_2 and X_3 to describe the thermodynamic state in the place of N_1, N_2 and N_3. However, they are not sufficient and must be supplemented by one more variable, which we choose to be $\xi = -N_1$, called the extent of reaction. With this expedient, the relation between the number of moles and the new variables becomes

$$N_1 = -\xi, \qquad N_2 = X_2 - 3\xi, \qquad N_3 = X_3 + 2\xi, \tag{19.8}$$

which replaced into the differential form of energy (19.4) gives

$$dU = TdS - pdV - Ad\xi + \mu_2 dX_2 + \mu_3 dX_3, \tag{19.9}$$

where

$$A = \mu_1 + 3\mu_2 - 2\mu_3, \tag{19.10}$$

is called affinity, a quantity introduced by De Donder and defined by

$$A = -\frac{\partial U}{\partial \xi}, \tag{19.11}$$

keeping S, V, X_2 and X_3 constant.

In this new description, the variables X_2 and X_3 do not have their values changed by the chemical reaction. In other words, they can be regarded as independent variables. This property however is not shared by the variable ξ that, in equilibrium, has its value dictated by the chemical reaction. In other terms, it should be understood as a dependent variable. The system should therefore be understood as a system of two components and not three and the state is now defined only by the variables X_2 and X_3, in addition to S and V. This means that, in equilibrium, U should be independent of ξ so that the derivative $(\partial U/\partial \xi)_{S,V,X_2,X_3} = A$ must vanish, which leads us to the result

$$\mu_1 + 3\mu_2 - 2\mu_3 = 0, \tag{19.12}$$

which is the condition for thermochemical equilibrium for the reaction represented by (19.3). At equilibrium, the chemical potentials are not independent but are related by (19.12) which is again the manifestation that, in equilibrium, the system in fact should be considered a two-component system rather than three.

Single Reaction

Here we extend the results above for a generic reaction among q chemical species. At the same time we highlight the role of two conjugate quantities, the extent of reaction and affinity, which are relevant for the description of the thermodynamics of chemical reactions.

The definition of the extent of reaction for a generic reaction is done as follows. Denote by B_i the chemical formula for each one of the q chemical species of a reactive system and suppose the occurrence of the following reaction among them

$$\sum_{i\,(\text{reag.})} \nu_i' B_i \rightleftharpoons \sum_{i\,(\text{prod.})} \nu_i' B_i, \tag{19.13}$$

where $\nu_i' > 0$ are the stoichiometric coefficients. It is convenient to write the chemical equation above in the equivalent form

$$\sum_{i=1}^{q} \nu_i B_i \rightleftharpoons 0, \tag{19.14}$$

where $\nu_i = -\nu_i' < 0$ for the reactants, $\nu_i = \nu_i' > 0$ for the products and $\nu_i = 0$ for the chemical species that do not participate in the reaction.

The extent of reaction ξ is defined by $\xi = N_1/\nu_1$ and the complementary variables X_i are defined such that they are invariant under the chemical reaction (19.14),

$$X_i = N_i - \frac{\nu_i}{\nu_1} N_1, \qquad i = 2, \ldots, q. \tag{19.15}$$

The inverse relation is given by

$$N_1 = \nu_1 \xi, \tag{19.16}$$

$$N_i = X_i + \nu_i \xi, \qquad i = 2, \ldots, q. \tag{19.17}$$

The differential form for the internal energy

$$dU = TdS - pdV + \sum_{i=1}^{q} \mu_i dN_i, \tag{19.18}$$

becomes, in terms of the new variables, the following form

$$dU = TdS - pdV - Ad\xi + \sum_{i=2}^{q} \mu_i dX_i, \tag{19.19}$$

where A is the affinity, given by

$$A = -\sum_{i=1}^{q} \mu_i \nu_i, \tag{19.20}$$

and, according to (19.19), defined by the derivative

$$A = -\frac{\partial U}{\partial \xi}, \tag{19.21}$$

keeping constant S, V and X_i. In thermochemical equilibrium we should have $A = 0$ which gives the condition for equilibrium

$$\sum_{i=1}^{q} \mu_i \nu_i = 0, \tag{19.22}$$

which is the generalization of the condition (19.12).

Multiple Reactions

Now consider the occurrence of r chemical reactions among the q chemical species, represented by

$$\sum_{i=1}^{q} \nu_{ij} B_i \rightleftharpoons 0, \qquad j = 1, 2, \ldots, r. \tag{19.23}$$

To each chemical reaction corresponds an extent of reaction ξ_j defined as follows. For each reaction j we choose one chemical species that participate in this reaction and attach to it the label j. The only requirement is that they are all distinct. The extents of reaction ξ_j are defined by the equation

$$N_i = \sum_{j=1}^{r} \nu_{ij}\xi_j, \qquad i = 1, 2, \ldots, r, \tag{19.24}$$

and the complementary variables X_i, by the equation

$$N_i = X_i + \sum_{j=1}^{r} \nu_{ij}\xi_j, \qquad i = r+1, \ldots, q. \tag{19.25}$$

In order that ξ_j actually describe the chemical reactions it is necessary that the variables X_i are invariant under each of the chemical reactions. From (19.24),

$$\xi_j = \sum_{i=1}^{r} a_{ji}N_i, \qquad j = 1, 2, \ldots, r, \tag{19.26}$$

where the coefficients a_{ij} are understood as the elements of the matrix that is the inverse of the matrix whose elements are the stoichiometric coefficients ν_{ij} and are related by

$$\sum_{i=1}^{r} a_{ji}\nu_{ik} = \delta_{jk}. \tag{19.27}$$

Let us suppose that the k-th reaction causes a change in N_i equal to $\Delta N_i = \nu_{ik}$. From (19.26) and from relation (19.27), we conclude that the change in ξ_j will be $\Delta\xi_j = \delta_{jk}$. Substituting these results in

$$\Delta N_i = \Delta X_i + \sum_{j=1}^{r} \nu_{ij}\Delta\xi_j, \qquad i = r+1, \ldots, q. \tag{19.28}$$

which is obtained from (19.25), we conclude that $\Delta X_i = 0$, a result that tell us that X_i is indeed invariant under the k-th reaction.

Replacing (19.26) into (19.25) and solving for X_i, we also get

$$X_i = N_i - \sum_{j=1}^{r}\sum_{k=1}^{r} \nu_{ij}a_{jk}N_k, \qquad i = r+1, \ldots, q, \tag{19.29}$$

which together with (19.26) comprises the inverse transformation of the transformation defined by (19.24) and (19.25).

Substituting the results (19.24) and (19.25) in (19.18), we obtain

$$dU = TdS - pdV - \sum_{j=1}^{r} A_j d\xi_j + \sum_{i=r+1}^{q} \mu_i dX_i, \tag{19.30}$$

where the affinities A_j are given by

$$A_j = -\sum_{i=1}^{q} \mu_i \nu_{ij}, \tag{19.31}$$

and, according to (19.30), defined by the derivative

$$A_j = -\frac{\partial U}{\partial \xi_j}, \tag{19.32}$$

keeping constant S, V, X_i and the other extents of reaction. In equilibrium $A_j = 0$ for each reaction and therefore the following condition must be fulfilled

$$\sum_{i=1}^{q} \mu_i \nu_{ij} = 0, \tag{19.33}$$

which is the condition for thermochemical equilibrium for the case of multiple reactions.

The equilibrium condition (19.33) tells us that the chemical potentials μ_i are not independent. The number of independent chemical potential equals the number of chemical species q minus the number of chemical reactions r, that is, $c = q - r$. A system of r chemical reactions and q chemical species must therefore be understood as a mixture of $c = q - r$ components. The chemical reactions (19.23) or equivalently (19.33), which connect the chemical potentials, should be linearly independent, which means that none can be obtained from a linear combination of the others.

Closed System and Thermodynamics Potentials

A closed system allows the exchange of heat and work but not the exchange of particles with the external environment. The only variations of the number of moles N_i are those dictated by the chemical reactions. As the quantities X_i are not affected by these variations, they remain constant. For a closed system the following differential form is valid for the internal energy U

$$dU = TdS - pdV - \sum_{j=1}^{r} A_j d\xi_j. \tag{19.34}$$

A thermodynamic system can be described in an equivalent manner by means of other thermodynamic potentials obtained from the internal energy U by Legendre transformations. From U, we introduce the Helmholtz free energy $F = U - TS$, the enthalpy $H = U + pV$ and the Gibbs free energy $G = H - TS$. The differential forms of these potentials differ from that related to the internal energy by the first two terms of (19.30) or (19.34). For closed systems with only one chemical reaction, described by the extension reaction ξ, the differential forms of the internal energy and other thermodynamic potentials are given by

$$dU = TdS - pdV - Ad\xi, \tag{19.35}$$

$$dF = -SdT - pdV - Ad\xi, \tag{19.36}$$

$$dH = TdS + Vdp - Ad\xi, \tag{19.37}$$

$$dG = -SdT + Vdp - Ad\xi. \tag{19.38}$$

The affinity A is given by the following equivalent forms

$$A = -\left(\frac{\partial U}{\partial \xi}\right)_{SV} = -\left(\frac{\partial F}{\partial \xi}\right)_{TV} = -\left(\frac{\partial H}{\partial \xi}\right)_{Sp} = -\left(\frac{\partial G}{\partial \xi}\right)_{Tp}. \tag{19.39}$$

It is worth emphasizing that the equilibrium properties are those obtained by taking the limit $A \to 0$.

19.2 Thermodynamic Coefficients

Heat of Reaction

We examine the thermochemical processes in a closed system with only one chemical reaction, described by the extent of reaction ξ, for which the differential forms (19.35)–(19.38) are valid. There are two basic thermodynamic processes as to how heat is transferred to a closed system. In one, the volume is kept constant, $dV = 0$. In this case there is no work done and heat introduced TdS becomes equal to the change of internal energy, $dU = TdS$, because $A = 0$. In the other case, the pressure p is kept constant. In this case, the system does work equal to pdV and the heat introduced becomes equal to the change in enthalpy. Indeed, taking into account that $A = 0$ and p is constant, then $TdS = dU + pdV = d(U + pV) = dH$.

The characterization of a chemical reaction as to the heat developed is made by means of a quantity known as heat of reaction, defined as the ratio of heat exchanged and variation in the extension reaction during a process at constant temperature. We define two types of heat of reaction. One is the isochoric heat of reaction, where the volume is kept constant in addition to the temperature and defined by

$$Q_V = T\left(\frac{\partial S}{\partial \xi}\right)_{TV} = \left(\frac{\partial U}{\partial \xi}\right)_{TV}. \tag{19.40}$$

The second equality is valid only when $A = 0$. The other is the isobaric heat of reaction where the pressure is kept constant in addition to the temperature and defined by

$$Q_p = T\left(\frac{\partial S}{\partial \xi}\right)_{Tp} = \left(\frac{\partial H}{\partial \xi}\right)_{Tp}. \tag{19.41}$$

The second equality is valid only when $A = 0$. Both heats of reaction are measured in units of energy per mole. When the heat of reaction is negative, the system releases heat to the environment and the reaction is exothermic. When the heat of reaction is positive, the system absorbs heat from the environment and the reaction is endothermic.

In a thermochemical process, in addition to the heat exchanged work may be developed. At constant temperature the work pdV performed by the system is equal to the decrease in Helmholtz free energy $F = U - TS$ because, taking into account that $A = 0$ and that T is constant, $pdV = -dU + TdS = -d(U - TS) = -dF$. At constant pressure, the determination of the work is reduced to determining the increase in volume.

To characterize a chemical reaction with respect to changes in volume, we introduce the volume of reaction defined as the ratio between the volume change at constant pressure and variation of the extent of reaction. There are two types of volume of reaction depending on which quantity, in addition to pressure, is kept constant: entropy or pressure. They are defined by

$$V_T = \left(\frac{\partial V}{\partial \xi}\right)_{Tp}, \qquad V_S = \left(\frac{\partial V}{\partial \xi}\right)_{Sp}. \tag{19.42}$$

The derivatives that define the heats of reaction and the volumes of reaction obey the Maxwell relations

$$\left(\frac{\partial S}{\partial \xi}\right)_{TV} = \left(\frac{\partial A}{\partial T}\right)_{V\xi}, \qquad \left(\frac{\partial S}{\partial \xi}\right)_{Tp} = \left(\frac{\partial A}{\partial T}\right)_{p\xi}, \tag{19.43}$$

$$\left(\frac{\partial V}{\partial \xi}\right)_{Tp} = -\left(\frac{\partial A}{\partial p}\right)_{T\xi}, \qquad \left(\frac{\partial V}{\partial \xi}\right)_{Sp} = -\left(\frac{\partial A}{\partial p}\right)_{S\xi}. \tag{19.44}$$

Heat Capacity and Compressibility

The heat capacities at constant volume C_v and at constant pressure C_p are defined in the usual form,

$$C_v = T\left(\frac{\partial S}{\partial T}\right)_{VA} = \left(\frac{\partial U}{\partial T}\right)_{VA}, \tag{19.45}$$

$$C_p = T\left(\frac{\partial S}{\partial T}\right)_{pA} = \left(\frac{\partial H}{\partial T}\right)_{pA}, \tag{19.46}$$

where A is kept constant. Again, the second equality in each equation is valid only when $A = 0$.

Other thermodynamic coefficients that can be used to characterize the thermochemical processes are the isothermal and adiabatic compressibilities

$$\kappa_T = -\frac{1}{V}\left(\frac{\partial V}{\partial p}\right)_{TA}, \qquad \kappa_S = -\frac{1}{V}\left(\frac{\partial V}{\partial p}\right)_{SA}, \tag{19.47}$$

and the coefficient of thermal expansion

$$\alpha = \frac{1}{V}\left(\frac{\partial V}{\partial T}\right)_{pA} = \frac{1}{V}\left(\frac{\partial S}{\partial p}\right)_{TA}. \tag{19.48}$$

In the three cases, the affinity A remains constant.

Deriving the heat of reaction Q_p with respect to temperature and using the definition of isobaric heat capacity, we obtain the Kirchhoff law

$$\frac{\partial Q_p}{\partial T} = \frac{\partial C_p}{\partial \xi}, \tag{19.49}$$

whose integrated form,

$$Q_p(T_1) - Q_p(T_0) = \int_{T_0}^{T_1} \frac{\partial C_p}{\partial \xi} dT, \tag{19.50}$$

allows the determination of the heat of reaction at a certain temperature when the heat of reaction is known at another temperature.

Deriving the volume of reaction V_T with respect to pressure and using the definition of isothermal compressibility, we get

$$\frac{\partial V_T}{\partial p} = -\frac{\partial}{\partial \xi}(V\kappa_T). \tag{19.51}$$

Thermochemical Coefficient

Here we examine the properties resulting from the thermochemical stability. We consider a closed system within which a single chemical reaction takes place which is described by the extension reaction ξ. We analyze the variations of this quantity as a result of changes in other thermodynamic variables such as temperature and pressure. We remark that a positive change, $\Delta\xi > 0$, means a displacement of the reaction to the right, that is, there is an increase of the products and a decrease of the reactants. A negative variation, $\Delta\xi < 0$, means a shift of the reaction to the left, that is, there is an increase of the reactants and a decrease of the products.

If a thermochemical system is moved from equilibrium by a disturbance in the extension reaction, the thermodynamic stability requires that the affinity resulting from this changes restores the equilibrium. That is, if $\Delta\xi > 0$ then $A < 0$ and if $\Delta\xi < 0$ then $A > 0$. In other terms, $\partial A/\partial\xi \leq 0$ and the quantity Ω, defined by

$$\Omega = -\left(\frac{\partial A}{\partial \xi}\right)_{Tp}, \tag{19.52}$$

and which we call thermochemical coefficient, holds the property

$$\Omega \geq 0, \tag{19.53}$$

which is the expression of the thermochemical stability.

We may equally rely on the convexity of the thermodynamic potentials to get the stability properties of thermochemical systems. We assume that the energy U is a convex function of all variables S, V and ξ. As a result of Legendre transformations, the other potentials are convex functions only of the set of extensive variable. The Helmholtz free energy F is convex function of V and ξ, the enthalpy H of S and ξ, and the Gibbs free energy of ξ alone. The convexity of any one of these potentials with respect to the extent of reaction leads to the result (19.53). For example, the convexity of the Gibbs free energy means that locally we have $\partial^2 G/\partial\xi^2 \geq 0$. Taking into account that the affinity $A = -(\partial G/\partial\xi)_{Tp}$ then $\Omega = -\partial A/\partial\xi \geq 0$.

Other thermochemical coefficients are defined by using other thermodynamic potentials. For example, we define

$$\Omega_V = -\left(\frac{\partial A}{\partial \xi}\right)_{TV}, \tag{19.54}$$

which is equal to $\Omega_V = \partial^2 F/\partial\xi^2$. Since F is a convex function of ξ then $\Omega_V \geq 0$.

Thermochemical Identities I

The most direct consequence of the stability conditions of a thermodynamic system has to do with the way a given thermodynamic variable changes as a result of changes in its conjugate. The introduction of heat, that is, of the entropy never leads to lowering the temperature, which means that the thermal capacity is positive or zero. A compression, that is, an increase in pressure never leads to an increase in volume, which means that the compressibility is positive or zero. As noted above, an increase of the extension reaction never leads to an increase in the affinity, which means that the thermochemical coefficient is positive or zero.

There are however other indirect consequences of stability, involving more than two thermodynamic quantities. The bodies that expand with increasing temperature, get warmer in compression. Those which contracts with increasing temperature, get cooler in a compression. This result can be shown by using the thermodynamic identity

$$\left(\frac{\partial p}{\partial T}\right)_V = -\frac{(\partial V/\partial T)_p}{(\partial V/\partial p)_T} = \frac{\alpha}{\kappa_T}. \tag{19.55}$$

Since $\kappa_T \geq 0$ then $\partial p/\partial T$ and α have the same sign. Therefore, the bodies that expand with increasing temperature at constant pressure ($\alpha > 0$), get warmer with increasing pressure ($\partial p/\partial T > 0$). The bodies that contracts with increasing temperature at constant pressure ($\alpha < 0$), get cooler with increasing pressure ($\partial p/\partial T < 0$). Using the identity

$$\left(\frac{\partial p}{\partial T}\right)_S = -\frac{(\partial S/\partial T)_p}{(\partial S/\partial p)_T} = \frac{(\partial S/\partial T)_p}{(\partial V/\partial T)_p} = \frac{C_p}{TV\alpha}, \tag{19.56}$$

we get the same rule for an adiabatic compression.

Next we present the extension of these results, due to De Donder, for situations where ξ and A are involved. The variation of the extent of reaction as a result of a variation in temperature at constant pressure is characterized by the derivative

$$\left(\frac{\partial \xi}{\partial T}\right)_{pA} = -\frac{(\partial A/\partial T)_{p\xi}}{(\partial A/\partial \xi)_{Tp}} = -\frac{(\partial S/\partial \xi)_{Tp}}{(\partial A/\partial \xi)_{Tp}}, \tag{19.57}$$

where we made use of a Maxwell relation. Therefore, from the definitions of Q_p and Ω, we reach the following result

$$\left(\frac{\partial \xi}{\partial T}\right)_{pA} = \frac{Q_p}{T\Omega}. \tag{19.58}$$

As the denominator is positive, since $\Omega \geq 0$, then $\partial \xi/\partial T$ and Q_p have the same sign. If the reaction is endothermic ($Q_p > 0$) then an increase in the temperature at constant pressure shifts the reaction to the right, toward the products ($\partial\xi/\partial T > 0$). If the reaction is exothermic ($Q_p < 0$) then an increase in temperature at constant pressure shifts the reaction to the left, toward reactants ($\partial\xi/\partial T < 0$). In both cases, the equilibrium is shifted in the direction where heat is absorbed, a rule established by van 't Hoff and by Le Chatelier. The same result can be achieved if the volume is maintained constant in the place of pressure. To this end it suffices to use the identity

$$\left(\frac{\partial\xi}{\partial T}\right)_{VA} = \frac{Q_V}{T\Omega_V}, \tag{19.59}$$

which is obtained in same manner.

Similarly, the variation in the extent of reaction as a result of changes in pressure made at constant temperature is characterized by the derivative

$$\left(\frac{\partial\xi}{\partial p}\right)_{TA} = -\frac{(\partial A/\partial p)_{T\xi}}{(\partial A/\partial\xi)_{Tp}} = -\frac{(\partial V/\partial\xi)_{Tp}}{(\partial A/\partial\xi)_{Tp}}, \tag{19.60}$$

where we have used a Maxwell relation. Using the definitions of V_T and Ω, we get the result

$$\left(\frac{\partial\xi}{\partial p}\right)_{T,A} = -\frac{V_T}{\Omega}. \tag{19.61}$$

Since the denominator is positive then ($\partial\xi/\partial p$) and V_T have opposite signs. In reactions that are accompanied by a decrease in the volume at constant temperature and pressure ($V_T < 0$), an increase in pressure shifts the reaction to the right, toward the products ($\partial\xi/\partial p > 0$). Those that are accompanied by an increase in volume at constant temperature and pressure ($V_T > 0$), an increase in pressure shifts the reaction to the left, toward the reactants ($\partial\xi/\partial p < 0$). In both cases, the equilibrium is shifted in the direction in which the volume decreases, a rule established by Le Chatelier. The same result is obtained if one considers the variation in the extent of reaction as a result of an adiabatic change in pressure.

Thermochemical Identities II

We examine now the identities analogous to that existing between the heat capacity at constant pressure and at constant volume,

$$C_p = C_V + \frac{TV\alpha}{\kappa_T}. \tag{19.62}$$

Suppose that we could insert or remove heat from a thermochemical system while preventing the reactions to occur. In this situation the extent of reaction ξ would remain constant while heat is exchanged. The thermal capacity of the system in this situation should be defined by

$$C_{p\xi} = T\left(\frac{\partial S}{\partial T}\right)_{p\xi} = \left(\frac{\partial H}{\partial T}\right)_{p\xi}, \tag{19.63}$$

where the derivative is performed keeping ξ constant instead of A constant. This heat capacity is distinct from C_p as defined by (19.46) where A is kept constant. We show that in fact $C_p \geq C_{p\xi}$. To this end we start from the identity

$$\left(\frac{\partial S}{\partial T}\right)_{pA} = \left(\frac{\partial S}{\partial T}\right)_{p\xi} + \left(\frac{\partial S}{\partial \xi}\right)_{Tp}\left(\frac{\partial \xi}{\partial T}\right)_{pA}. \tag{19.64}$$

Using the results above, we find

$$C_p = C_{p\xi} + \frac{Q_p^2}{T\Omega}. \tag{19.65}$$

Since $\Omega \geq 0$, then $C_p \geq C_{p\xi}$, that is, the heat capacity of a reactive system would be greater if the reaction did not occur. This result tells us in addition that an increase in the temperature induces an endothermic reaction such that the heat input is greater than the heat required in the absence the reaction.

Analogously one can imagine the change in volume in the situation where ξ is taken as constant. This situation is characterized by the compressibility at ξ constant in the place of A constant,

$$\kappa_{T\xi} = -\frac{1}{V}\left(\frac{\partial V}{\partial p}\right)_{T\xi}. \tag{19.66}$$

Using the results above, we get

$$\left(\frac{\partial V}{\partial p}\right)_{TA} = \left(\frac{\partial V}{\partial p}\right)_{T\xi} + \left(\frac{\partial V}{\partial \xi}\right)_{Tp}\left(\frac{\partial \xi}{\partial p}\right)_{TA}, \tag{19.67}$$

$$\kappa_T = \kappa_{T\xi} + \frac{V_T^2}{V\Omega}. \tag{19.68}$$

Since $\Omega \geq 0$, then $\kappa_T \geq \kappa_{T\xi}$, that is, the compressibility of a reaction system is greater than the compressibility in the absence of reaction. This result also says that an increase in the volume causes a decrease in pressure smaller than that which would be obtained in the absence of the reaction.

The two results above can be summarized as follows. The variation of an extensive quantity X due to a variation in its conjugate x is greater than the variation that would result in the absence of reaction, $(\partial X/\partial x)_A > (\partial X/\partial x)_\xi$, which is the statement of the Le Chatelier-Braun principle due to Ehrenfest.

19.3 Thermochemical Equilibrium

Spontaneous Processes

Imagine an isolated system consisting of a vessel within which several chemical species take place. The species are inert and the system is in equilibrium in a given state. At a certain moment they cease to be inert and begin to react. We can imagine, for example, that the reactants are separated and that at a particular instant the walls that separate them are removed. From that moment the system evolves spontaneously to a new equilibrium state whose entropy is greater than that of the initial state. The entropy change obeys the inequality $\Delta S \geq 0$, which can therefore be used as a criterion for occurrence of spontaneous reactions in isolated systems.

In systems that are not isolated, spontaneous processes may occur with decreasing entropy. In these situations we must use other criteria. If the temperature T and volume V are kept constant, the system evolves to a state in which the Helmholtz free energy is smaller. The change in the Helmholtz free energy obeys the inequality

$$\Delta F = \Delta U - T\Delta S \leq 0, \tag{19.69}$$

which is the criterion for the occurrence of spontaneous reactions systems maintained at T and V constant. If the temperature T and pressure p are constant, the system evolves to a state in which the Gibbs free energy is smaller. The change in the Gibbs free energy obeys the inequality

$$\Delta G = \Delta H - T\Delta S \leq 0, \tag{19.70}$$

which is the criterion for the occurrence of spontaneous reactions in systems maintained at T and p constant.

The exothermicity of a chemical reaction, characterized by $\Delta H < 0$, is not sufficient to ensure spontaneity in processes at T and p constant, unless $\Delta S > 0$ or the temperature is sufficiently low. In this regime, according to the Nernst principle, ΔS becomes very small, and vanishes when $T \to 0$ so that ΔH approaches ΔG. The criterion $\Delta H < 0$ was recognized by Berthelot and Thomsen, but it alone is valid only at sufficiently low temperatures. Endothermic reactions, characterized by $\Delta H > 0$, may indeed occur spontaneously at high temperatures, regime in which the second portion can prevail over the first

The obedience to the above criteria do not guarantee that the reaction will actually occur because the criteria do not include information on reaction rates. According to Table 19.1, the formation of water from hydrogen and oxygen is very favorable but the reaction rate is very low frustrating the reactive process.

Table 19.1 Standard enthalpy Δh, standard entropy Δs and standard Gibbs free energy Δg of some reactions at the pressure of 1 atm and temperature of 25 °C. *Source*: CRC

Reactants	Products	Δh (kJ/mol)	Δs (J/mol)	Δg (kJ/K mol)
$2H_2 + O_2$	$2H_2O(l)$	−571.6	−326.8	−474.2
$N_2 + 3H_2$	$2NH_3$	−91.8	−198.3	−32.8
$2CO + O_2$	$2CO_2$	−566.0	−173.1	−514.4
$2NO + O_2$	$2NO_2$	−116.2	−146.6	−72.6
$2N_2O$	$2N_2 + O_2$	−163.2	148.5	−207.4
$CH_4 + 2O_2$	$CO_2 + 2H_2O(l)$	−889.5	−243.3	−818.1
$C_2H_6O(l) + 3O_2$	$2CO_2 + 3H_2O(l)$	−1368	−138.9	−1326

Standard Enthalpy of Reaction

In the study of thermochemical processes it is essential to determine the variation of thermodynamic potentials in cases in which the reactants and products are separated in their pure states. More precisely, in the initial state there are only reactants, which are separated in their pure states, at a certain temperature and pressure, and whose quantities in moles are equal to the stoichiometric coefficients of reaction. In the final stage there are only products, which are separated in their pure states, at the same temperature and pressure of the reactants, and whose quantities in moles are equal to the stoichiometric coefficients of reaction. As the reactants and products are separated, the initial thermodynamic potential is the sum of the potential of the individual reactants and the final potential is the sum of potential of the products individually. Taking as reference the reaction (19.13), the variations of thermodynamic potentials, Δu, Δf, Δh and Δg, are given by

$$\Delta u = \sum_i \nu_i u_i, \qquad \Delta f = \sum_i \nu_i f_i, \tag{19.71}$$

$$\Delta h = \sum_i \nu_i h_i, \qquad \Delta g = \sum_i \nu_i g_i, \tag{19.72}$$

and are called standard energy of reaction, standard Helmholtz free energy of reaction, standard enthalpy of reaction and standard Gibbs free energy of reaction. The quantities u_i, f_i, h_i and g_i are respectively the molar quantities of each of the chemical species alone. Note that the four expressions above are in fact differences between the thermodynamic potentials of the products and reactants of the reaction (19.13) because, according to the convention used the stoichiometric coefficients are positive for products and negative for the reactants.

It is worth mentioning that the standard enthalpy of reaction as well as other standard potentials have the additivity property. If we consider that a certain chemical reaction, as the reaction (19.14), is equivalent to several intermediate reactions, which we consider to be those given by (19.23), that is, such that

$\nu_i = \sum_j \nu_{ij}$ then the standard enthalpy of reaction Δh is equal to the sum of the standard enthalpies Δh^j of the intermediate reactions

$$\Delta h = \sum_j \Delta h^j, \tag{19.73}$$

which is the Hess law. The demonstration is as follows

$$\sum_j \Delta h^j = \sum_j \sum_i \nu_{ij} h_i = \sum_i \sum_j \nu_{ij} h_i = \sum_i \nu_i h_i = \Delta h. \tag{19.74}$$

The standard entropy of reaction and the standard volume of reaction are given by

$$\Delta s = \sum_i \nu_i s_i, \qquad \Delta v = \sum_i \nu_i v_i, \tag{19.75}$$

where s_i and v_i are the molar entropy and the molar volume, respectively of the chemical species alone. For ideal gases $v_i = v = RT/p$ is independent of the gas and in this case $\Delta v = v \sum_i \nu_i$ and the work done is $p\Delta v = RT \sum_i \nu_i$. Therefore, if the sum of the stoichiometric coefficients of the products is greater than those of the reactants, the volume will increase and there will be work done. The standard potentials of reaction are related by

$$\Delta h = \Delta u + p\Delta v, \tag{19.76}$$

$$\Delta f = \Delta u - T\Delta s, \tag{19.77}$$

$$\Delta g = \Delta h - T\Delta s. \tag{19.78}$$

Table 19.1 shows the values of Δh, Δs and Δg for some reactions. The values for the reverse reactions are obtained by changing the signs of these values.

The standard enthalpy of reaction corresponding to the synthesis of one mole of a substance from its constituents is called enthalpy of formation. By convention, constituents are simple chemical substances stable at the pressure of 1 atm and temperature of 25 °C. The enthalpy of formation water is the standard enthalpy of the reaction H_2+(1/2)O_2 $\rightarrow$ H_2O. Note that hydrogen and oxygen are gases at the standard state while the water is liquid at the standard state. The heat of combustion is defined as the standard enthalpy of reaction of one mole of a substance with oxygen having as products CO_2 and H_2O.

Equilibrium Constant

We have seen that in thermochemical equilibrium the chemical potentials of the various chemical species are not independent but are related in accordance with (19.22),

$$\sum_{i=1}^{q} \mu_i \nu_i = 0, \tag{19.79}$$

valid for a system within which a single chemical reaction occurs, represented by (19.14). An equivalent way of describing the thermochemical equilibrium is through the activity of each chemical species defined as follows. For each chemical species, we introduce a reference chemical potential μ_i° which depends only on T and p. We define the relative activity of a_i, or simply activity by

$$a_i = e^{(\mu_i - \mu_i^\circ)/RT}, \tag{19.80}$$

and therefore

$$\mu_i = \mu_i^\circ + RT \ln a_i. \tag{19.81}$$

Replacing (19.81) into (19.79), we get the condition for equilibrium in terms of activities

$$\sum_i \nu_i \{\mu_i^\circ + RT \ln a_i\} = 0, \tag{19.82}$$

which we write in the form

$$a_1^{\nu_1} a_2^{\nu_2} \ldots a_q^{\nu_q} = K, \tag{19.83}$$

where K, called equilibrium constant, is given by

$$RT \ln K = -\sum_i \nu_i \mu_i^\circ, \tag{19.84}$$

and depends only on T and p. Since the sum at right is the standard Gibbs free energy we reach the following relation

$$RT \ln K = -\Delta g, \tag{19.85}$$

which can be written as

$$K = e^{-\Delta g/RT}. \tag{19.86}$$

Guldberg-Waage Law

The equilibrium equation (19.83) is as general as the condition (19.79), being valid for any system. To use any of these two conditions for specific cases, we must know the dependence of the potential chemicals or activities with T, p and N_i. Next we consider the case of dilute solutions and of ideal solutions, which include the mixtures of ideal gases. For these systems the reference chemical potential can be chosen such that the relative activity a_i of a chemical species coincides with its mole fraction x_i, that is,

$$a_i = x_i, \tag{19.87}$$

or equivalently

$$\mu_i = \mu_i^\circ + RT \ln x_i. \tag{19.88}$$

The mole fractions are defined by $x_i = N_i/N$, where $N = \sum_i N_i$. The condition for equilibrium in terms of the mole fraction reads therefore

$$x_1^{\nu_1} x_2^{\nu_2} \dots x_q^{\nu_q} = K. \tag{19.89}$$

Recall that K depends only on temperature T and pressure p. Note that the expression at left is actually a ratio because some stoichiometric coefficients are negative and others positive. For the reaction (19.3), $\nu_1 = -1$, $\nu_2 = -3$ and $\nu_3 = 2$ so that the left side equals $x_3^2/x_1x_2^3$. We remark that an increase in the equilibrium constant means an increase of the products at the expense of reactants: the reaction shifts to the right or toward the products. A decrease in the constant means a decrease of the products and an increase of the reactants: the reaction shifts to the left or towards the reactants.

The chemical potential μ_i of a component of an ideal gas mixture is given by

$$\mu_i = g_i + RT \ln x_i, \tag{19.90}$$

where g_i is the molar Gibbs free energy of component i alone and is given by

$$g_i = \alpha_i - RT \ln \frac{RT}{p}. \tag{19.91}$$

The first term α_i only depends on temperature and is given by $\alpha_i = -b_i T - c_i T \ln RT$, where b_i and c_i are constants specific to each gas. We can therefore identify g_i with the reference chemical potential μ_i° and write the equilibrium constant (19.84) as

$$\ln K = \ln K_c + \nu \ln \frac{RT}{p}, \tag{19.92}$$

where $\nu = \sum_i \nu_i$ and K_c is defined by

$$RT \ln K_c = -\sum_i \nu_i \alpha_i, \tag{19.93}$$

and depends only on temperature. We see that the equilibrium constant K corresponding to a mixture of ideal gases varies with pressure according to

$$K = K_c \left(\frac{RT}{p}\right)^{\nu}. \tag{19.94}$$

Notice that K_c depends only on temperature and therefore an increased in pressure may increase the equilibrium constant as long as $\nu < 0$ which occurs when the sum of the stoichiometric coefficients of the products is less than that of the reactants. In other words, when the reaction occurs with a decrease in number of molecules.

The equilibrium condition can be written in terms of concentrations, defined by $\rho_i = N_i/V$. Since $x_i = N_i/N$ then $x_i = \rho_i V/N = \rho_i RT/p$. Replacing this result (19.89) and using (19.94) we arrive at the equilibrium equation in terms of the concentrations,

$$\rho_1^{\nu_1} \rho_2^{\nu_2} \dots \rho_q^{\nu_q} = K_c, \tag{19.95}$$

valid for a mixture of ideal gases. Notice that the equilibrium constant K_c depends only on temperature. Equation (19.95) is known as the law of mass action or Guldberg-Waage law of chemical equilibrium.

It is also possible to write the condition for equilibrium in terms of the partial pressures. Using relation $p_i = \rho_i RT$ between the partial pressure p_i of each gas and the concentration ρ_i, we get the relation

$$p_1^{\nu_1} p_2^{\nu_2} \dots p_q^{\nu_q} = K_p, \tag{19.96}$$

valid for a mixture of ideal gases where $K_p = K_c(RT)^{\nu}$ and depends only on temperature.

Van 't Hoff Equation

The behavior of the equilibrium constant with temperature can be obtained from standard enthalpy through the van 't Hoff equation, obtained as follows. From (19.85),

$$R \ln K = -\frac{\Delta g}{T}. \tag{19.97}$$

Taking the derivative with respect to temperature

$$R\frac{\partial \ln K}{\partial T} = \frac{\Delta s}{T} + \frac{\Delta g}{T^2}, \tag{19.98}$$

where we used the relation $\Delta s = -\partial \Delta g/\partial T$. Using in addition the relation $\Delta h = \Delta g + T\Delta s$, we arrive at the van 't Hoff equation

$$\left(\frac{\partial \ln K}{\partial T}\right)_p = \frac{\Delta h}{RT^2}, \tag{19.99}$$

which relates the isobaric variations of the equilibrium constant with temperature and the standard enthalpy Δh. If the reaction is endothermic ($\Delta h > 0$) then an increase in temperature at constant pressure shifts the reaction to the right, toward the products ($\partial K/\partial T > 0$). If the reaction is exothermic ($\Delta h < 0$) then an increase in temperature at constant pressure shifts the reaction to the left, toward the reactants ($\partial K/\partial T < 0$). In both cases the equilibrium is displaced in the direction in which heat is absorbed, a rule introduced originally by van 't Hoff.

Similarly, we obtain the variations of the equilibrium constant with pressure, at constant temperature. Deriving (19.85) with respect to pressure, and taking into account that $\partial \Delta g/\partial p = \Delta v$ we reach the result

$$\left(\frac{\partial \ln K}{\partial p}\right)_T = -\frac{\Delta v}{RT}. \tag{19.100}$$

In reactions that are accompanied by a decrease in the volume at constant temperature and pressure ($\Delta v < 0$), an increase in pressure shifts the reaction to the right, towards the products ($\partial K/\partial p > 0$). Those that are accompanied by an increase in volume at constant temperature and pressure ($\Delta v > 0$), an increase in pressure shifts the reaction to the left, towards the reactants ($\partial K/\partial p < 0$). In both cases, the equilibrium is shifted in the direction where the volume decreases.

Ideal Gas Mixture

Here we determine the Gibbs free energy of a ideal gas mixture enclosed in a closed vessel at a given temperature and pressure, given by

$$G = \sum_i N_i(g_i + RT \ln x_i), \tag{19.101}$$

where $x_i = N_i/N$, $N = \sum_i N_i$ and g_i is the molar Gibbs free energy of the i-th gas alone and depends only on of T and p. We consider a single chemical reaction, represented by (19.14) and described by the extent of reaction ξ. The number of moles depends on ξ according to (19.16) and (19.17). Here we examine the situation in which at the minimum value of ξ, which we consider to be zero, the vessel contains only reactants, whose numbers of moles we denote by N_i^R. In addition the number of moles is equal to the stoichiometric coefficients, that is, $N_i^R = (|\nu_i| - \nu_i)/2$. At the maximum value of ξ, which we consider to be the unity, there are only products whose number of moles we denote by N_i^P. In addition the number of moles is equal to the stoichiometric coefficient, that is, $N_i^P = (|\nu_i| + \nu_i)/2$. From these conditions we find the relation between N_i and ξ, given by

$$N_i = N_i^R(1-\xi) + N_i^P\xi. \tag{19.102}$$

From these relation we find

$$\sum_i N_i g_i = g_R(1-\xi) + g_P\,\xi, \tag{19.103}$$

where $g_R = \sum_i N_i^R g_i$ is the free energy of the reactants alone and $g_P = \sum_i N_i^P g_i$ is the free energy of the products alone. The mole fraction is given by

$$x_i = \frac{N_i^R(1-\xi) + N_i^P\xi}{N_R(1-\xi) + N_P\xi}, \tag{19.104}$$

where $N_R = \sum_i N_i^R$ and $N_P = \sum_i N_i^P$. The equilibrium is determined by $\partial G/\partial\xi = 0$ and is given by

$$\Delta g + RT\sum_i \nu_i \ln x_i = 0, \tag{19.105}$$

where $\Delta g = g_P - g_R$, which can be written in the form

$$\prod x_i^{\nu_i} = e^{-\Delta g/RT} = K. \tag{19.106}$$

Let us apply the above results for the synthesis of ammonia from its constituents whose reaction is given by

$$^1/_2\,\mathrm{N_2} + {}^3/_2\,\mathrm{H_2} \rightleftharpoons \mathrm{NH_3}. \tag{19.107}$$

In this case $\nu_1 = -1/2$, $\nu_2 = -3/2$ and $\nu_3 = 1$ so that $N_R = 2$ and $N_p = 1$ and therefore

$$x_1 = \frac{1-\xi}{2(2-\xi)}, \qquad x_2 = \frac{3(1-\xi)}{2(2-\xi)}, \qquad x_3 = \frac{\xi}{2-\xi}, \tag{19.108}$$

which replaced in the equilibrium equation

$$\frac{x_3}{x_1^{1/2} x_2^{3/2}} = K, \tag{19.109}$$

leads us to the equation that determines ξ

$$\frac{4\xi(2-\xi)}{3\sqrt{3}(1-\xi)^2} = K, \tag{19.110}$$

whose solution is

$$\xi = 1 - \frac{2}{\sqrt{3\sqrt{3}K + 4}}. \tag{19.111}$$

Table 19.1 gives the value $\Delta g = -16.4\,\mathrm{kJ/mol}$ for the reaction (19.107), at the pressure of 1 atm and temperature of 25 °C, from which we can calculate K by means of (19.106) with the following result $K = 747$ which gives $\xi = 0.968$. This value for the production of ammonia is large but the rate of reaction at room temperature is very small. At high temperatures the rate increases and the process becomes feasible as long as ξ is appreciable. At the pressure of 1 atm and the temperature of 500 °C we use the result of problem 19.2 to obtain the following value for the equilibrium constant $K = 0.0034$ from which we get $\xi = 0.0022$, which is very small. Increasing the pressure it is possible to increase ξ. Indeed, according to (19.94) the equilibrium constant is proportional to $p^{-\nu}$. In the present case $\nu = \nu_1 + \nu_2 + \nu_3 = -1$ and thus K is proportional to the pressure p. Therefore, for a pressure of 100 atm, and temperature of 500 °C, $K = 0.34$ which gives $\xi = 0.16$, a reasonable value. The use of high pressures ant high temperatures, together with the use of a catalyst, comprises the Haber method for the production of ammonia.

Consistency of the Equilibrium Constants

We have seen that a system consisting of various chemical reactions, represented by (19.23), the equilibrium conditions must take into account each one of the reactions. These conditions are given by (19.33), valid for each one of the chemical equations. From them we get the equilibrium conditions for each equation in terms of the activities, mole fractions, and concentrations. The forms of these conditions are analogous to those valid for a single chemical equation. For example, the equilibrium condition in terms of the activities with respect to the j-th chemical equation is given by

$$a_1^{\nu_{j1}} a_2^{\nu_{j2}} \dots a_q^{\nu_{jq}} = K_j, \tag{19.112}$$

where K_j is the equilibrium constant associated to the j-th chemical equation.

A chemical reaction such as that represented by the chemical equation (19.14) may be composed by a set of intermediate reactions. We assume that they form a set of r reactions represented by the chemical equations (19.23). As the intermediate reactions should result in the overall reaction then the stoichiometric coefficients are related by

$$\nu_i = \sum_{j=1}^{r} \nu_{ji}. \tag{19.113}$$

In the thermodynamic equilibrium, the chemical equilibrium equations must be valid for the overall reaction and for each of the intermediate reactions.

The consistency of the equilibrium equations must be such that the equilibrium equations of intermediate reactions should result in the equilibrium equation of the overall reaction. In fact, using the result (19.113) we see that (19.33), which describe the thermochemical equilibrium of the intermediate reactions in terms of chemical potentials, imply (19.22), which represent the thermochemical equilibrium of the overall reaction in terms of the chemical potentials. The same consistency exists for the equilibrium equation in terms of activities, and also in terms of mole fractions and concentrations. For each intermediate reaction the condition (19.112) is valid. Multiplying (19.112) for $j = 1, 2, \dots, r$ and using the result (19.113), the product results in the equilibrium equation (19.89) with K given by

$$K = K_1 K_2 \dots K_r, \tag{19.114}$$

or equivalently by

$$\ln K = \ln K_1 + \ln K_2 + \dots + \ln K_r. \tag{19.115}$$

The validity of the Guldberg-Waage law for each intermediate reaction results in the validity of the law for the overall reaction or any reaction that can be considered as a combination of intermediate reactions. The equilibrium constants of the intermediate reactions are related to the equilibrium constant of the overall reaction by (19.114) or equivalently (19.115).

Problems

19.1 The derivative of any function ϕ that depends on the number of moles N_i, with respect to the extent of reaction ξ can be carried out as follows

$$\frac{\partial \phi}{\partial \xi} = \sum_i \frac{\partial \phi}{\partial N_i} \nu_i,$$

because $\partial N_i/\partial\xi = \nu_i$. Use this result to show that for a ideal mixture the following relations are valid

$$Q_p = \frac{\partial H}{\partial \xi} = \sum_i \nu_i h_i = \Delta h,$$

$$V_T = \frac{\partial V}{\partial \xi} = \sum_i \nu_i v_i = \Delta v, \qquad \frac{\partial C_p}{\partial \xi} = \sum_i \nu_i c_i,$$

where c_i are the molar heat capacities of the chemical species alone.

19.2 Use the results of the previous problem to show that the Kirchhoff law for an ideal mixture is given by $\partial \Delta h/\partial T = c$, where $c = \sum_i \nu_i c_i$. Supposing that c is constant, $\Delta h = \Delta h_0 + c(T - T_0)$. Use this result to obtain the following expression for the equilibrium constant

$$\ln \frac{K}{K_0} = \frac{c}{R} \ln \frac{T}{T_0} - \frac{1}{R}(\Delta h_0 - cT_0)(\frac{1}{T} - \frac{1}{T_0}).$$

Use this expression to determine K at 500 °C for th reaction (19.107) from the data at 25 °C. The heat capacities of N_2, H_2 and NH_3 are 29.1, 28.8 and 35.1 J/mol K at 25 °C.

19.3 Show that the thermochemical coefficient $\Omega = -(\partial A/\partial\xi)_{Tp}$ can be obtained explicitly for systems described by the chemical potential (19.88) and is

$$\Omega = \frac{RT}{N} \sum_i \left(\frac{\nu_i}{x_i} - \nu \right)^2 x_i,$$

where ν is defined by $\nu = \sum_i \nu_i$, which shows clearly that $\Omega \geq 0$.

19.4 Determine ξ as a function of the equilibrium constant for the reaction

$$CO + {}^1/_2\, O_2 \rightleftharpoons CO_2,$$

where the reactants and products are considered to be ideal gases. Calculate ξ at 1 atm and 25 °C.

19.5 Show that the following van 't Hoff equation is valid for an ideal gas mixture

$$\left(\frac{\partial \ln K_c}{\partial T} \right)_p = \frac{\Delta u}{RT^2},$$

where K_c is the equilibrium constant in terms of the concentrations and $\Delta u = \sum_i \nu_i u_i$, where u_i is the molar energy of the i-th component of the mixture.

References

General

1. M. Bailyn, *A Survey of Thermodynamics* (American Institute of Physics, New York, 1994)
2. H.B. Callen, *Thermodynamics* (Wiley, New York, 1960); *Thermodynamics and an Introduction to Thermostatistics*, 2nd edn. (Wiley, New York, 1985)
3. S. Carnot, *Réflexions sur la Puissance Motrice du Feu et sur les Machines propes à Developper cette Puissance* (Bachelier, Paris, 1824)
4. É. Clapeyron, Mémoire sur la puissance motrice de la chaleur. J. de l'Ecole Royale Polytechnique **14**, 153 (1834)
5. R. Clausius, Über die bewegende Kraft der Wärme und die Gesetze welche sich daraus für die Wärmelehre selbst ableiten lassen. Ann. der Phys. und Chem. **79**, 368, 500 (1850)
6. R. Clausius, Über verschiedene für die Anwendung bequeme Formen der Hauptgleichungen der mechanischen Wärmetheorie". Ann. der Phys. und Chem. **125**, 353 (1865)
7. P. Duhem, *Thermodynamique et Chimie* (Hermann, Paris, 1902); 2de édition (1910)
8. P. Duhem, *Traité d'Énergétique ou de Thermodynamique Générale*, 2 vols. (Gauthier-Villars, Paris, 1911)
9. P.S. Epstein, *Textbook of Thermodynamics* (Wiley, New York, 1937)
10. E. Fermi, *Thermodynamics* (Prentice Hall, New York, 1937); (Dover, New York, 1956)
11. J.W. Gibbs, A method of geometrical representation of the thermodynamic properties of substances by means of surfaces. Trans. Connecticut Acad. **2**, 382 (1873)
12. J.W. Gibbs, On the equilibrium of heterogeneous substances. Trans. Connecticut Acad. **3**, 108 (1876); **3**, 343 (1878)
13. E.A. Guggenheim, *Thermodynamics* (North Holland, Amsterdam, 1949); 2nd edn. (1950); 3rd edn. (1957); 4th edn. (1959); 5th edn. (1967); 6th edn. (1977); 7th edn. (1985)
14. J.P. Joule, On the calorific effects of magneto-electricity, and on the mechanical value of heat. Philos. Mag. **23**, 263, 347, 435 (1843)
15. J.P. Joule, On the mechanical equivalent of heat. Philos. Trans. R. Soc. **140**, 61 (1850)
16. C. Kittel, *Thermal Physics* (Wiley, New York, 1969)
17. D. Kondepudi, I. Prigogine, *Modern Thermodynamics* (Wiley, New York, 1998)
18. R. Kubo, *Thermodynamics* (North-Holland, Amsterdam, 1966)
19. L.D. Landau, E.M. Lifshitz, *Statistical Physics* (Clarendon Press, Oxford, 1938); (Pergamon, Oxford, 1958); 2nd edn. (1969)
20. J.C. Maxwell, *Theory of Heat* (Longmans, London, 1871); 5th edn. (1877); 9th edn. (1888)

M.J. de Oliveira, *Equilibrium Thermodynamics*, Graduate Texts in Physics,
DOI 10.1007/978-3-662-53207-2

21. J.R. Mayer, Bemerkungen über die Kräfte der unbelebten Natur. Ann. der Chem. und Pharm. **42**, 233 (1842)
22. W. Nernst, Ueber die Berechnung chemischer Gleichgewichte aus thermischen Messungen. Kgl. Ges. d. Wiss. Gött. **1906**, 1–40 (1906)
23. W. Nernst, *Die theoretischen und experimentellen Grundlagen des neuen Wärmesatzes* (Knapp, Halle, 1918); 2te Auflage (1924)
24. A.B. Pippard, *The Elements of Classical Thermodynamics* (Cambridge University Press, London, 1957)
25. M. Planck, *Vorlesungen über Thermodynamik* (Veit, Leipzig, 1897); 2te Auflage (1905); 3te Auflage (1911); Walter de Gruyter, Berlin, 7te Auflage (1922); 9te Auflage (1930)
26. F. Reif, *Fundamentals of Statistical and Thermal Physics* (McGraw-Hill, New York, 1965)
27. D. Ruelle, *Thermodynamic Formalism* (Addison-Wesley, Reading, 1978)
28. Yu.B. Rumer, M.Sh. Ryvkin, *Thermodynamics, Statistical Physics, and Kinetics* (Mir, Moscow, 1980)
29. F.W. Sears, *Thermodynamics, the Kinetic Theory of Gases, and Statistical Mechanics* (Addison Wesley, Reading, 1950); 2nd edn. (1953); F.W. Sears, G.L. Salinger, *Thermodynamics, Kinetic Theory, and Statistical Thermodynamics*, 3rd edn. (Addison-Wesley, Reading, 1975)
30. A. Sommerfeld, *Thermodynamik und Statistik* (Dieterich'sche Verlagsbuchhandlung, Wiesbaden, 1952)
31. R.E. Sonntag, G.J. Van Wylen, *Introduction to Thermodynamics: Classical and Statistical* (Wiley, New York, 1971)
32. W. Thomson (Lord Kelvin), On an absolute thermodynamic scale, founded on Carnot's theory of the motive power of heat, and calculated from Regnault's observations. Philos. Mag. **33**, 313 (1848)
33. W. Thomson (Lord Kelvin), On the dynamical theory of heat, with numerical results from Mr. Joule's equivalent of a thermal unit, and M. Regnault's observations on steam. Philos. Mag. **4**, 8, 105, 168 (1852)
34. L. Tisza, *Generalized Thermodynamics* (MIT Press, Cambridge, 1966)
35. J.D. van der Waals, *Over de Continuiteit van den Gas- en Vloeistoftoestand* (Sijthoff, Leiden, 1873)
36. H.L.F. von Helmholtz, *Über die Erhaltung der Kraft* (Reimer, Berlin, 1847)
37. M.W. Zemansky, *Heat and Thermodynamics* (McGraw-Hill, New York, 1937); 2nd edn. (1943); 3rd edn. (1951); 4th edn. (1957); 5th edn. (1968); M.W. Zemansky, R.H. Dittman, *Heat and Thermodynamics, An Intermediate Textbook*, 6th edn. (McGraw-Hill, New York, 1981); 7th edn. (1996)

Liquids, Mixtures and Thermochemistry

38. P.W. Atkins, *Physical Chemistry* (Oxford University Press, Oxford, 1978); 2nd edn. (1982); 3rd edn. (1986); 4th edn. (1990); 5th edn. (1994); 6th edn. (1998)
39. T. De Donder, *L'Affinité* (Lamertin, Bruxelles, 1923)
40. A. Findlay, *The Phase Rule and its Application* (Longmans, London, 1904); 5th edn. (1923); 9th edn. (Dover, New York, 1951)
41. E.A. Guggenheim, *Mixtures* (Clarendon Press, Oxford, 1952)
42. W. Heitler, Zwei Beiträge zur Theorie konzentrierter Lösungen. Ann. Physik **80**, 629 (1926)
43. J.H. Hildebrand, Solubility. XII. Regular solutions. J. Am. Chem. Soc. **51**, 66 (1929)
44. J.H. Hildebrand, *Solubility of Nonelectrolytes*, 2nd edn. (Reinhold, New York, 1936); J.H. Hildebrand, R.L. Scott, *Solubility of Nonelectrolytes*, 3rd edn. (Reinhold, New York, 1950)
45. J.G. Kirkwood, I. Oppenheim, *Chemical Thermodynamics* (McGraw-Hill, New York, 1961)
46. A.N. Krestóvnikov, V.N. Vigdoróvich, *Termodinámica Química* (Editorial Mir, Moscú, 1980)
47. G.N. Lewis, M. Randall, *Thermodynamics and the Free Energy of Chemical Substances* (McGraw-Hill, New York, 1923); 2nd edn., 1961, revised by K.S. Pitzer and L. Brewer

48. W.J. Moore, *Physical Chemistry* (Prentice-Hall, New York, 1950); 2nd edn. (1955); 3rd edn. (Prentice-Hall, Englewood Cliffs, NJ, 1962); 4th edn. (1972)
49. L.K. Nash, *Elements of Chemical Thermodynamics* (Addison-Wesley, Reading, 1962)
50. I. Prigogine, R. Defay, *Thermodynamique chimique* (Dunot, Paris, 1944); (Desoer, Liège, 1950)
51. J.S. Rowlinson, *Liquids and Liquid Mixtures* (Butterworths, London, 1959); J.S. Rowlinson, F.L. Swinton, *Liquids and Liquid Mixtures*, 3rd edn. (Butterworth Scientific, London, 1982)
52. J. Waser, *Basic Chemical Thermodynamics* (Benjamin, New York, 1966)
53. E.N. Yeremin, *Fundamentals of Chemical Thermodynamics* (Mir, Moscow, 1981)

Solids and Alloys

54. N.W. Ashcroft, N.D. Mermin, *Solid State Physics* (Holt, Rinehart and Winston, New York, 1976)
55. W.L. Bragg, E.J. Williams, The effect of thermal agitation on atomic arrangement in alloys. Proc. R. Soc. A **145**, 699 (1934)
56. A.D. Bruce, R.A. Cowley, *Structural Phase Transitions* (Taylor and Francis, London, 1981)
57. P. Gordon, *Principles of Phase Diagrams in Materials Systems* (McGraw-Hill, New York, 1968)
58. W. Gorsky, Röntgenographische Untersuchung von Umwandlungen in der CuAu. Z. Physik **50**, 64 (1928)
59. C. Kittel, *Introduction to Solid State Physics* (Wiley, New York, 1953); 4th edn. (1971); 7th edn. (1996)
60. M.A. Krivoglaz, A.A. Smirnov, *The Theory of Order-Disorder in Alloys* (Elsevier, New York, 1965)
61. R. Kubo, T. Nagamiya, *Solid State Physics* (MacGraw-Hill, New York, 1969)
62. L.D. Landau, Zh. Éksp. Teor. Fiz. **7** 19, 627 (1937)
63. L.D. Landau, E.M. Lifshitz, *Theory of Elasticity* (Pergamon, Oxford, 1959); 2nd edn. (1970)
64. T. Muto, Y. Takagi, The theory of order-disorder transitions in alloys. Solid State **1**, 193 (1955)
65. D.C. Wallace, *Thermodynamics of Crystals* (Wiley, New York, 1972)

Magnetic and Ferromagnetic Materials

66. M.F. Collins, *Magnetic Critical Scattering* (Oxford University Press, New York, 1989)
67. P. Curie, Propriétés magnétiques des corps a diverses températures. Ann. de Chim. et de Phys. **5**, 289 (1895)
68. L.J. de Jongh, A.R. Miedema, Experiments on simple magnetic model systems. Adv. Phys. **23**, 1 (1974)
69. D. Jiles, *Introduction to Magnetism and Magnetic Materials* (Chapman and Hall, London, 1991)
70. L. Landau, Eine mögliche erklärung der feldabhängigkeit der suszeptibilität bei niedrigen temperaturen. Phys. Z. Sowjetunion **4**, 675 (1933)
71. R.A. McCurrie, *Ferromagnetic Materials* (Academic, London, 1994)
72. L. Néel, Propriétés magnétiques des ferrites; ferrimagnétisme et antiferromagnétisme. Ann. Phys. **3**, 137 (1948)
73. J.S. Smart, *Effective Field Theories of Magnetism* (Saunders, Philadelphia, 1966)
74. J.H. Van Vleck, *The Theory of Electric and Magnetic Susceptibilities* (Oxford University Press, London, 1932)

75. P. Weiss, L'hypothèse du champ moléculaire et la propriété ferromagétique. J. Phys. **6**, 661 (1907)
76. P. Weiss, G. Foëx, *Le Magnétisme* (Librairie Armand Colin, Paris, 1926)

Dielectrics and Ferroelectrics

77. J.C. Burfoot, *Ferroelectrics* (D. van Nostrand, London, 1967)
78. P. Debye, Einige Resultate einer kinetischen Theorie der Isolatoren. Phys. Z. **13**, 97 (1912)
79. P. Debye, *Polare Molekeln* (Hirzel, Leipzig, 1929)
80. A.F. Devonshire, Theory of barium titanate. Part I. Philos. Mag. **40**, 1040 (1949); Theory of barium titanate. Part II. Philos. Mag. **42**, 1065 (1951)
81. E. Fatuzzo, W.J. Merz, *Ferroeletricity* (North-Holland, Amsterdam, 1967)
82. W. Känzig, *Ferroelectrics and Antiferroelectrics* (Academic, New York, 1957)
83. M.E. Lines, A.M. Glass, *Principles and Applications of Ferroelectrics and Related Materials* (Clarendon Press, Oxford, 1979)
84. J.C. Slater, Theory of the transition in KH_2PO_4. J. Chem. Phys. **9**, 16–33 (1941)

Liquid Crystals

85. P.J. Collings, *Liquid Crystals* (Princeton University Press, Princeton, 1990)
86. P.G. de Gennes, *The Physics of Liquid Crystals* (Clarendon Press, Oxford, 1974); P.G. de Gennes, J. Prost, *The Physics of Liquid Crystals*, 2nd edn. (Clarendon Press, Oxford, 1993)
87. A.M. Figueiredo Neto, S.R.A. Salinas, *The Physics of Lyotropic Liquid Crystals* (Oxford University Press, Oxford, 2005)
88. E.B. Priestley, P.J. Wojtowicz, P. Sheng (eds.), *Introduction to Liquid Crystals* (Plenum, New York, 1974)

Thermal Radiation

89. M. Planck, Ueber eine Verbesserung der Wien'schen Spectralgleichung. Verh. Dtsch. Phys. Ges. **2**, 202 (1900)
90. M. Planck, *Vorlesungen über Theorie der Wärmestrahlung* (Barth, Leipzig, 1906); 2te auflage (1913)

Phase Transitions and Critical Phenomena

91. C. Domb, *The Critical Point* (Taylor and Francis, London, 1996)
92. M.E. Fisher, The theory of equilibrium critical phenomena. Rep. Prog. Phys. **30**, 615 (1967)
93. R.B. Griffiths, Thermodynamic functions for fluids and ferromagnets near the critical point. Phys. Rev. **158**, 176 (1967)
94. R.B. Griffiths, Thermodynamic model for tricritical points in ternary and quaternary fluid mixtures. J. Chem. Phys. **60**, 195 (1974)

95. R.B. Griffiths, Phase diagrams and higher-order critical points. Phys. Rev. B **12**, 345 (1975)
96. R.B. Griffiths, J.C. Wheeler, Critical points in multicomponent systems. Phys. Rev. A **2**, 1047 (1970)
97. R.B. Griffiths, J.C. Wheeler, *The Thermodynamics of Phase Transitions* (1976)
98. P. Heller, Experimental investigations of critical phenomena. Rep. Prog. Phys. **30**, 731 (1967)
99. H.E. Stanley, *Introduction to Phase Transitions and Critical Phenomena* (Oxford University Press, New York, 1971)
100. J.-C. Tolédano, P. Tolédano, *The Landau Theory of Phase Transitions* (World Scientific, Singapore, 1987)
101. B. Widom, Equation of state in the neighborhood of the critical point. J. Chem. Phys. **43**, 3898 (1965)
102. J. Zernike, General considerations concerning the number of virtual phases. Rec. Trav. Chim. **68**, 585 (1949)

Collectanea

103. J. Kestin (ed.), *The Second Law* (Dowden, Hutchinson and Ross, Stroudsburg, Pennsylvania, 1976)
104. R.B. Lindsay (ed.), *Energy: Historical Development of the Concept* (Dowden, Hutchinson and Ross, Stroudsburg, Pennsylvania, 1975)
105. W.F. Magie, *A Source Book in Physics* (McGraw-Hill, New York, 1935)

Chapter 5

106. A. Michels, W. de Graaff, T. Wassenaar, J.M.H. Levelt, P. Louwerse, Physica **25**, 25 (1959)
107. J.R. Roebuck, H. Osterberg, Phys. Rev. **48**, 450 (1935)

Chapter 6

108. K. Clusius, L. Riccoboni, Z. Phys. Chem. B **38**, 81 (1937)
109. W.S. Corak, M.P. Garfunkel, C.B. Satterthwaite, A. Wexler, Phys. Rev. **98**, 1699 (1955)
110. P. Debye, Ann. Phys. **39**, 789 (1912)
111. E.D. Eastman, W.C. McGavock, J. Am. Chem. Soc. **59**, 145 (1937)
112. A. Einstein, Ann. Phys. **22**, 180 (1907)
113. W.F. Giauque, J.O. Clayton, J. Am. Chem. Soc. **55**, 4875 (1933)
114. W.F. Giauque, J.W. Stout, J. Am. Chem. Soc. **58**, 1144 (1936)
115. P.H. Keesom, N. Pearlman, Phys. Rev. **91**, 1354 (1953)
116. J.F. Nagle, J. Math. Phys. **7**, 1484 (1966)
117. W. Nernst, Kgl. Ges. d. Wiss. Gött. **1906**, 1–40 (1906)
118. L. Pauling, J. Am. Chem. Soc. **57**, 2680 (1935)
119. M. Planck, *Vorlesungen über Thermodynamik* (Veit, Leipzig, 1911); 3te Auflage, p. 266
120. O. Sackur, Ann. Phys. **36**, 958 (1911)
121. A. Sommerfeld, Z. Phys. **47**, 1 (1928)
122. H. Tetrode, Ann. Phys. **38**, 434 (1912)
123. G.K. White, J.G. Collins, J. Low Temp. Phys. **7**, 43 (1972)

Chapter 8

124. L. Beck, G. Ernst, J. Gürtner, J. Chem. Thermodyn. **34**, 277 (2002)
125. Table 8.2. R.D. Goodwin, L.A. Weber, J. Res. NBS A **73**, 1 (1969); M. Barmatz, Phys. Rev. Lett. **24**, 651 (1970); H.D. Bale, B.C. Dobbs, J.S. Lin, P.W. Schmidt, Phys. Rev. Lett. **25**, 1556 (1970); L.A. Weber, Phys. Rev. A **2**, 2379 (1970); G.R. Brown, H. Meyer, Phys. Rev. A **6**, 364 (1972); A. Tominaga, Y. Narahara, Phys. Lett. A **41**, 353 (1972); C.E. Chase, G.O. Zimmerman, J. Low Temp. Phys. **11**, 551 (1973); H.A. Kierstead, Phys. Rev. A **7**, 242 (1973); A.V. Voronel, V.G. Gorbunova, V.A. Smirnov, N.G. Shmakov, V.V. Shchekochikhina, Sov. Phys. JETP **36**, 505 (1973); J.M.H. Levelt-Sengers, Physica **73**, 73 (1974); M. Barmatz, P.C. Hohenberg, A. Kornblit, Phys. Rev. B **12**, 1947 (1975); R. Hocken, M.R. Moldover, Phys. Rev. Lett. **37**, 29 (1976); A.V. Voronel, in *Phase Transitions and Critical Phenomena*, vol. 5b, p. 343, ed. by C. Domb, M.S. Green (Academic, New York, 1976); D.R. Douslin, R.H. Harrison, J. Chem. Thermodyn. **8**, 301 (1976); D. Balzarini, M. Burton, Can. J. Phys. **57**, 1516 (1979); M.W. Pestak, M.H.W. Chan, Phys. Rev. B **30**, 274 (1984); J.R. de Bruyn, D.A. Balzarini, Phys. Rev. A **36**, 5677 (1987); J.R. de Bruyn, D.A. Balzarini, Phys. Rev. B **39**, 9243 (1989); L. Beck, G. Ernst, J. Gürtner, J. Chem. Thermodyn. **34**, 277 (2002)
126. R.B. Griffiths, Phys. Rev. **158**, 176 (1967)
127. A. Michels, B. Blaisse, C. Michels, Proc. R. Soc. A **160**, 358 (1937)
128. A.V. Voronel, V.G. Gorbunova, V.A. Smirnov, N.G. Shmakov, V.V. Shchekochikhina, Sov. Phys. JETP **36**, 505 (1973)

Chapter 10

129. A.S. Darling, R.A. Mintern, J.C. Chaston, J. Inst. Met. **81**, 125 (1952–1953)
130. B.F. Dodge, A.K. Dunbar, J. Am. Chem. Soc. **49**, 591 (1927)
131. J.L. Murray, Metall. Trans. A **15**, 261 (1984)
132. H. Stöhr, W. Klemm, Z. Anorg. Allgem. Chem. **241**, 305 (1939)
133. Table 10.1. D.R. Thompson, O.K. Rice. J. Am. Chem. Soc. **86**, 3547 (1964); B. Chu, F.J. Schoenes, W.P. Kao, J. Am. Chem. Soc. **90**, 3042 (1968); A.M. Wims, D. McIntyre, F. Hynne, J. Chem. Phys. **50**, 616 (1969); P.N. Pusey, W.I. Goldburg, Phys. Rev. A **3**, 766 (1971); H.K. Schurmann, R.D. Parks, Phys. Rev. Lett. **26**, 367 (1971); D. Balzarini, Can. J. Phys. **52**, 499 (1974); E.S.R. Gopal et al., Phys. Rev. Lett. **32**, 284 (1974); S.C. Greer, R. Hocken, J. Chem. Phys. **63**, 5067 (1975); J.I. Lataille, T.S. Venkataraman, L.M. Narducci, Phys. Lett. A **53**, 359 (1975); S.C. Greer, Phys. Rev. A **14**, 1770 (1976); D.T. Jacobs, et al., Chem. Phys. **20**, 219 (1977); D. Beysens, A. Bourgon, Phys. Rev. A **19**, 2407 (1979); M.A. Anisimov et al., Sov. Phys. JETP **49**, 844 (1979); D. Beysens, J. Chem. Phys. **71**, 2557 (1979); A. Sivaraman et al., Ber. Bunsen Phys. Chem. **84**, 196 (1980); J. Shelton, D. Balzarini, Can. J. Phys. **59**, 934 (1981); M. Nakata, T. Dobashi, N. Kuwahara, M. Kaneko, B. Chu, Phys. Rev. A **18**, 2683 (1987)

Chapter 12

134. W.L. Bragg, E.J. Williams, Proc. R. Soc. A **145**, 699 (1934)
135. Table 12.1. O.W. Dietrich, J. Als-Nielsen, Phys. Rev. **153**, 711 (1967); J. Als-Nielsen, O.W. Dietrich, Phys. Rev. **153**, 717 (1967); J. Ashman, P. Handler, Phys. Rev. Lett. **23**, 642 (1969); M.B. Salamon, F.L. Lederman, Phys. Rev. B **10**, 4492 (1974); M.F. Collins, *Magnetic Critical Scattering* (Oxford University Press, Oxford, 1989)

136. C. Franz, M. Gantois, J. Appl. Crystallogr. **4**, 387 (1971)
137. W. Gorsky, Z. Phys. **50**, 64 (1928)
138. L.D. Landau, Zh. Éksp. Teor. Fiz. **7** 19, 627 (1937)
139. W. Shockley, J. Chem. Phys. **6**, 130 (1938)
140. Figure 12.2. C. Sykes, H. Wilkinson, J. Inst. Met. **61**, 223 (1937); O. Rathmann, J. Als-Nielsen, Phys. Rev. B **9**, 3921 (1974); P.K. Kumar, L. Muldawer, Phys. Rev. B **14**, 1972 (1976); M. Hansen, *Constitution of Binary Alloys*, 2nd edn. (McGraw-Hill, New York, 1958)

Chapter 13

141. L. Brillouin, J. Phys. **8**, 74 (1927)
142. D.L. Connelly, J.S. Loomis, D.E. Mapother, Phys. Rev. B **3**, 924 (1971)
143. P. Curie, Ann. Chim. Phys. **5**, 289 (1895)
144. Table 13.3. P. Heller, Rep. Prog. Phys. **30**, 731 (1967); W. Rocker, R. Kohlhass, Z. Naturforsch. A **22**, 291(1967); J.S. Kouvel, J.B. Comly, J. Appl. Phys. **20**, 1237 (1968); J.T. Ho, J.D. Litster, Phys. Rev. B **2**, 4523 (1970); Phys. Rev. Lett. 22, 603 (1969); N. Menyuk, K. Dwight, T.B. Reed, Phys. Rev. B **3**, 1689 (1971); G. Ahlers, A. Kornblit, Phys. Rev. B **12**, 1938 (1975); A. Kornblit, G. Ahlers, E. Buehler, Phys. Rev. B **17**, 282 (1978); C.J. Glinka, V.J. Minikiewicz, PRB **16**, 4084 (1977); M.F. Collins, *Magnetic Critical Scattering* (Oxford University Press, New York, 1989); T. Tanaka, K. Miyatani, J. Appl. Phys. **82**, 5658 (1997)
145. W.E. Henry, Phys. Rev. **88**, 559 (1952)
146. P. Langevin, Ann. Chim. Phys. **5**, 70 (1905); J. Phys. **4**, 678 (1905)
147. P. Weiss, J. Phys. **6**, 661 (1907)
148. P. Weiss, R. Forrer, Ann. Phys. **5**, 153 (1926)

Chapter 14

149. M.E. Fisher, Physica **26**, 618 (1960); Philos. Mag. **7**, 1731 (1962)
150. Table 14.2. A. Kornblit, G. Ahlers, Phys. Rev. B **8**, 5163 (1973); L.J. de Jongh, A.R. Miedema, Adv. Phys. **23**, 1 (1974); G. Ahlers, A. Kornblit, Phys. Rev. B **12**, 1938 (1975); M. Barmatz, P.C. Hohenberg, A.Kornblit, Phys. Rev. B **12**, 1947 (1975); M.F. Collins, *Magnetic Critical Scattering* (Oxford University Press, New York, 1989)
151. L. Landau, Phys. Z. Sowjetunion **4**, 675 (1933)
152. D.P. Landau, B.E. Keen, B. Schneider, W.P. Wolf, Phys. Rev. B **3**, 2310 (1970)
153. M.A. Lasher, J. van den Broek, C.J. Gorter, Physica **24**, 1061, 1076 (1958)
154. L. Néel, Ann. Phys. **3**, 137 (1948)
155. N.F. Oliveira Jr., A.P. Filho, S.R.A. Salinas, in *AIP Conference Proceedings*, vol. 29 (1975), p. 463
156. R. Pauthenet, Ann. Phys. **3**, 424 (1958)

Chapter 15

157. R. Blinc, J. Phys. Chem. Solids **13**, 204 (1960)
158. P.G. de Gennes, Solid State Commun. **1**, 132 (1963)
159. P. Debye, Phys. Z. **13**, 97 (1912)

160. A.F. Devonshire, Philos. Mag. **40**, 1040 (1949); Philos. Mag. **42**, 1065 (1951)
161. S. Hoshino, T. Mitsui, F. Jona, R. Pepinsky, Phys. Rev. **107**, 1255 (1957)
162. Y. Ishibashi, J. Phys. Soc. Jpn. **56**, 2089 (1987)
163. W.J. Merz, Phys. Rev. **91**, 513 (1953)
164. R. Sänger, O. Steiger, Gächter, Helv. Phys. Acta **5**, 200 (1932)
165. J.C. Slater, J. Chem. Phys. **9**, 16–33 (1941)
166. Y. Takagi, J. Phys. Soc. Jpn. **3**, 271, 273 (1948)
167. C.T. Zahn, Phys. Rev. **24**, 400 (1924); Phys. Rev. **27**, 455 (1926); C.T. Zahn, J.B. Miles, Jr., Phys. Rev. **32**, 497 (1928)

Chapter 16

168. A.D. Bruce, R.A. Cowley, *Structural Phase Transitions* (Taylor and Francis, London, 1981)
169. A.F. Devonshire, Philos. Mag. **40**, 1040 (1949); Philos. Mag. **40**, 1065 (1949)
170. L.D. Landau, E.M. Lifshitz, *Theory of Elasticity* (Pergamon Press, Oxford, 1959); 2nd edn. (1970)
171. P. Toledano, M.M. Fejer, B.A. Auld, Phys. Rev. B **27**, 5717 (1983)

Chapter 17

172. P.J. Collings, *Liquid Crystals* (Princeton University Press, Princeton, 1990)
173. P.G. de Gennes, J. Prost, *The Physics of Liquid Crystals*, 2nd edn. (Clarendon Press, Oxford, 1993)
174. I. Lelides, G. Durand, Phys. Rev. Lett. **73**, 672 (1994)

Chapter 18

175. D.J. Fixsen, E.S. Cheng, J.M. Gales, J.C. Mather, R.A. Shafer, E.L. Wright, Astrophys. J. **473**, 576 (1996)
176. M.F. Modest, *Radiate Heat Transfer* (McGraw-Hill, New York, 1992); 2nd edn. (Academic, Amsterdam, 2003)
177. M. Planck, Verh. Dtsch. Phys. Ges. **2**, 202 (1900)

Chapter 19

178. J. Heer, The principle of Le Chatelier and Braun. J. Chem. Educ. **34**(8), 375–380 (1957)
179. L.K. Nash, *Elements of Chemical Thermodynamics* (Addison-Wesley, Reading, 1962)
180. J. Waser, *Basic Chemical Thermodynamics* (Benjamin, New York, 1966)

Tables e Handbooks

[EG] *Encyclopedie de Gaz, L'Air Liquide, Division Scientifique* (Elsevier, Amsterdam, 1976)
[AIP] D.E. Gray (coordinating editor), *American Institute of Physics Handbook*, 3rd edn. (McGraw-Hill, New York, 1972)
[TT] J. Hilsenrath (ed.), *Tables of Thermodynamic and Transport Properties of Air, Argon, Carbon Dioxide, Carbon Monoxide, Hydrogen, Nitrogen, and Steam* (Pergamon, New York, 1960)
[LB] Landolt-Börnstein, *Zahlenwerte und Funktionen aus Physik, Chemie, Astronomie, Geophysik und Technik*, 6te Auflage (Springer, Berlin, 1950–1980); Neue Series (1961–1985)
[CRC] D.R. Lide (editor-in-chief), *CRC Handbook of Chemistry and Physics*, 78th edn. (CRC Press, Boca Raton, FL, 1997–1998)
[ICT] E.W. Washburn (editor-in-chief), *International Critical Tables* (McGraw-Hill, New York, 1926–1930)

Index

M.J. de Oliveira, *Equilibrium Thermodynamics*, Graduate Texts in Physics,
DOI 10.1007/978-3-662-53207-2

Printed in the United States
By Bookmasters